BLOOD IMMUNITY

AND

BLOOD RELATIONSHIP

BLOOD IMMUNITY

AND

BLOOD RELATIONSHIP

A DEMONSTRATION OF CERTAIN BLOOD-RELATIONSHIPS
AMONGST ANIMALS BY MEANS OF

THE PRECIPITIN TEST FOR BLOOD

by

GEORGE H. F. NUTTALL, M.A., M.D., Ph.D.

University Lecturer in Bacteriology and Preventive Medicine, Cambridge.

Including

Original Researches by

G. S. GRAHAM-SMITH, M.A., M.B., D.P.H. (Camb.)

and

T. S. P. STRANGEWAYS, M.A., M.R.C.S.

CAMBRIDGE:
at the University Press
1904

CAMBRIDGE
UNIVERSITY PRESS

32 Avenue of the Americas, New York NY 10013-2473, USA

Cambridge University Press is part of the University of Cambridge.

It furthers the University's mission by disseminating knowledge in the pursuit of
education, learning and research at the highest international levels of excellence.

www.cambridge.org
Information on this title: www.cambridge.org/9781107492899

First published 1904
First paperback edition 2015

A catalogue record for this publication is available from the British Library

ISBN 978-1-107-49289-9 Paperback

THIS VOLUME IS DEDICATED

TO

PAUL EHRLICH AND ÉLIE METCHNIKOFF

WHOSE GENIUS AND INFLUENCE

HAVE SO GREATLY ADVANCED AND STIMULATED

THE SEARCH AFTER TRUTH AMID

THE COMPLEX PROBLEMS OF IMMUNITY.

"Blut ist ein ganz besondrer Saft."

GOETHE's *Faust*, Part I

PREFACE.

THE investigations recorded in this volume were carried out in the Pathological Laboratory of the University of Cambridge, chiefly during the year 1902.

My original intention was to publish a volume on the demonstration of certain blood-relationships amongst animals as indicated by 16,000 tests made by myself with precipitating antisera upon 900 specimens of blood obtained from various sources, followed by a critical review of the literature on antibodies in blood.

Subsequently it was deemed advisable to add a section dealing with the practical applications of the precipitin test for blood in legal medicine, together with some investigations carried out at my suggestion by Messrs G. S. Graham-Smith and F. Sanger. The latter were published in the *Journal of Hygiene*, vol. III., 1903, but are added for the reason that they bear directly on the subjects treated in this volume. Finally two new sections were added dealing with further original researches conducted in this Laboratory. The latter are entitled "The Results of 500 Quantitative Precipitin-Tests upon the Bloods of Primates, Insectivora, Carnivora, Ungulata, Cetacea, Marsupialia and Aves" by myself and Mr T. S. P. Strangeways, and "Blood-Relationship amongst the Lower Vertebrata and Arthropoda, etc." by Mr G. S. Graham-Smith.

The results recorded in these papers should be of interest not only to zoologists, physiologists, and those engaged in practical medico-legal work, but also to those interested in the complex problems of immunity.

Although the source from which each specimen of blood was obtained has been fully acknowledged in the text, I here wish to place on record my indebtedness to the many gentlemen who, often at much inconvenience to themselves, collected and forwarded these specimens, and especially my appreciation of the scientific spirit which led them to give, wherever possible, the accurate nomenclature of the species, without which the value of the work would have been considerably impaired.

The publication of these researches has been rendered possible through a grant made for the purpose by the Royal Society, and the generous interest of the Syndics of the University Press, Cambridge, to whom I am greatly indebted.

G. H. F. N.

Pathological Laboratory, Cambridge.
January, 1904.

ERRATA.

Beginning page 298 (Tables) the following names of animals should be corrected the numbers referring to those printed in black letters on the left-hand margin :

797	*read* Euplocamus		846	*read* Bit*is*
827	„ Python sebae		857	„ Xenopus laevis
836	„ Zamenis *ravergieri*		860	„ Megalobatrachus.
844	„ Naja *h*aje			

p. 260, top of first column, delete "3. Suborder Hyracoidea, Fam. Hyracidae" and substitute

    1. Suborder
        MYSTACOCETI.
    2. Suborder
        ODONTOCETI.

CONTENTS.

PART II.

THE PRECIPITINS.

SECTION IX. ON THE PRACTICAL APPLICATION OF THE PRECIPITIN REACTIONS
IN LEGAL MEDICINE, ETC.

INTRODUCTION.

In the absence of palaeontological evidence the question of the interrelationship amongst animals is based upon similarities of structure in existing forms. In judging of these similarities the subjective element may largely enter, in evidence of which we need but look at the history of the classification of the Primates.

Linnaeus placed Man, the Apes, Lemurs and Bats in the division Primates. Blumenbach first placed Man in a special order, the Bimana, including the Apes and Semi-Apes under Quadrumana, this classification being retained by Cuvier and others. Huxley (1863) showed that all true Apes are as genuinely "Bimana" as Man, and gave comparative anatomical proof that the differences between Man and the higher Apes are less than between these and the lowest Apes. Huxley therefore separated Primates into Anthropoidae (Man), Simiidae (Apes), and Lemuridae (Semi-Apes). Haeckel (1866)[1] did not think that Man should be placed in a separate order. Zoologists to-day agree in placing Man and Apes in one order, the Anthropoidea.

The question as to the degree of relationship between the Anthropoidea is one upon which there is some disagreement. Haeckel[2] (1899) has recently brought together all the evidence speaking for the descent of man from Old World apes, whose recent ancestors belonged to the tailless Anthropoids, whose older ancestors belonged to Cynopithecidae. Years ago[1], he pointed out that the African man-like apes, the Gorilla and Chimpanzee, are black in colour, and "like their countrymen the Negroes, have the head long from back to front (dolichocephalic). The Asiatic man-apes are on the contrary mostly of a brown, or yellowish-brown colour and have the head short from back to front (brachycephalic), like their countrymen, the Malays and Mongols." A closer

---

[1] Haeckel (1879), *The Evolution of Man*, vol. II. (New York, D. Appleton & Co.).

[2] Haeckel, "Ueber unsere gegenwärtige Kenntniss vom Ursprung des Menschen," (Bonn, 1899).

relationship was claimed to exist between man and the Old World apes than between man and the New World apes, by Darwin[1]. Selenká[2] has pointed out that the placenta in Simiidae possesses a similar structure (Placenta discoidalis capsularis) to that in man, as distinguished from Old World monkeys, where the placenta bidiscoidalis prevails. The very interesting observations upon the eye made by Johnson (1901)[3] also demonstrated the close relationship between the Old World forms and man, the macula lutea tending to disappear as we descend in the scale of New World monkeys, and being absent in the Lemurs. The results which I published upon my tests with precipitins directly supported this evidence, for the reactions obtained with the bloods of Simiidae closely resembled those obtained with human blood, the bloods of Cercopithecidae came next, followed by those of the Cebidae and Hapalidae, which gave but slight reactions with antihuman serum, whilst the blood of Lemuroidea gave no indication of blood-relationship.

According to Dubois (1896)[4] the relationships amongst the Anthropoidea are represented by the accompanying genealogical tree, based upon that of Haeckel (1895). In the paper by Dubois, the bones of *Pithecanthropus erectus* are described as those of a probably ancestral form of man, these having been found in early pliocene deposits in Java. He places Dryopithecus between the Cercopithecidae and Simiidae, after Gaudry (1890), and considers Prothylobates to represent a generalized hypothetical form to the common ancestor of all the man-like apes. He regards Palaeopithecus as the direct ancestor of Pithecanthropus. However this may be, the fact remains, that the degrees of reaction obtained by me in my blood tests are in strict accord with this genealogy, as pointing to the more remote relationship of the Cercopithecidae, but especially of the New World monkeys, as indicated in the tree. And we shall see that the study of the haemolysins has given results in accord with what has been observed for the precipitins.

A perusal of the pages relating to the tests made upon the many bloods I have examined by means of precipitating antisera, will very clearly show that this method of investigation permits of our drawing

[1] See Darwin, *The Descent of Man.*

[2] Selenká, cited by Friedenthal (I. 1900).

[3] Johnson, G. L. (1901), "Contributions to the Comparative Anatomy of the Mammalian Eye, etc." *Philos. Trans. of the Royal Society*, B. vol. cxciv. pp. 1—82.

[4] Dubois, E. (14, iv. 1896), " Pithecanthropus erectus, eine Stammform des Menschen." *Anatom. Anzeiger*, Bd. xii. pp. 1—22. 3 figures.

certain definite conclusions. It is a remarkable fact, as I stated on a former occasion (16, XII. 1901) with regard to my results with the Anthropoidea, and this applies as well to other groups of animals, that

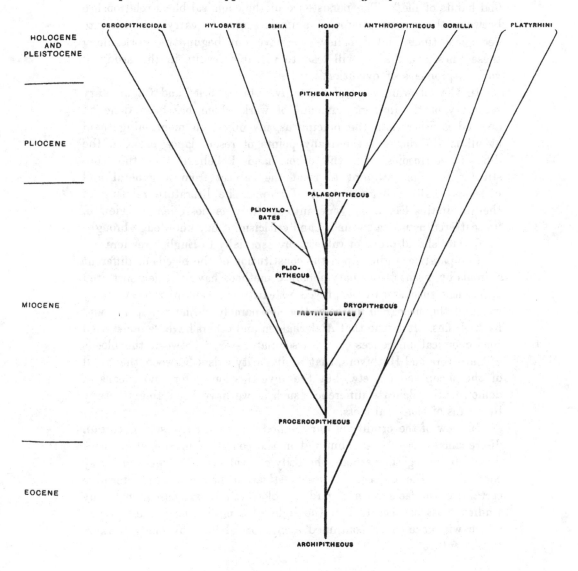

GENEALOGY OF ANTHROPOIDEA
AFTER DUBOIS 1896

Fig. 1.

1—2

a common property has persisted in the bloods of certain groups of animals "throughout the ages which have elapsed during their evolution from a common ancestor, and this in spite of differences of food and habits of life." The persistence of the chemical blood-relationship between the various groups of animals serves to carry us back into geological times, and I believe we have but begun the work along these lines, and that it will lead to valuable results in the study of various problems of evolution.

In the following pages I have given but a brief and fragmentary summary of the immense amount of work which has been done on antibodies other than the precipitins, my object in mentioning them at all is to bring out the many points of resemblance between the different antibodies. On the other hand, I believe that this constitutes the first attempt to treat the subject from a general and more especially zoological point of view. The literature relating to the precipitins has been gone into as fully as possible. In view of its scattered nature, a summary and criticism of our knowledge brought up to date should prove of value, more especially to English readers.

Comparative studies upon the constitution of the bloods of different animals by means of ordinary chemical methods have not demonstrated differences such as can be proved to exist by means of what we may well call the biological methods, the use namely of the precipitins and haemolysins. It is true that Abderhalden and others have demonstrated that chemical differences of a coarser nature exist between the bloods of Carnivora and Herbivora, that a similarity exists between the blood of the sheep and ox, etc., but the investigations afford no means of demonstrating delicate differences such as we have been able to study by means of biological tests.

In view of the crudity of our methods, it is not surprising if certain discrepancies may be encountered in the course of investigations conducted by biological methods, the body of evidence is however perfectly conclusive. The object of my investigation has been to determine certain broad facts with regard to blood-affinities, consequently my studies must be regarded in the light of a preliminary investigation, which will have to be continued along special lines by many workers in the future.

PART I.

RELATING TO ANTIBODIES IN GENERAL, AND ANTI-BODIES OTHER THAN THE PRECIPITINS.

SECTION I.

EHRLICH'S THEORY REGARDING THE FORMATION OF ANTIBODIES, TOGETHER WITH A BRIEF CONSIDERATION OF TOXINS AND ANTITOXINS.

IN 1890 Behring discovered that the serum of an animal immunified against diphtheria was capable, when injected into a fresh animal, of conferring immunity upon the latter, which, failing the use of the immune serum, died from the effects of the diphtherial toxin it received. Soon after, Behring and Kitasato obtained similar results with tetanus. Ehrlich (1891) next treated animals with increasing doses of ricin and abrin (the toxic substances contained in the castor-oil bean and the seeds of jequirity) and found that the toxin was neutralized in vitro when added to the treated animal's serum, proof of neutralization being afforded by the fact that when certain proportions of toxin and immune serum were mixed in vitro, these mixtures were innocuous when injected into animals. He proved that the neutralizing action of immune serum upon each of these toxins was specific, that is, the antiserum for abrin did not neutralize ricin, and vice versa. Immunization against the toxins of snake venom had already been practised by Sewall (1887), and subsequently Calmette (1894) and Frazer (1895), but it was reserved for Ehrlich to throw more light upon the nature of the acquired immunity to toxins through the formation of antitoxins in the bodies of toxin-treated animals. He and Madsen showed that toxin and antitoxin combined in definitely measurable quantities in vitro, more rapidly in concentrated solutions, the union being retarded by cold, hastened by heat. In other words, the union took place in a manner similar to that observed with regard to known chemical bodies.

Ehrlich concluded that "the power of toxins to combine with antibodies must depend upon a specific atom-group in the toxin-complex possessing a maximal specific relation to definite atom-groups of the antitoxin-complex, so that it rapidly unites therewith, like a lock and key," a figure borrowed from Emil Fischer in describing the action of specific ferments.

Ehrlich's theory (1897) had its origin in an hypothesis advanced by Weigert in 1896, to the effect that a condition of physiological equilibrium is maintained in the body by virtue of mutually restraining influences exerted upon each other by the cells which compose it. Assuming that a cell or a group of cells are destroyed through some agency, the equilibrium is upset, an element of restraint will be removed, and there will be an overproduction, comparable to the compensating hyperplasia observed when an organ is removed, say a kidney, a double function being thrown upon the one which remains. According to Weigert, there is always hypercompensation where there has been destruction of cells. The above conception is narrowed down by Ehrlich to what takes place in a single cell.

The processes which lead to the formation of antibodies such as antitoxins are, according to Ehrlich, essentially similar to those taking place normally in the process of assimilation. We now know that normal serum contains a number of antibodies having similar actions to those artificially produced as a result of immunization with this or that substance, we know of normal agglutinins, haemolysins, bacteriolysins, antitoxins, antiferments, etc., all of which go to prove the correctness of Ehrlich's views in this respect. In normal processes of assimilation food substances are taken up by the cells with whose substance they enter into chemical combination. That a toxin also enters into chemical union with certain cells of the body was indicated by the experiments of Wassermann, who found that when he mixed tetanus toxin with the brain substance of the susceptible guinea-pig, that the mixture was no longer toxic for the guinea-pig. There was evidently a special affinity between the brain substance and the toxin, for the reason that emulsions of other guinea-pig organs when brought in contact with tetanus toxin exerted no such effect. It would appear from this experiment that a toxin may have a special affinity for certain tissue cells, and this appeared to explain the neuro-toxic character of the symptoms which are observed in tetanus.

Antitoxins are not formed when any of the chemically defined poisons are introduced into an animal. The effect of toxins is also different from these with respect to the period of incubation which precedes their apparent action in corpore whatever the dose may be, if we except snake venom and eel-serum, where we however have antitoxins formed. According to Ehrlich (1901) toxins enter into specific chemical combination with the protoplasm of certain cell groups, other poisons like alkaloids do not. Substances which enter into chemical combination are assimilable, it being immaterial whether they belong to the class of substances we regard as foods or as toxins. The toxins of vegetable and animal origin possess the same characters as the albumens or their derivatives. Antibodies are formed not only for toxins, but also for food substances, such as milk, serum, etc., which exert no injurious action.

The food-stuff or toxin enters into combination with the cell or antibody by means of its haptophorous group (see below), and the protoplasm which is capable of combining with these bodies, which in other words, is receptive, possesses corresponding "receptors," which unite with the haptophorous groups.

Diphtheria antitoxin was found to enter into chemical combination with diphtheria toxin, combining in definite proportions according to what is known as the law of multiples. When a culture-filtrate of the diphtheria bacillus is allowed to stand for some time it is found to lose its toxicity as far as its immediate effects upon experimental animals are concerned. A larger dose of old filtrate is required to kill an animal in 24—48 hours than when a fresh filtrate is used. Neverthe-less the filtrate continues to combine with antitoxin in the same manner as before. In other words, the combining power remains, the toxic power is lowered. This is due to the conversion of toxin into "toxoid." Ehrlich concluded from the above observation that the toxin-molecule contains two independent atom-complexes, the one haptophorous (which persists), which combines with the antitoxin or the corresponding cell receptors, the other toxophorous (labile), being the cause of the specific toxic effect. The haptophorous group serves as an intermediary in binding the toxophorous group to the cell.

When an animal is treated with culture-filtrate containing toxoid it is rendered immune to toxin, and its serum is found to contain antitoxin. According to Ehrlich, the haptophorous (non-toxic) group, combines with the cell receptors, or "side chains," because of a special affinity between the groups. At first the combination may be effected within the cell substance, but owing to an increased demand upon the cell for receptors fitted to the haptophors, receptors are produced in increasing quantity, and are finally thrown off as free receptors into the circulation. It is the free receptors which, circulating in the blood, lead to its antitoxic character. The serum of an animal treated with toxoid is antitoxic, for the reason that the receptors disarm the toxin of its haptophorous group, and consequently do not permit the toxophorous group to combine with the cell protoplasm.

Toxoid is therefore to be regarded as a toxin molecule, the toxophorous group of which has been destroyed. Similar or analogous observations have been made upon other antibodies, such as tetanolysin (Madsen), snake venom (Myers), milk-curdling ferment (Morgenroth), etc.

Immunity to the effects of a toxin would therefore appear to depend upon the *absence* of certain haptophorous groups in the toxin, for failing these the toxin does not become anchored to the cell, and the latter remains uninjured. Whereas the haptophorous group immediately enters into combination with its corresponding receptor, the action of the toxophorous group may be delayed for weeks. If tetanus toxin is injected into frogs, the animals being maintained at 20° C. (Courmont), large amounts may be injected with impunity. Morgenroth has shown that the haptophorous group of the tetanus toxin enters into combination with the nervous substance of the frog, but that the toxophorous group does not do so unless the temperature is raised.

Antitoxins are not of themselves toxic. As stated, the evidence of their combining with toxins is based upon animal experiment, a mixture of suitable proportions of homologous toxin and antitoxin being without effect upon an animal which is susceptible to the toxin alone. Much more favourable are the conditions of experiments with other antibodies, such as haemolysins, agglutinins, and precipitins for the reason that the interaction of the substances can be observed in vitro.

The study of the specific bacteriolysins[1], begun by Behring and Nissen (1890) on animals immunified with *Vibrio Metchnikovi*, but brought especially into prominence by Pfeiffer (1894) in his studies upon

[1] That a difference between the bacteriolytic power of normal and immune serum might exist was already indicated by me (1888, p. 388). The serum of a sheep immunified against anthrax was found to be more bacteriolytic for anthrax bacilli than normal sheep's serum. The desirability of further experiments in this direction was indicated by me at the time.

immunity to cholera, led to results of great importance. It is however especially the studies upon the specific haemolysins which have helped to further our knowledge concerning immunity. The results with the haemolysins show that these and the bacteriolysins are both strictly comparable in their mode of production and constitution. In both the injection of cellular elements is followed by the appearance of specific lysins in the serum of the treated animal. The cytolytic serum is rendered inactive by heat (56° C.) and can be reactivated by fresh serum which of itself is without effect. The action between cytolysin and cell has been compared by Ehrlich to that which takes place between toxin and antitoxin. In the case of the cytolysins we have to deal with a stable immune-body and a labile complement, which cannot act of itself but requires the intermediation of the immune or intermediary body. When this is secured, the complement produces changes of a digestive character in the cells subject to its action, and for this reason it may be safely considered to possess the character of a digestive ferment. The similarities observed in these antibodies led to the extension of the theory of Ehrlich to other antibodies such as the agglutinins and precipitins.

It appears therefore that all antibodies are formed on the same general principles, although they may possess different properties. Wherever they are formed the substance must be assimilable which gives rise to their formation. Toxins are relatively simple bodies, they are highly soluble and readily enter into combination with the protoplasmic molecule through the intermediation of the haptophorous group. Food-stuffs on the other hand are more complicated, they require to be simplified for assimilation, and in consequence the antibodies to which they give rise must necessarily be more complicated. In the latter case the albuminous molecule anchors the food-stuff by means of receptors, which act as intermediaries, permitting a digestive, fermentlike group (complement or the like) to attack the complicated food molecule.

Before proceeding to consider antibodies in general I will add a few facts regarding the antitoxins in particular. When an animal is being immunified against a toxin, say tetanus or diphtheria toxin, the antitoxin content of its serum and milk gradually rises, as periodic injections of increasing doses of toxin are being practised. Brieger and Ehrlich (1893, p. 341) noted however that the curve of antitoxin content pursued a wavy course (wellenförmiger Verlauf) in animals treated with tetanus toxin, the milk containing antitoxin in lesser quantity after each toxin injection, the fall in antitoxin being followed in due course by a rise to a point higher than before. Salomonsen and Madsen (IV. 1897, p. 326) also observed this in horses treated with diphtheria toxin. The antitoxin content of the serum or milk always fell after a fresh toxin injection, gradually rising afterwards, the maximum amount being usually reached 9—10 days after the last toxin injection. A similar observation has been made with regard to the precipitins, and the phenomenon will probably be observable in relation to all antibodies. (See Appendix, Note 1.) The decrease of antitoxin appears to denote that the toxin and antitoxin are combining in corpore.

The question of the regeneration of antitoxins in corpore has been the subject of some investigation. Thus Roux and Vaillard (II. 1893, p. 82) found that they could remove (by repeated bleedings in a few days) as much blood from a tetanus-immune rabbit as the animal originally possessed, but the newly-formed blood was apparently as antitoxic as that removed at first. Salomonsen and Madsen (XI. 1898, p. 763) made similar observations with regard to diphtherial antitoxin in the serum of goats and horses, the animals being in a condition of antitoxic equilibrium. These observations, as also the fact recorded by the last authors mentioned (*Compt. rend. Acad. des Sc.*, Paris, 1898) that the administration of pilocarpine increased the amount of antitoxin present in the serum of immunified animals, have been brought forward in evidence as to the secretive nature of the antibodies.

It has been claimed that normal antitoxins are present in the serum of animals. Thus Meade Bolton, in Philadelphia, and Cobbett, in Cambridge, found a certain percentage of normal horse sera to possess slight but distinct antitoxic properties for diphtheria toxin. Cobbett has however recently had occasion to observe diphtheria in the horse[1]. Consequently a doubt arises as to the former observation bringing strict evidence as to the existence of normal antitoxins. On the other hand their existence can scarcely be doubted for the reason that substances fulfilling the function of different antibodies have frequently been noticed in normal sera. Quite recently, moreover, von Dungern (1902, p. 37) has observed that rabbit serum contained a "normal antitoxin" for the toxin contained in the eggs of starfish (*Asterias glacialis*, and *Astropecten aurantiacus*), the toxin in question acting upon the spermatozoa of sea-urchins.

---

[1] Cobbett, C. L. (1900), *Diphtheria occurring spontaneously in the horse*, Lancet, vol. II. p. 573.

The chemical nature of antitoxins has been the subject of considerable investigation without much light being thrown on the subject. Brieger and Ehrlich (1893, p. 345) precipitated tetanus antitoxin from the milk of immunified animals by means of ammonium sulphate, finding that the antitoxin was included in the first precipitate obtained by adding 27—30% of the salt. The remaining filtrate still contained much albumin, but very little antitoxin. Freund and Sternberg (1899, p. 432) precipitated diphtheria antitoxin from antidiphtherial horse serum by means of 50% ammonium sulphate. Seng used magnesium sulphate to saturation. Jacoby (cited by Michaëlis 1902, p. 41) found antiricin in rabbit serum to be present in the fraction precipitated through the addition of 25—33% ammonium sulphate. The antitoxins are therefore precipitated under the same conditions as the globulins, but not the albumins. We shall see that other antibodies behave similarly in this respect. Some authors claim that the antitoxins are identical with the globulins, others that they are only entangled with the globulins, being precipitated under the same conditions. Pick (1902, p. 5) refers the disagreement to differences in the chemical methods used by different authors, and discusses the subject at length. Pick found diphtheria antitoxin in horse serum to be present in the pseudoglobulin fraction, whereas in the goat it was present in the euglobulin fraction.

Ehrlich (1901, Schlussbetrachtungen, etc.) has sought by means of schematic figures to render our conception clearer regarding the way in which the various antibodies enter into chemical union with different substances, by means of what he terms receptors. Before proceeding, the reader is referred to the figures and description on the succeeding page.

In normal assimilative processes, as also in the earlier stages of immunization (either artificial or in consequence of disease), the various receptors are attached to the molecule of the cell which is receptive either for the foodstuff, toxin or the like. When the receptors are produced in excess they are thrown off into the circulation. Accepting the above diagrammatic representation of the receptors attached to the cell-molecules as a base, then the freed receptors would be represented similarly, only detached at their base. Such freed receptors have been termed " haptins "; they possess the same structure as the attached receptors. The haptins corresponding to receptors of the Orders I. and II. have been styled " uniceptors," those corresponding to receptors of the III. Order as " amboceptors," by Ehrlich.

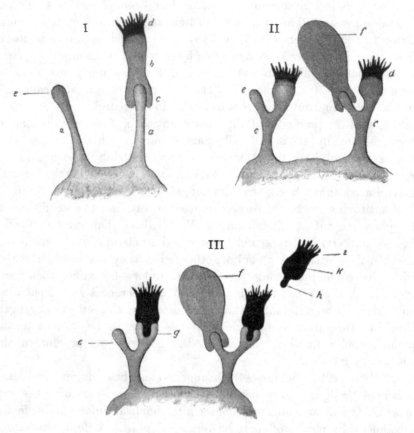

Fig. 2.

Receptors of the First Order, are represented in Fig. 2, I., at *a*, being attached to the cell-molecule beneath. The portion *e* represents the haptophorous complex, whilst *b* represents a toxin molecule, which possesses a haptophorous *c* and toxophorous *d* group. This represents the union of toxin and antitoxin, or ferment and antiferment, the latter possessing but one haptophorous group, the union between antibody and the toxin or ferment being direct.

Receptors of the Second Order, are represented at *c*, in Fig. 2, II., in which *e* represents the haptophorous, and *d* the zymophorous group of the complex, *f* being the food-molecule with which it enters into combination. Receptors of this order are possessed by agglutinins and precipitins. The digestive cell, or the antibody present in the serum in this case possesses one haptophorous and one zymophorous group.

Receptors of the Third Order, are represented in Fig. 2, III., *e* being the haptophorous group, *g* the complementophile group of the complex, *k* the complement with its haptophorous *h* and zymotoxic *z* group; whilst *f* represents the food-molecule which has become linked to the receptor. Such receptors are found in haemolysins, bacteriolysins and other cytolysins, the union with these cellular elements being effected by means of the immune-body, this permitting the complement to act as a digestive ferment.

It appears from the foregoing that it is immaterial whether the receptors or haptins and the haptophorous groups are attached to cellular elements or in solution, the diagrammatic representation indicates *the mode of chemical union*. A molecule provided with haptophoric groups is as we have seen styled a haptin, the other groups with which it is provided being named in accordance with the action produced by the antibody. Thus, a toxic haptin possesses haptophorous and *toxo*phorous groups; a ferment haptin possesses haptophorous and *zymo*phorous groups; precipitins a haptophorous and *ergo*phorous group, the latter term being suggested by Michaëlis and Oppenheimer (1902, p. 360).

Antibodies resemble each other in a number of points. Analogous bodies to toxoid (Ehrlich), are complementoid (Ehrlich and Morgenroth), agglutinoid (Eisenberg and Volk), precipitoid (Kraus and Pirquet). The chemical examination of various antibodies by fractional precipitation (summarized by Pick, 1902, p. 34) shows that all observers agree in their not being bound up with fibrino-globulin and serum-albumin. This has been found to be the case with diphtheria and tetanus antitoxins, with cholera-lysin, with cholera and typhoid agglutinins, with precipitins. Anti-antibodies for anti-spermotoxin, anti-hæmolysin, and precipitins have been obtained, but not anti-antitoxins and anti-bacterio-agglutinins. As Kraus and Eisenberg (27. II. 1902, p. 211) point out, this is explainable by Ehrlich's theory. The diphtheria antitoxins and typhoid agglutinins only possess affinities for diphtheria toxin and typhoid bacilli respectively, and consequently if they are introduced into an animal the immune-bodies they contain find no substances to which they can become anchored, and it follows that they will not lead to the formation of new substances such as anti-antitoxin and anti-agglutinin. On the other hand, other immune-substances introduced into animals of a corresponding species, will lead to the formation of anti-immune bodies. All the antibodies have the property of entering into chemical union with the bodies to which they owe their origin, specific affinities existing between them. The names which have been applied to various antibodies are in accordance with their action in corpore or in vitro.

I do not think it expedient here to enter further into the theories of Ehrlich with regard to the more intimate nature of toxins, the so-called "toxin-spectra" and the like, regarding which the reader is referred to the papers by Ehrlich. A useful summary on the subject will be found in the collective review by Aschoff (1902). A number of points of a critical nature, directed against the acceptance of some of Ehrlich's views, have been brought forward by Baumgarten, Gruber, and by Emmerich and Loew. It is difficult to enter upon these questions without considerable detail which would be out of place here, the reader is therefore referred to these authors in the bibliography.

That a number of different antibodies may coexist in a serum has been amply proved, for example, Bordet (1900) after injecting fowl's blood into rabbits, observed the formation of agglutinins, haemolysins, and precipitins in the animals' serum, this having also been seen by Tchistovitch (1899) in animals treated with horse blood. Similarly agglutinins may coexist with antitoxins and with antiferments etc.

Normal Sera, as we shall see, may contain *Antitoxin,*
Antiferments,
Cytotoxins,
Agglutinins,
Precipitins,

the amount of antibody present being usually very slight as compared to what is observable in sera which contain specific antibodies.

An immunizing substance may produce
$\left\{\begin{array}{l} \textit{Antitoxin} \\ \textit{Antiferment} \\ \textit{Cytotoxin} \text{ including Haemolysins,} \\ \text{Spermotoxins, Nephrotoxins, etc.} \\ \textit{Bacteriolysins} \\ \textit{Agglutinins} \\ \textit{Precipitins} \end{array}\right.$ consisting of $\left\{\begin{array}{l} \text{Complement} \\ \text{Immune-body or} \\ \text{Amboceptor} \end{array}\right.$

The Antibodies :

$\left\{\begin{array}{l} \textit{Cytotoxins} \\ \textit{Antiferment} \\ \textit{Agglutinins} \\ \textit{Precipitins} \end{array}\right.$ give rise to anti-antibodies $\left\{\begin{array}{l} \textit{Anti-cytotoxins}\text{—consisting of} \left\{\begin{array}{l} \text{Anti-complement} \\ \text{Anti-immune-} \\ \text{\quad body} \end{array}\right. \\ \textit{Anti-antiferment} \\ \textit{Anti-agglutinins} \\ \textit{Anti-precipitins} \end{array}\right.$

Synonymous Terms.

Cytotoxin is used for any substance in serum, venom or bacterial cultures, or of plant origin, which destroys cellular elements, either animal or vegetable. The haemolysins and other toxic substances which kill but do not dissolve cellular elements are included under Cytotoxins, also the bacteriolysins (bactericidal substance, alexin).

Precipitin is used to include the antibodies contained in lactosera, called "coagulins" by some authors. The term "serotoxin" (corresponding to cytotoxin) suggested by Nedriagailoff (11. VIII. 1901) will scarcely receive acceptance. Sera containing precipitins for blood are referred to frequently as haematosera.

Complement is used synonymously with addiment (Ehrlich), alexine, cytase (of French authors).

Immune-body is used synonymously with Amboceptor, Immunkörper, Präparator, Copula, Desmon, Hilfskörper, Zwischenkörper (of German authors), Fixateur, substance sensibilisatrice or préventive spécifique (of French authors), Fixer, and Go-between (of English authors).

Intermediary-body is confined in its use to the corresponding substance to the foregoing, which is found in normal sera, and corresponds to Ehrlich's "Zwischenkörper."

SECTION II.

ANTIFERMENTS AND FERMENTS.

THE discovery of antiferments dates from Hildebrand (1893, p. 32) who found an *anti-emulsin* in the serum of rabbits subjected to repeated emulsin injections. Fermi and Pernossi (1894) found that when trypsin was injected into guinea-pigs it disappeared within 24 hours, its presence being determined by the action of the animals' blood and organs upon gelatin. The experiments, both in corpore and in vitro, showed that the trypsin was destroyed (see Achalme, 1902). Von Dungern (1898) obtained an *antidiastatic ferment* by treating animals with the proteolytic ferment contained in certain bacterial cultures. Morgenroth (1899 and 1900), experimenting with goats, obtained *anti-rennet* ferment through injecting animals with rennet. He next obtained *anti-cynarase* by injecting cynarase ferment (from the blossoms of *Cynara cardunculus*), and proved that this and the preceding antiferment were distinct. Briot (1900) working independently to Morgenroth, discovered that he could obtain anti-rennet in the serum of rabbits treated with rennet. Delezenne (cited by Metchnikoff, 1901, p. 115) whilst finding that the normal serum of animals exerted scarcely any effect on gelatin, that of animals treated with gelatin did. He injected animals with fluid gelatin, and observed that their serum soon acquired the power of rapidly dissolving gelatin. We may therefore speak of this as a *gelatin* ferment. The ferment, or anti-gelatin serum, resembles the precipitins in that it resists heating to 56° C. Whereas Landsteiner (23. III. 1900) states that the action of *anti-trypsin* is non-specific, being bound up with serum albumin, Glässner (unpublished research, cited by his colleague Rostoski, 1902 b., p. 60) came to the opposite conclusion, experimenting with anti-trypsin *normally* present in the serum of the horse and ox. The actions of these anti-trypsins were most marked against their homologous trypsins. The anti-trypsin is bound up with pseudoglobulin. Achalme (1902, p. 744) cites several authors who have observed the presence of anti-trypsin in normal serums. Metchnikoff (1901, p. 117) cites Röden as having found normal horse serum to retard or prevent rennet action, and states that still others have found normal sera to more or less impede the digestion of albuminoids through trypsin[1]. Moll (1902) immunified rabbits

---

[1] See also S. Korschun, "Ueber Lab und Antilab," *Zeitschr. f. physiol. Chem.* XXXVI. p. 141, who studied the anti-rennet in normal horse serum, and by injecting horse serum into goats obtained anti-anti-rennet. He considers anti-rennet to act on rennet as does antitoxin on toxin. He also found what appears to be a pseudo-anti-rennet. The reader is referred to the original.

against the urea-splitting ferment of *Micrococcus ureae*, demonstrating the presence of *anti-urease* in their serum. Although normal serum and urine antagonize urease, the normal antagonistic substance is distinct from anti-urease. I will further mention that Sachs (1902) has in a similar manner succeeded in obtaining *anti-pepsin*, and Gessard (1902, cited above) *anti-tyrosin*.

Anti-coagulins have been discovered by Bordet and Gengou (III. 1901). These investigators injected rabbit serum, or plasma, into guinea-pigs and found that the serum of the guinea-pigs acquired the property of preventing the coagulation of rabbits' blood. They attribute the greater part of the anti-coagulating action to the neutralization of the rabbit's fibrin-ferment. The anti-coagulin is specific, or nearly so, in its action, and the experiments made therewith indicate that the fibrin-ferments of different animals possess a different constitution, although capable, all of them, of producing coagulation of the same fibrinogen. The anti-coagulin resists heating to 58·5° C. The authors cite Camus (1901) as having also worked upon anti-coagulins.

The anti-coagulating action of leech-extract studied by Haycroft, then by Dickinson (1870) and others, can be counteracted by a *coagulin*. Thus Wendelstadt (1901) found that if he injected leech-extract into rabbits he obtained an anti-body which impeded the anti-coagulating action of the leech extract. According to this author, the coagulin is chiefly formed in the pancreas, then in the liver and kidneys. The source of these different anti-ferments will doubtless vary considerably, depending on their nature.

SECTION III.

THE CYTOTOXINS OF BLOOD SERUM.

THE discovery of the cytotoxins dates back to the time when blood-transfusion was first practised, it being noticed that the bloods of different animals transfused into man were more or less directly injurious, and not capable of replacing human blood for this purpose. The transfusion of foreign blood led to the formation of clots, thrombi, serous exudation, and more or less haemolysis. A *résumé* of the earlier work on this subject will be found in Ziemssen's *Klinische Vorträge*, 1887, wherein the observations of Panum, Ponfick, Hayem, Landois and others are recorded.

Especially important for our subject are the investigations of Landois (1875) on blood-transfusion. He found that the transfusion of foreign blood might prove fatal to an animal. The transfusion was followed by haemoglobinuria due to the haemoglobin derived from the injected blood corpuscles, but in addition to the dissolution of the treated animal's corpuscles. Where the blood transfused emanated from a closely related species Landois observed no ill effects to follow its transfusion, this being the case for instance when transfusion was practised between the dog and wolf, horse and donkey, hare and rabbit. He concluded that large transfusions could only be practised between closely allied species. The first to study the phenomena of haemolysis in heterologous blood serum was Creite (1869), whose description of the appearances observed leaves little to be desired.

The term cytotoxin has been proposed by Metchnikoff (VI. 1900, p. 369) for such animal poisons as affect cellular elements, whether they be animal or vegetable. For the cytotoxins which dissolve the blood cells the term haemolysin has gained general usage, and for this reason I have retained the term in the following pages. Bordet (V. 1900, p. 257) prefers the term haemotoxin to haemolysin, but I see no advantage in abandoning the term lysins, which certainly suggests something more as to their effects. Corresponding terms are spermotoxin for toxin in serum which acts more especially on spermatozoa, leucotoxin for serum which destroys leucocytes. Flexner and Noguchi (II. 1902) separate the haemolysins into erythrolysins (dissolve red corpuscles) and leukolysins (dissolve leucocytes).

Besides the serum of animals the secretion of certain glands may be cytotoxic, as is seen with snake venoms. Cytotoxins may also be of vegetable origin, being either derived from bacteria or higher plants. We shall refer to these elsewhere (see p. 28). The haemolysins act by separating the haemoglobin from the stroma of the blood corpuscles, causing the blood with which they come in contact to "lake." Haemolysis may be effected also by hypotonic salt solutions, or by destructive chemicals, but this apparently purely physical phenomenon does not concern us.

It is true that Baumgarten (16, XII. 1901; and 27, X. 1902) would explain haemolysis as due to altered physical conditions brought about by haemolytic sera, as seen in plasmolysis due to non-isotonic solutions. He however found that this did not offer a full explanation, for a haemolytic serum acted more rapidly than saline solution of equal specific gravity. He therefore supposes that the agglutinin present in haemolytic serum causes this increased action, claiming moreover that the agglutinin is identical with Ehrlich's immune body. It appears to him *a priori* probable that the serum of different animals may contain different quantities of osmotic substance. He does not believe that the action of complement is similar to that of digestive ferment, as considered by Buchner, Bordet, and Ehrlich and his school.

Haemolysins may be present in certain normal sera, or they may be artificially produced by injections of the blood of other animals than that treated, if we except the observations upon isolysins. In their fundamental characters the haemolysins, both natural and artificial, correspond to the bacteriolysins; they offer the advantage over bacteriolysins, that it is possible to experiment with them in the test-tube. The discovery of the artificial haemolysins dates from Belfanti and Carbone (1898).

The Mode of Action of Haemolysins and Bacteriolysins.

Bordet and Ehrlich have shown that there are at least two bodies concerned in the haemolytic or bacteriolytic action of a serum. The one is the specific immune-body, which is thermostable; the other is the thermolabile complement, which is present in normal sera. The artificially produced haemolysins, namely those

2—2

produced in animals by the injection of foreign blood corpuscles, appear to have the same constitution as the normal haemolysins.

Bordet (1895—1900) demonstrated the existence of the two bodies referred to above by heating immune-serum to 55° C., or allowing the serum to stand for some hours. Such serum was no longer cytolytic, but if he added fresh normal serum to it, the normal serum of itself being inactive, the heated immune-serum regained its haemolytic power, that is, it was reactivated. The immune-body persists in stored sera, and withstands a temperature of 65—70° C.; the thermolabile body as we have seen is destroyed at 55° C. and is soon lost. According to Bordet the immune-body or "matière sensibilisatrice," as he terms it, becomes anchored to a red blood cell or bacterium, rendering it highly susceptible to the influence of the complement, to the action of which it was previously insusceptible. It is the complement, or "alexine" as he terms it, which reactivates the immune-body. Ehrlich and Morgenroth (1899) showed by most ingenious experiments that the immune-body became fixed to the susceptible cells.

By injecting complement, Bordet (v. 1900) produced anti-complement, which prevented the action of the immune-body by neutralizing the complement, the anti-complement not having any specific affinity for the immune-body. Bordet and Gruber consider the complement or "alexine" identical in all cases, the former and Gengou (1902, p. 738) found fresh dog serum added to fresh rabbit corpuscles lost its bactericidal properties. Ehrlich and Morgenroth consider that normal sera contain a *multiplicity* of complements. Metchnikoff (1901, p. 123) points out that there is agreement as to the action of anti-complement upon complement being direct. Bordet found that heating a mixture of these bodies did not liberate the anti-complement, for the heated mixture does not regain its anti-haemolytic power, as we might expect it to do if anti-complement were freed by heating. Heating to 55°, as we have seen, affects complement, but not anti-complement. The conclusion is therefore justified, that complement and anti-complement enter into chemical combination.. Ehrlich and Morgenroth have found moreover that anti-complement only prevents the action of complement, that it exerts no action either on the immune-body or susceptible cell.

Ehrlich (1900) and Morgenroth, by the method of "elective absorption," found that a normally haemolytic serum contains a variety of intermediary bodies (analogous to the immune-bodies in artificially haemolytic sera). They found that a serum which had been treated with one species of corpuscles until it ceased to act on these still had the power to haemolyse other species of corpuscles. In the same way a multiplicity of complements appropriately adapted to the various intermediary bodies have been found by them to exist in normal sera. The union of cellular elements and the intermediary body is effected, according to Ehrlich, by means of combining or haptophoric groups, common to both cell and intermediary body, the complement being similarly linked to the intermediary body. In other words, the intermediary body possesses two haptophoric groups, the one (complementophilic) combining with the complement, the other with the cell. The complement has also two haptophores, the one combining with the intermediary body, the other possessing a fermentative action (zymotoxic, or toxo-phoric group) which dissolves the cell.

Flexner and Noguchi (1902) find that there are a multiplicity of intermediary bodies in snake venoms, for venom solution treated with the washed blood corpuscles of the dog, rabbit, and guinea-pig in succession was found to give up to each a part of its intermediary bodies, no one kind of corpuscle being capable of fixing the entire content of intermediary bodies. It therefore appears as if there might be an indefinite number of intermediary bodies in venom. When a complement foreign to the corpuscles is used it never causes complete haemolysis. It is evident that these results with venom are very similar to those of Ehrlich with haemolytic sera.

That different immune-bodies and complements may coexist in a serum has been shown by Wendelstadt (16, IV. 1902) who treated animals with different bloods simultaneously, for instance with the blood of the ox, sheep, and pig, finding three immune-bodies and three complements in their serum; the complement for the pig being more resistant to heat than the others.

Ehrlich and Morgenroth found that they could not obtain anti-complement when they treated an animal with complement derived from a *closely related species*, but they obtained it readily in many cases where the relationship was distant. The injection of goat serum into the sheep and *vice versa* did not give rise to anti-complement. This observation is quite in accord with what has been observed with regard to the formation of precipitins.

An excess of immune-body may impede haemolysis upon the addition of complement, as Nolf observed in haemolysing fowl corpuscles with the immune serum of the rabbit; moreover Neisser and Döring are cited by Eisenberg (v. 1902, p. 303) as having made an analogous observation when haemolysing rabbit corpuscles with human serum from a case of chronic nephritis. Ide (27, VII. 1902, p. 273) has confirmed this. For an explanation of this interesting phenomenon of "Komplementablenkung" see Neisser and Wechsberg (1901), who observed that an excess of immune-body prevented the action of a bacteriolytic serum; as did also Lipstein (16, IV. 1902), and Walker (I. 1903).

That the normal haemolytic action of eel serum upon rabbit corpuscles can be prevented by *specific anti-haemolytic serum* has been shown by Camus and Gley (29, I. 1898) and Kossel (14, II. 1898) experimenting with the serum of rabbits immunified against the serum of the eel. The rabbits were treated with increasing doses of eel serum, for which they gradually developed increasing quantities of antitoxin (otherwise anti-haemolysin), this being confirmed by Tchistovitch (1899). Bordet (1900) also found that the serum of a guinea-pig immunified with rabbit corpuscles, when injected in carefully graded doses into a rabbit (being highly haemolytic), caused the formation of anti-haemolysin in the rabbit, the antibody preventing haemolysis. The experiment was reversed by Schütze (5, VII. 1901) with similar result. Schütze thought he could trace the formation of the anti-haemolysin to the action of the immune-body, for he treated the guinea-pigs with complement-free (heated) rabbit serum. Müller (21, II. 1901) also treated rabbits with heated fowl serum, the antiserum obtained being anti-haemolytic. Ehrlich and Morgenroth (1901), however, consider that Schütze, and Müller, by heating their sera, converted the complement into *complementoid* (analogous to toxoid) and that the anti-haemolytic action they observed depended upon anti-complement; for the complementoids, although incapable of acting upon corpuscles, give rise, when injected, to anti-complements. Müller (21, II. 1901, p. 185) failed to produce anti-haemolysin in guinea-pigs treated with normal rabbit serum which had been heated. Ehrlich and Morgenroth (III. 1900) obtained *anti-isolysin* by injecting isolysin (see later) into a goat, noting that there were individual differences in the action of both isolysin and anti-isolysin depending upon the animal from which they were derived.

Wassermann (1901, *Zeitschr. f. Hyg.*, Bd 37) injected washed rabbit leucocytes into guinea-pigs, obtaining weak but evident "anti-complement" to the complements of rabbit serum, the haemolysis of goat blood by rabbit serum being prevented. Donath and Landsteiner (25, VII. 1901) obtained a similar result in rabbits by injecting dog leucocytes, crushed lymph-glands, red blood corpuscles and milk severally, but they do not conclude that the antisera obtained in these different ways are identical in character.

Müller (24, VI. 1901) finds that a number of *normal* sera possess anti-haemolytic properties, several for instance protected rabbit's corpuscles against duck haemolysin. This action appears to be due to anti-complement. When he treated (12, VIII. 1902) animals with such anti-haemolytic sera, these being inactivated, their serum acquired the property of haemolysing the corpuscles of the animal which possessed the anti-haemolytic serum.

Besredka (25, X. 1901) also claims to have observed the existence of anti-haemolysins in the normal sera of man, rabbit, guinea-pig, fowl, and goose. He considers that they represent anti-auto-cytotoxins, whose formation is brought about by the constantly occurring death of certain cells in the body (red blood corpuscles for example), the assimilation of these cells bringing about the formation of this antibody. This view receives support from what has been observed with regard to isolysins in disease. The presence of anti-haemolysin is to be considered as evidence of a reaction on the part of the body, its object being to combat the continual destruction of bodily cells, although at the same time haemolysins are produced. We know that fresh rabbit serum; for instance, haemolyses guinea-pig corpuscles. When heated to 55° C. it ceases to be haemolytic, whereas this heated serum when added to fresh rabbit serum in certain proportions will prevent its exerting a haemolytic action on guinea-pig corpuscles. This does not happen when the heated sera of other animals (man, ox, fowl, goose, etc.) are added instead of rabbit serum. There would therefore appear to be evidence as to the existence of both anti-haemolysin and haemolysin in the one serum.

The action of normal haemolysin is not confined to red blood corpuscles. For instance, Delezenne (x. 1900) found that it affected other cells of the same animal as well, to a greater or less extent. Haemolysins act both on the red blood corpuscles and leucocytes, as can be seen in experiments conducted *in vitro*. To see the effects on other cells the experiments must be conducted to a large extent *in corpore*. Taking eel serum as a type of a haemolytic serum, we find that it is destructive to red blood corpuscles (Mosso, Camus and Gley, Kossel, etc.), to leucocytes (Delezenne, 1898), nerve cells (Kossel and Westphal), renal epithelium (Pettit, 1898). The effects on the nervous system were already observed by Mosso (1888) and Delezenne (x. 1900, p. 693), the latter finding eel serum 200 times as toxic when introduced into the nerve centres as when administered subcutaneously to the same animal.

Metchnikoff (x. 1899, p. 762) produced "*leucotoxins*" by injecting an emulsion of rat spleen into the guinea-pig. After 47 days the treated animal's serum was found to agglutinate and dissolve rat leucocytes suspended in rat's abdominal lymph. The mononuclears were first attacked, then the polynuclears, and finally the Mastzellen of Ehrlich. He also obtained leucotoxin by injecting an emulsion of rabbits' mesenteric glands into guinea-pigs. Although these glands only contained mononuclears, the serum was destructive both to rabbit mononuclears and polynuclears. The leucotoxins were markedly specific, that for the rat scarcely affected mouse leucocytes, that for the rabbit had no effect on rat leucocytes and *vice versâ*. Funck (26, v. 1900), and Besredka (vi. 1900, p. 391) have experimented also upon these lines, the latter preparing leucotoxins for different animals (horse, ox, sheep, goat, dog) by injecting their lymph glands into other animals. He found that none of the leucotoxins affected human leucocytes, but this was an exception, for ox leucotoxin, obtained from the guinea-pig, was leucotoxic for rabbits; and human leucotoxin was leucotoxic for guinea-pigs. Besredka found leucotoxic sera less stable than haemolytic sera, and no leucotoxin was formed when lymph glands heated to 55° C. were injected. As with other toxins, the leucotoxins, when injected into animals, produce illness and death; he found for instance (p. 393) that an intraperitoneal injection of 3 c.c. of leucotoxic serum for the guinea-pig, obtained from a rabbit, killed a guinea-pig in 3 4 hours. Besredka (p. 397) obtained *anti-leucotoxin* in the serum of guinea-pigs treated with guinea-pig leucotoxin (from rabbit), the antibody preventing leucolysis when mixed with leucolytic serum.

In their studies on snake venom, Flexner and Noguchi (1902) found the venoms to dissolve and agglutinate rabbit's leucocytes. Whereas the agglutinating principle seemed to be identical for both leucocytes and red corpuscles, the lysin for each was distinct. Venomized leucocytes required a complement for their destruction just as do the red blood corpuscles. The different varieties of leucocytes were found to react differently to venom.

The presence of leucotoxins in fresh normal dog serum was already observed by Buchner (1893, p. 120), who stated that the fresh serum killed but did not dissolve the leucocytes of man and rabbit. When dog serum was heated to 55° C. it was no longer leucotoxic.

The spermotoxins discovered by Landsteiner (29, iv. 1899, p. 549) have proved of considerable interest. He obtained spermotoxin by injecting ox spermatozoa into guinea-pigs. Whereas ox spermatozoa introduced into the peritoneum of the normal guinea-pig remained motile for a considerable time, those introduced into the peritoneum of a guinea-pig which had received several injections of spermatozoa were immobilized very quickly. This was confirmed by Metchnikoff (x. 1899, p. 739) who injected ox and human spermatozoa into guinea-pigs. Whereas the spermatozoa remained alive for 30 hours in normal guinea-pig's serum, they were immobilized in a few minutes in that of a treated guinea-pig. The antiserum only immobilized the spermatozoa, it did not dissolve them. Moxter (25, i. 1900) found that spermotoxic serum for the sheep also haemolysed sheep's blood corpuscles, and that the immune-body combined with sheep spermatozoa. He denied the specificity of spermotoxin, considering, however, that it did possess a special affinity for spermatozoa. The blood cells appeared to him to be more susceptible to a variety of antibodies than are the other cells. Metchnikoff (vii. 1900, p. 373) considers that the haemolytic action of Moxter's spermotoxin may have been due to haemolysin produced in his treated animals through injecting blood contained in the testicular emulsion he used. He argues that if the action of these antibodies is not specific, then an artificial haemolytic serum should destroy the spermatozoa, whereas it does not. He found that a rabbit serum which was haemolytic for sheep corpuscles did not exert any more action upon sheep spermatozoa than did normal rabbit serum, he claims in fact that they lived longer in the haemolytic serum. He concludes therefrom that the antisera may be specific for certain cells as distinct from others belonging to the same animal. To further test this point, Metalnikoff (IX. 1900, p. 581) treated a rabbit with a mixture of blood and spermatozoa from sheep, obtaining an antiserum which haemolysed corpuscles and immobilised spermatozoa. In confirmation of Metchnikoff's results, he found that he could remove the haemolysin by adding corpuscles to the serum, the remaining fluid remaining spermotoxic. On the other hand spermatozoa fixed both the intermediary bodies of spermotoxin and haemolysin, and he concludes that this must have been what misled Moxter. Metalnikoff (p. 578) obtained spermotoxin from rabbits treated with guinea-pig spermatozoa. He noted that normal rabbit serum was scarcely less spermotoxic than that of treated rabbits, but there was an essential difference, namely, that the normal spermotoxin heated to 56° C. could not be reactivated, whereas the artificial spermotoxin could be reactivated by adding fresh normal guinea-pig serum thereto, the latter having no influence of itself. Consequently artificial spermotoxins, as is the case with haemolysins, are composed of an immune or intermediary body and of complement.

The existence of normal spermotoxins, as observed by Metalnikoff, has also been proved by Weichardt (25, XI. 1901, p. 835). Two out of three normal rat sera were markedly spermotoxic, the third feebly so, for rabbit spermatozoa. Pigeon serum was found to be spermotoxic for rabbit, dog, and goose spermatozoa. He observed the normal spermotoxic action of a goose's serum to vary widely in one and the same animal at various times.

Following along the lines of Ehrlich and Morgenroth's work with haemolysins, Metalnikoff (p. 585) succeeded in producing *auto-spermotoxin* by injecting guinea-pig spermatozoa into guinea-pigs. The auto-spermotoxin, which was active *in vitro*, could be reactivated by fresh normal guinea-pig serum after having been inactivated by heat.

Halban and Landsteiner (25, III. 1902, p. 475) observed that spermotoxic serum obtained by injecting rabbits with ox spermatozoa agglutinated ox corpuscles much more powerfully than did normal rabbit serum.

The reaction of different guinea-pigs to foreign spermatozoa injections was found to vary considerably by Weichardt (p. 833), some producing a highly spermotoxic serum, others little or no spermotoxin.

Anti-spermotoxin, comparable to anti-haemolysin, etc., was produced by Metalnikoff (IX. 1900, p. 583) by injecting guinea-pig spermotoxin (from rabbit) into normal guinea-pigs. Serum, containing anti-spermotoxin, exerted but a slight effect when fresh, but its effect was marked when it was inactivated, this being due to the fresh serum containing a preponderance of anti-complement. Consequently anti-spermotoxic serum consists of anti-intermediary body and anti-complement. Weichardt (25, XI. 1901) reaches the same conclusion, adding that anti-spermotoxin also contains an anti-agglutinin.

Neurotoxins are another class of antibodies obtained by injecting the brain-substance of one species into another. Metchnikoff succeeded in obtaining a neurotoxin for pigeons in the serum of rats treated with pigeon-brain emulsion. His pupil Delezenne (x. 1900, p. 696), who cites the unpublished experiment, subsequently obtained dog neurotoxin by treating a duck with dog-brain substance. He (p. 703) tested this neurotoxin on rabbits with negative result, whereas it affected cats, though less than dogs. This indicates a relationship between these Carnivores, such as I have been able in a measure to demonstrate by means of precipitins. Again as with the precipitins (see later) he was unable to obtain positive results by repeatedly injecting rabbit-brain substance into guinea-pigs (p. 695). Boeri (28, x. 1902) on the other hand states that he injected rabbit-brain emulsion intraperitoneally into guinea-pigs, and found their serum when injected subdurally into rabbits to produce prolonged nervous excitation. When heated to 55° C. the serum had no such effect.

Neurotoxins are normally present in venoms of serpents, as found by Flexner and Noguchi (1902). The neurotoxin is distinct from the haemolytic substance in venom, for a venom, robbed of its haemolysin by the addition of corpuscles, was neurotoxic, and *vice versâ*. This agrees with Ehrlich's view, as supported, amongst others, by Wassermann and Takaki's experiments upon the fixation of tetanus toxin in certain centres, etc.

Trichotoxins were first obtained by von Dungern (1899), the antibody, as the name implies, exerting a special effect upon ciliated epithelium. He injected rabbits and guinea-pigs with the tracheal epithelium of the ox, and found their serum to immobilise the ox ciliated epithelium. He however found that the antiserum also haemolysed ox corpuscles, but it had more affinity for the epithelium than did an antiserum obtained by corresponding blood injections. Rabbits injected with cows' milk gave an antiserum which also immobilized the epithelium, and it was also haemolytic to a considerable degree. The three antisera (for milk, corpuscles, epithelium) were subjected to comparative tests on 5 % blood corpuscle suspensions, the antisera being inactivated and then reactivated by the addition of complement. The results showed differences with regard to the haemolytic properties, those of the antisera for milk and epithelium being distinct from the third, but indistinguishable from each other.

Other authors have finally claimed to have produced a variety of toxic substances in animals treated with emulsions of different glands, this being at present the subject of much active investigation. It is only necessary to mention a few of these so as to indicate the general drift of the work. Lindenmann (II. 1900, p. 57) treated guinea-pigs with rabbit kidney-emulsion, obtaining, he claims, a highly potent "*nephrotoxin*" which caused necrosis and profound disintegration of the epithelium of the convoluted tubules in the rabbit's kidney, the glomeruli remaining unaffected. Néfédieff (25, I. 1901, p. 18) repeated the experiment, and reversed it, confirming the result. He found the nephrotoxin to be also haemolytic.

Delezenne (VIII. 1900) obtained antisera for liver cells, or "*hepatotoxins*," and these have also been studied by Deutsch (cited by Néfédieff). Mankowski (1902) treated cats with intraperitoneal injections of dog thyroid gland emulsion, obtaining a serum having toxic properties for the gland in dogs, the action being in the main thyreolytic. The serum containing the "*thyreotoxin*" could be inactivated and reactivated, consequently it contained an intermediary body and complement. Goutscharnkow (1901) also observed toxic effects in dogs injected with a similar antiserum obtained from sheep. He however found that the thyroids of sheep were affected by injecting these animals with dogs' thyroids, general systemic disturbance being produced.

Von Dungern (1902, p. 51) treated rabbits with emulsions of the eggs of starfish and sea-urchins, and found the antisera to agglutinate the corresponding spermatozoa of these animals.

From the work of Metchnikoff it would appear as if the antibodies for special cells of one animal exerted no influence on others. Delezenne (x. 1900, p. 704) also found that his artificial neurotoxin for the dog did not haemolyse dog corpuscles, and that his artificial haemolysin exerted no neurotoxic action. We have seen that other workers have noted special effects on the cells, emulsions of which had been used for the production of antibodies, but that they acted besides on other elements. I would add that this has been the case in the investigations of Boeri (28, x. 1902) who found a neurotoxic serum to be also haemolytic, and a haemolytic serum to be slightly neurotoxic. The evidence is certainly unanimous with regard to the different antibodies exerting a *special* action.

Regarding the Source of Haemolysins and Bacteriolysins.

Gengou (1901, *Ann. de l'Inst. Pasteur*, xv.) concluded that the haemolysins are derived from leucocytes, for the reason that plasma separated from fresh blood, kept cold throughout, by centrifugalization, was less haemolytic than serum. G. Ascoli (9, x. 1902, p. 736) came to the opposite conclusion. Ascoli obtained plasma in the same manner as Gengou, from a dog which had been immunified with rabbit corpuscles. When 2—5 c.c. of this dog's plasma were injected into a normal rabbit the latter developed haemoglobinuria lasting 24—48 hours. Ascoli concludes that plasma possesses the same haemolytic properties as serum both *in corpore* and *in vitro*. Whereas Nolf claimed that serum injections did not lead to the formation of haemolysins, the contrary has been found to be the case by von Dungern (1899) and Morgenroth (1902). Schattenfroh (1901) was unable to obtain haemolytic sera by goat serum injections, but did so by injecting goat and human urine, the latter observation being confirmed for human urine by Ruffer and Crendiropoulo (24, I. 1903). Ide (27, VII. 1902, p. 269) considers that such results are to be explained by the presence of haemoglobin in serum and urine.

Ide produced anti-pseudoglobulin and anti-serum-albumin and found the first not to haemolyse corresponding corpuscles, whereas the second did so, although its action was but one-tenth as strong as that of anti-haemoglobulin. He concludes therefrom that intact serum and serum albumin possess receptors in common with those of the red blood corpuscles, and these receptors, Ide thinks, are to be found in the haemoglobulin which remains present in serum and serum albumin.

Shibayama (5, XII. 1901) found haemolysins for dog corpuscles in the spleen and lymphatic glands of normal guinea-pigs, not elsewhere. He considers that when dog blood is injected into such animals these organs are stimulated to increased production of haemolysin in the sense of Ehrlich's theory.

The seat of origin of bacteriolysins *in corpore* has been studied by a number of observers, and has been the subject of considerable discussion, for which reason I shall not enter upon it here (see Metchnikoff, 1901; Pfeiffer and Marx, 1898; Deutsch, 1899; Wassermann, 1899; Römer, 1902; and the reviews of Ritchie, 1902 and Aschoff, 1902). According to Pick (1902, pp. 15—21) bacteriolysins are contained in the euglobulin fraction of serum, this being contrary to Pfeiffer and Proskauer (1896).

That there is *a difference in the specific bacteriolytic sera obtained from different animals* appears to be the case from the investigations of Sobernheim (1899), who found that anti-anthrax serum obtained from immunified sheep was able to protect sheep but not rabbits. In this case we are dealing, not with an antitoxic serum but with one whose action is antibacterial. As Ehrlich (Croonian Lect. 1900) notes, Kitt had a precisely similar experience with symptomatic anthrax. It will be noted that an analogous observation has been made with regard to the precipitins. This would appear to be due to the complexity of bacteriolysins and precipitins as compared with antitoxins, where no such differences have been noted. It may be supposed that in reacting to a highly complex body the organism impresses more of its own character upon the antibody which it evolves.

Depending upon *the age of an animal*, its blood corpuscles will behave differently both to natural and artificial haemolysins. Thus Camus and Gley (1899, p. 779) found the normal haemolysin of eel serum to be markedly resisted by the corpuscles of young rabbits, and Delezenne (x. 1900, p. 702) found the corpuscles of young dogs to be much more resistant to the artificial haemolysins (for dog) than those of adult animals. We shall see that the precipitins also give different reactions with foetal and adult blood.

The bacteriolysins to which reference has been made in the preceding pages constitute a class of cytotoxins essentially similar to the haemolysins. They act upon bacteria. Their presence in normal blood and various body fluids was demonstrated by Nuttall (1888); whereas artificial or specific bacteriolysins were first demonstrated by Pfeiffer in animals immunified with the cholera vibrio. The work done on the bacteriolysins has been very extensive, but it is impossible to enter into the subject here at all in detail. It will suffice to say that the specific bacteriolysins behave essentially as do the specific haemolysins, and that they possess the same constitution. They are composed of a labile complement (destroyed at 55° C., as I first showed for normal bacteriolysins) and of a stable immune-body. The latter is capable of conferring immunity, as was first shown by Fraenkel and Sobernheim (1894, p. 154), on heated anti-cholera serum which had lost its bactericidal properties. Wassermann (3, I. 1901) was able to neutralize the complement by means of anti-complement, the latter being specific and only capable of binding one sort of complement. As in haemolytic sera, agglutinins may be present together with lysins. Pfeiffer observed the agglutinin to disappear in stored immune-sera (of goat and dog treated with *B. typhosus*) and the immune-body to persist therein. Mertens (13, VI. 1901), who cites this unpublished observation of Pfeiffer, was able to confirm it on cholera immune-serum 5 years old. The existence of immune-bodies in anti-microbic sera has been also proved by Bordet and Gengou (v. 1901) in experiments upon the blood of animals treated with the germs of plague, typhoid, anthrax, rouget du porc, and with Proteus vulgaris. Immune-bodies were found in the serum of typhoid convalescents.

The occurrence of substances corresponding to immune-bodies in normal sera has been denied, although Malvoz (25, VIII. 1902) has found normal adult dog serum to act like immune-serum towards *B. anthracis*, that is, when inactivated by heat its "immune-body" combined with the bacilli, and rendered them susceptible to the bactericidal action of the complement in the serum of other animals, such as the rabbit, guinea-pig and rat. It is worthy of note in this connection, that the adult dog is most refractory to anthrax infection. The immune-body is absent in young dogs, as also in the guinea-pig, ox, and rat, all of which are susceptible, whereas it may or may not be present in the relatively resistant rabbit.

The rapid decomposition of bodies of animals which have died from the effects of snake venom is due to the loss of bactericidal power of the blood, as has been shown by Welch and Ewing, and by Flexner and Noguchi (1902), the venom combining with the serum complement of bacteriolytic serum. Antivenin (Calmette) neutralized both the bacteriolytic and haemolytic action of venom *in vitro*.

Eisenberg (v. 1902) cites Kraus and Clairmont as finding that heated bacteria are not dissolved, although, as Bail and Wilde found, they are capable of absorbing the bacteriolysin.

Bacterial Haemolysins.

The bacterial haemolysins have received a considerable amount of attention. Ehrlich showed that tetanus toxin is haemolytic, Madsen finding that it contained a toxin which produced convulsions, a tetanospasmin, as distinguished from the tetanolysin. Kraus and Clairmont (17, x. 1901) found tetanus toxin, as also the products of *Staphylococcus pyogenes*, *Streptococcus*, *Vibrios*, and putrefactive bacteria, to be haemolytic. Neisser and Wechsberg, as also Bulloch and Hunter, have studied the haemolytic action of *Bacillus pyocyaneus* and of the *Staphylococcus*; Levy, Castellani, and Lubenau of *B. typhosus*, *B. dysenteriae*, and *Micrococcus tetragenus*. Madsen found tetanolysin very unstable; Kraus and Clairmont found the haemolysin to be destroyed after 15 minutes' exposure at 60° C. Neisser and Wechsberg found staphylolysin to be injured at 48°, and destroyed at 58° C. in 20 minutes. That the bacteriohaemolysins possess a constitution similar to toxins was indicated by Ehrlich and Madsen, who found that they could neutralize them by means of anti-haemolysin. They moreover found that normal horse serum possessed anti-haemolytic action. They were unable to establish any relation between the antitoxic value of a serum and its haemolytic power. The conclusion reached as the result of comparative experiments is that the bacterio-haemolysins and anti-bacterio-haemolysins are specific in character.

The Complement.

As I have stated elsewhere, cytolytic sera are inactivated by being heated to 55° C., this being due to the destruction of the complement. A normal serum thus treated cannot be reactivated, whereas an immune-serum can. The fermentlike body, or complement, is not specific, for the reason that complement can be supplied from another animal which has not been treated. (The toxophorous group of the toxin molecule is also regarded as non-specific by Ehrlich.) Some authors consider that there is but one complement in cytolytic sera, this being the alexin of some writers, whereas Ehrlich and his school claim that any serum contains a multiplicity of complements. A number of points in connection with complements have been considered in the preceding pages, and it is not my object to enter at all fully into a consideration of them here. It will suffice to mention a few facts indicative of their importance.

The amount of complement in serum appears to vary considerably. Thus Weichardt (25, XI. 1901, p. 834), using normal rabbit sera to reactivate spermotoxic guinea-pig serum, found that only 5 out of 11 rabbits' sera contained complement suitable for reactivation. Serum obtained from a man contained complement at one time, none when tested after a period of 8 weeks. Sweet (XII. 1902) has collected the data contained in the literature relating to the variation in the complement-content of cytolytic sera. A decrease in complement was observed by Abbott and Bergey (1902) in the serum of rabbits treated with ox corpuscles, the animals also receiving alcohol. Ehrlich and Morgenroth (1900) found a decrease in complement in rabbits which had received a dose of phosphorus sufficient to kill them after the lapse of 3 days ; the haemolytic power of their serum had disappeared on the second day in a manner corresponding to what is seen in sera inactivated by heat. Metalnikoff (1900) observed a decrease of complement in a rabbit treated with guinea-pig spermatozoa, an enormous abscess having formed, which subsequently burst and healed. When fresh guinea-pig serum was added to this rabbit's serum it was spermotoxic, not otherwise. On the other hand, Nolf (1900) claimed to observe an increase in complement in animals treated with fowl serum ; Müller (1901) made observations which show that the complement-content may be increased by injecting peptone solution, bouillon, aleuronat solution, in other words indifferent fluids. Sweet, from whom I have made these citations, studied the question in rabbits immunified with ox corpuscles. He found that he could increase the complement-content by the injection of substances having a positive chemotactic action on leucocytes, using *Staphylococcus pyogenes aureus*, sterile oil of turpentine, sterile aleuronat suspension. He found the haemolytic complement in a free state in the blood plasma, and in the serous part of an exudate, not being contained in leucocytes, nor being set free by the process of coagulation.

The relation of complement in a serum containing blood corpuscles subject to the action of snake venom has been very clearly brought out by Flexner and Noguchi (1902). These authors experimented with the haemolysins contained in the venoms of *Crotalus adamanteus* (rattlesnake), *Ancistrodon piscivorus* (water-moccasin), *Naja tripudians* (cobra), and *Ancistrodon contortrix* (copperhead). They found that when they had thoroughly washed the blood corpuscles of the dog, sheep, ox, pig, rabbit, and guinea-pig, that the corpuscles were unaffected by contact with these venoms. On the other hand if the corpuscles were bathed in their respective complement-containing sera, haemolysis took place promptly upon the addition of the venoms. An excess of serum increased the haemolytic power. They found :

Venom intermediary body, added to blood corpuscles, had no effect

„ „ „   „   „ „ „   „   and serum complement gave
haemolysis.

Serum complement alone   „ „ „   „   had no effect.

This result is comparable to what has been observed with regard to the non-haemolytic action of blood serum from normal animals upon the washed corpuscles of certain other species, and directly supports a contention of Friedenthal's (mentioned later), that tests made with normal sera upon the washed corpuscles of other animals do not afford a means of studying blood relationships with haemolysins.

The importance of a further study of the complement is well shown by the investigations of Moro (31, x. 1901) upon serum, and its bactericidal properties under different conditions. Using the methods devised by me (1888), but expressing the bactericidal effect in terms of per cent. of bacteria destroyed by contact with the serum, he found that fresh

Placental blood serum killed			$58 \cdot 9\,^0/_0$
Older children's	„	„	$46 \cdot 3\,^0/_0$
Bottle-fed infant's	„	„	$33 \cdot 4\,^0/_0$
Breast-fed	„	„	„	$77 \cdot 0\,^0/_0$

An experiment conducted upon a single infant gave :

Placental blood of mother killed	$56 \cdot 0\,^0/_0$
Breast-fed infant 2 weeks old, its serum killed	$72 \cdot 9\,^0/_0$
Serum of the same child after bottle-feeding for 2 weeks killed	$40 \cdot 7\,^0/_0$.

The blood serum of a breast-fed child had more haemolytic action on rabbit corpuscles than did that of a bottle-fed infant's; even the sickliest breast-fed infant had more bactericidal substance in its serum than the healthiest bottle-fed infant. The serum of a new-born infant only killed 59 $^0/_0$ of the bacteria.

That a fundamental interest attaches to the ferment-like complements in relation to normal physiological processes is moreover very clear from a suggestive paper by Wassermann (1, I. 1903), whose investigations were stimulated by those of Moro just quoted. It is a well-established fact that breast-fed infants thrive as a rule better than those fed upon cows' milk. Heubner, using Rubner's method, has studied the process of assimilation in such children comparatively, and found that, reckoned in calories, the child receiving mother's milk showed the most growth; in other words, that homologous mother's milk achieves as much with a few calories as does cow's milk with a very large number of calories. The infant fed on cow's milk has to expend energy in the form of glandular and digestive activity to assimilate the heterologous milk. That there is such an increased activity is evident from the following.

If heterologous food-stuffs, such as goat serum, etc., are injected, say into a guinea-pig's peritoneum, the peritoneal exudation is capable, soon after, of destroying and dissolving large numbers of bacteria (*B. typhosus*), whereas the contrary is the case in a normal animal. The peritoneal exudation in the serum-treated guinea-pig is rich in digestive complements which have appeared for the purpose of acting upon the serum injected, and it is these digestive substances (complements) which destroy the bacteria.

If we introduce an homologous serum, instead of the foregoing, that is guinea-pig serum into the peritoneum of another guinea-pig, the bacteria subsequently introduced into the peritoneum are not destroyed, no complements or ferments being required for the assimilation of the homologous serum.

We have noted above that Moro found the serum of bottle-fed infants markedly less bactericidal than that of breast-fed infants. He was unable to find complements in milk. The conclusion appears therefore justified, that the reason for there being less complement in the serum of the bottle-fed infant is to be traced to the complement being used up in the process of assimilating the cow's milk.

The conclusion reached above appears all the more justified when we view the results obtained in the study of other antibodies *in corpore*, such as the fall in the amount of antitoxin or of precipitin in a serum when a substance capable of being acted upon by the antibodies is introduced into the economy.

The treatment of animals for the production of Specific Haemolysins.

As it is not my object to enter at all fully into this subject, it will suffice to cite the methods pursued by but a few observers. Bordet (x. 1898, p. 692) injected 10 c.c. of defibrinated rabbit blood intraperitoneally into guinea-pigs and after 5–6 injections noted the presence of haemolysins. Deutsch (15, v. 1901, p. 662) defibrinates the blood (say human) he wishes to inject, allows it to settle 24 hours in the ice chest, then removes the serum and injects 10 c.c. of corpuscles subcutaneously into rabbits, a second and third injection being made at intervals of 7 days. He bleeds the rabbits for anti-serum 21 days after the first injection. He says that haemolysins, agglutinins and even precipitins (?) appear one week after the first injection, as with the antibodies for *B. typhosus* (Deutsch, 1899).

That injections of urine may lead to the formation of haemolysins was discovered by Schattenfroh (30, I. 1901). He injected rabbits subcutaneously with human, goat, and horse urine, the animals receiving 120—150 c.c. intervals of 2–3 days elapsing between injections. The serum of the rabbits treated with human and goat urine acquired powerfully haemolytic and agglutinating properties for the red blood corpuscles of the animals which had yielded the urine. This special action was very marked in the serum of the human-urine-treated rabbits; normal rabbit sera, or those of rabbits treated with other urines, possessing little or no globulicidal effect on human corpuscles. The serum of a rabbit treated with horse urine gave no marked result. The serum of the rabbits treated with goat urine did not contain precipitin, nor anti-complement for goat serum, this being in marked contrast to the serum of rabbits treated either with active or inactive goat serum. Schattenfroh suggests that the method of urine injection may be of use in the production of specific haemolysins for blood-testing by the method recommended by Deutsch.

Ruffer and Crendiropoulo (24, I. 1903) have confirmed this observation, finding that the injection of human urine two or three times into rabbits produced a haemolytic serum for human corpuscles, the haemolysin acting however slightly also upon the corpuscles of the guinea-pig.

As has been found for the precipitins, Metalnikoff (18, IV. 1901, pp. 532, 533) has observed the formation of haemolysins in the serum of rats *fed* for 1 to 8 weeks on horse blood. An exposure to 55° C. inactivated their serum, it being reactivatable through fresh normal serum; it consequently contained immune-body and complement. Similarly rats fed on rabbit blood, and rabbits fed on horse blood, yielded specific haemolysins.

Again, as has been found for the precipitins, Calmette and Breton (1, XII. 1902) observed a decrease in the amount of haemolysin in animals treated for longer periods. They found, however, that when such over-treated animals are allowed to rest for some months they produce a much more active antiserum after but two injections.

SECTION IV.

THE ACTION OF DIFFERENT SERA UPON THE BLOOD CORPUSCLES OF CERTAIN ANIMALS *IN VITRO* AND *IN CORPORE.*

I. The Effects of Normal Haemolysins.

Primates.

Human Serum: There is no evidence that the sera of any human races are haemolytic for the blood corpuscles of other races of man, in proof of which we have a large experience with regard to transfusion. Transfusion of the blood from a member of one race into that of another has not been followed by any ill effects.

Friedenthal (1900, p. 505) transfused 25 c.c. of defibrinated human blood into a chimpanzee, the animal remaining healthy, its urine showing no trace of haemoglobin. He also transfused 10 to 20 c.c. of defibrinated human blood into *Macacus sinicus* and *M. cynomolgus* without ill effects, a slight transitory haemoglobinuria being observed in consequence, which was attributed to the haemoglobin in the defibrinated blood injected, the blood corpuscles contained therein having undergone haemolysis.

The effects of human serum *in vitro* have been studied by Friedenthal, upon the blood corpuscles of other Primates, these corpuscles not being washed (see p. 34). He found the corpuscles of the Ourang-outang and Gibbon to remain unaffected, whereas the corpuscles of *Macacus sinicus, M. cynomolgus, Rhesus nemestrinus* (Old World Monkeys), *Pithesciurus sciureus,* and *Ateles ater* were haemolysed. Human serum also haemolysed the corpuscles of *Lemur varius.* He does not state that he observed any differences with regard to the rate at which haemolysis took place, a matter worth noting when we consider the results I have obtained with precipitins.

In his Huxley Lecture, Welch (11, x. 1902, p. 1108) refers to an unpublished investigation by H. T. Marshall (working under Ehrlich and Morgenroth) upon the receptors of the red blood corpuscles of man and of two species of monkey. It appears that Marshall has found human and monkey corpuscles to have a large number of receptors in common, but also ones peculiar to each. Welch points out that this is in harmony with my published results upon the precipitins in relation to the phylogenetic relationships between animals.

The action of human serum upon the blood of other animals than Primates has been considerably studied; it will suffice however if I cite but a few data. Cl. Bernard[1] stated that 10 c.c. of human blood was toxic for the rabbit. Friedenthal, experimenting with unwashed corpuscles of the eel, frog, snake, pigeon, fowl, night heron, horse, pig, ox, rabbit, guinea-pig, dog, cat, hedgehog, *in vitro*, has found all of these to be haemolysed by human serum.

Macacus Serum was found to haemolyse some human blood corpuscles, not others, there being evidently differences in the resisting power of the corpuscles of different individuals (Friedenthal).

Few as these observations are, they are certainly in remarkable accord with what I have observed upon the precipitin reactions.

Insectivora.

Erinaceus Serum haemolyses the blood corpuscles of the cat and rabbit (Friedenthal).

Carnivora.

Canine Sera : The older experiments demonstrated that transfusion could be practised with impunity between different Canidae : the dog, wolf, and fox. On the other hand Cl. Bernard (*loc. cit.*) found rabbits to be killed by the injection of 10 c.c. of dog serum. Friedenthal and Lewandowsky (1899, p. 532) killed rabbits of 1500 grammes weight usually in a few minutes by injecting 7—14 c.c. of dog's serum ; the lethal dose required being larger when the rabbits received subcutaneous injections instead of intraperitoneal injections.

Experiments with dog serum *in vitro*, have been mostly conducted upon washed corpuscles, that is corpuscles from which the serum had been removed, if we except those of older date. Ehrlich and Morgenroth (1899, ii. p. 13 repr.) found dog serum at times to powerfully

---

[1] "Leçons sur les propriétés physiologiques et les altérations pathologiques des liquides de l'organisme," ii. p. 459 ; cited by Friedenthal and Lewandowsky (1899, p. 532).

haemolyse *cat* corpuscles, at other times it exerted no effect, whilst London (1902, p. 56) states they are not haemolysed. Daremberg (1891) found dog serum to haemolyse *human* corpuscles in 9 minutes. The corpuscles of the *rabbit* and *guinea-pig* are haemolysed in $2\frac{1}{2}$ minutes (Daremberg); Buchner (1893, p. 122), Ehrlich and Morgenroth (III. 1900), Flexner and Noguchi (1902) state that rabbit corpuscles are rapidly haemolysed. Buchner stated that the corpuscles of the guinea-pig were more slowly haemolysed than those of the rabbit, but this does not appear to have been noted by Ehrlich and Morgenroth, and Flexner and Noguchi. Rat corpuscles are haemolysed (Ehrlich and Morgenroth).

The blood corpuscles of the Ungulata are haemolysed by dog serum, the bloods tested being those of the horse, pig, and sheep (haemolysed in 1 minute, Daremberg), ox (haemolysed in 9 minutes, Daremberg), sheep and goat (Ehrlich and Morgenroth).

Feline Sera: Friedenthal (1900) cites an experiment in which he transfused the blood of the domestic Angora cat into the vessels of *Felis ocelot*, and from the latter into the cat. He chose animals of similar size, put their blood circulations into communication, and allowed their bloods to mix until about one-half of their blood volume had been exchanged. The cat remained perfectly well after the operation, the ocelot unfortunately died from the effects of prolonged narcosis. He pertinently remarks to these experiments " Also getrennte Familien, gesondertes Blut." Friedenthal and Lewandowsky found the lethal dose of cat serum for rabbits to be the same as that of the dog (see above).

Experiments with cat serum *in vitro* have shown that it does not haemolyse the blood corpuscles of *Felis ocelot* and *Felis jaguarundi*, and *vice versâ* (Friedenthal, 1900), nor of other cats (Friedenthal and Lewandowsky, 1899). Cat serum was found to haemolyse the corpuscles of the rabbit (F. and L. 1899, London, 1902, p. 56), hedgehog, and frog (Friedenthal, 1900).

Rodentia.

Rabbit Serum: Transfusion effected between two rabbits or between the rabbit and hare is not followed by ill effects (Friedenthal, 1900, and Friedenthal and Lewandowsky, 1899).

The experiments with rabbit serum *in vitro* are not concordant, apparently for the reason that some observers experimented with washed, others with unwashed corpuscles. Working with unwashed corpuscles, Friedenthal found rabbit serum to rapidly haemolyse the corpuscles of the horse, pig, ox, monkey and man, whereas it exerted no effect on hare's corpuscles. Other observers, experimenting with washed corpuscles, state that rabbit serum does not haemolyse the corpuscles of the *cat* (London, 1902, p. 56), that its action is very variable upon the corpuscles of the guinea-pig (Ehrlich and Morgenroth, 1899, ii, p. 13 repr.), and only slight upon dog's corpuscles (Flexner and Noguchi, 1902).

Guinea-pig Serum: Experiments *in vitro*, made apparently under comparable conditions with washed corpuscles, have shown guinea-pig serum to exert practically no haemolytic action on rabbit corpuscles (Bordet, x. 1898, p. 692; Flexner and Noguchi, 1902); to rapidly haemolyse those of the rat and mouse (Bordet, 1898); to haemolyse those of the hedgehog (Friedenthal, 1900); and to exert a feeble action on the amoeboid cells of the crayfish (Szczawinski, 28, xi. 1902), the artificial haemolysin acted more powerfully.

Rat Serum: Injections of rat blood into mice and *vice versâ* are not injurious (Friedenthal, 1900). Metalnikoff (18, iv. 1901, p. 532) experimenting with the serum of the white rat found that it did not haemolyse nor agglutinate the corpuscles of the horse. I have not found further data with regard to the action of these sera.

Ungulata.

Pig Serum: The haemolytic action of pig serum upon *human* corpuscles was already observed by Landois in transfusion experiments upon man. Friedenthal (1900) also found it to haemolyse the corpuscles of the ox, horse, dog, cat and rabbit, besides those of man.

Uhlenhuth (1897) found 12 c.c. of pig serum to be fatal to 1 kilo of rabbit when injected intravenously; he cites Weiss (1896) as having found 35 c.c. to be the fatal dose for rabbits (subcutaneous injection?). Pig serum produced infiltration and necrosis when subcutaneously injected into guinea-pigs.

Ox Serum was found to be injurious to man in the transfusion experiments of Landois, because it haemolysed human corpuscles. Friedenthal found it to haemolyse the corpuscles of the pig, horse, rabbit, dog, cat, and also man. The lethal dose by intravenous injection in rabbits was found to be 6 c.c. per kilo of rabbit by Uhlenhuth. According to Rumno and Bordoni, and Weiss (1896), the lethal dose of ox serum is 8 c.c., whilst Guinard and Dunarest give 9 c.c. as fatal for rabbits, these authors being cited by Uhlenhuth. Friedenthal and Lewandowsky found the lethal dose of calf serum for rabbits to be the same as that of dog serum (see above, p. 34). Subcutaneously injected into guinea-pigs it produced the same effects as pig serum.

Gruber (3, XII. 1901) dwells on the importance of concentration in haemolytic experiments, stating that ox serum, even when highly diluted, acts on rabbit and guinea-pig corpuscles, whilst *it only acts on sheep corpuscles when concentrated.* This observation, we see, shows the converse of what has been observed with the precipitins, which act on higher dilutions of related bloods.

Sheep Serum was also found injurious to man in Landois' transfusion experiments. According to Uhlenhuth (1897) the lethal dose per kilo of rabbit is 11 c.c., Rumno and Bordoni give it at 12 c.c., Weiss at 20 c.c. It produces the same effects as pig serum when subcutaneously injected into guinea-pigs, according to Uhlenhuth, who cites the authors named.

Goat Serum was found to exert a very slight haemolytic action or no action upon the corpuscles of the sheep by Ehrlich and Morgenroth (1899).

Equine Serum : The transfusion of horse serum into man was found to be injurious to man in the experiments of Landois. On the other hand he found transfusions between the horse and donkey not to be injurious. Friedenthal (1900, p. 503) found the sera of the horse and donkey without action on the corpuscles of these species, whereas they haemolysed those of man, ox, sheep, guinea-pig, and rabbit. According to Morgenroth and Sachs (7, VII. 1902, p. 633) the haemolytic action of horse serum varies greatly, compared to the constant haemolytic power of other sera, such as those of the dog and goat. The haemolysin of the horse appears to be very labile, for they often observed it to be active for the corpuscles of the rabbit and guinea-pig when fresh, but inactive, or but slightly active, after being kept 24 hours on ice. Uhlenhuth cites Weiss as giving the lethal dose of horse serum at 44 c.c., and Guinard and Dunarest at 324 c.c., for rabbits, this great variation being in accord with the statement of Morgenroth and Sachs just cited. Friedenthal and Lewandowsky (1899) found it necessary to inject 70 c.c. of horse serum intravenously into a rabbit of 1300 g. to kill it. Uhlenhuth (1897, p. 391) found horse serum not to exert effects such as do the sera of man, sheep, pig, rabbit, and ox, when injected subcutaneously into guinea-pigs. (See under pig serum, p. 36.)

Aves.

Experiments conducted *in vitro* by Friedenthal (1900) showed the serum of the fowl to haemolyse the corpuscles of several vertebrates including those of some birds, namely those of *Falco tinnunculus* (Accipitres) and *Nyctocorax* (Ciconiiformes). The serum of the night heron haemolysed the corpuscles of the fowl. The serum of the gull (*Larus argentatus*) haemolysed the corpuscles of the frog.

Reptilia.

The toxicity of snake sera for Mammalia is well-marked. Friedenthal found the intravenous injection of 2 c.c. of the serum of the "Kreuzotter" (presumably *Pelias berus* Merr.) to be fatal to rabbits of average size. Dr Graham-Smith in trying to produce precipitins in rabbits by injecting the serum of *Tropidonotus natrix*, found a similar dose to be lethal for guinea-pigs by intraperitoneal injection. Friedenthal found ·5 c.c. of his snake serum to kill frogs when injected into the posterior lymph-sac. He also found the serum of *Tropidonotus* ("Ringelnatter") to be highly haemolytic.

The agreement between Aves and Reptilia in possessing a highly toxic serum, may be added to the other points of similarity, as to anatomical structure (skeleton, urinary apparatus, scales, eggs, etc.) and the microscopic appearances of their blood, which have led systematists to group both classes under Sauropsida. It will be seen, however, that more striking proof has been obtained by means of the precipitins with regard to the relationship between these classes.

Amphibia.

Frog Serum possesses haemolytic properties for the corpuscles of the dog, guinea-pig and rabbit, although it acts less rapidly than the serum of *Necturus*, whose corpuscles it slightly haemolyses (Flexner and Noguchi, 1902). Friedenthal states that the haemolytic power decreases in starved frogs.

Necturus Serum: The serum of this amphibian quickly haemolyses the corpuscles of the dog, guinea-pig, and rabbit, but slightly affecting those of the frog (Flexner and Noguchi).

Pisces.

Eel Serum: The highly toxic properties of the serum of *Anguilla* were discovered by Mosso (1888, p. 144), who found that ·5 c.c. thereof injected intravenously into a dog weighing 15,206 g., produced death in 7 minutes, even a dose of ·02 c.c. per kilo being fatal. He referred to the toxic body as "ichthiotoxin." The serum of *Muraena* proved also highly toxic, ·66 c.c. killing a dog of 6,160 g. in 5 minutes; ·3 c.c. killing a rabbit of 1,030 g. in 2½ minutes. A toxic serum is also possessed by *Conger myrus* and *C. vulgaris.* Pigeons are susceptible to large doses according to Camus and Gley (1899, p. 785). These authors (24, VII. and X. 1899, p. 786) found the blood corpuscles of *new-born rabbits* to be highly resistant to the haemolytic action of eel serum. They considered this due to a resisting power inherent in the cell, not to the existence of normal anti-haemolysin. (See p. 22.)

The study of the haemolytic effects of eel serum *in vitro*, begun by Mosso, were extended by others. Tchistovitch (25, V. 1899, p. 407) and Camus and Gley (X. 1899, p. 784) found the corpuscles of the fowl very resistant. The latter found the corpuscles of the pigeon very resistant, whilst Tchistovitch states they are susceptible, though almost insoluble. Friedenthal, working with unwashed hedgehog corpuscles, found them haemolysed by eel serum; whereas Camus and Gley state that they resist haemolysis. Camus and Gley state that the corpuscles of the guinea-pig and rabbit are haemolysed by eel serum, whereas those of the Bat, *Bufo vulgaris, Rana temporaria* and *R. esculenta,* and *Testudo graeca* are very resistant. They attribute this resistance to the corpuscles, not to anti-haemolysin. Friedenthal, working with unwashed frog corpuscles, states that they were haemolysed by the eel serum. These somewhat contradictory results may be explained by the fact observed by Ehrlich and Morgenroth (1899, II. p. 13 repr.) that the haemolytic power of eel serum may vary greatly, at times haemolysing powerfully, at other times not at all.

In Friedenthal's experiments, the blood of *Acanthias vulgaris* was rapidly haemolysed by eel serum, and that of *Labrus maculatus* haemolysed teleostian blood. The serum of *A. vulgaris* haemolysed the corpuscles of a bird (*Larus argentatus*), a rat (*Mus decumanus*), as well as that of other Teleostea (*Labrus maculatus* and *Anguilla vulgaris*), and the serum of the shark was "not indifferent" to the corpuscles of other Elasmobranchs, viz. *Raja batis.*

Lower Animals.

Friedenthal was unable to demonstrate the existence of any haemo-lytic action on the part of the blood of Crustacea (*Cancer pagurus*), Oligochaeta (*Arenicola piscatorum*), or Sea-urchin, when these were brought in contact with the blood corpuscles of the gull (*Larus argentatus*) and rat.

II. The Effects of Artificial Haemolysins upon the Red Blood Corpuscles of different Animals in corpore and in vitro.

The demonstration of artificial haemolysins dates from the discovery of Belfanti and Carbone (VIII. 1898) who found that they were formed in the serum of animals treated with a foreign blood. They injected large amounts of rabbit blood intraperitoneally into the horse and found the serum of the horse to become highly toxic to rabbits. Similarly, the serum of a dog, treated with rabbit blood, became toxic for rabbits. Such sera were not toxic for the animals yielding the antisera, very slightly so for other animals than those which yielded the blood injected. Bordet subsequently showed that the toxic action of the serum of treated animals corresponded with its haemolytic power *in vitro* upon the particular species of corpuscles with which the animal had been treated.

Ehrlich and Morgenroth (1899) treated a goat with sheep's corpus-cles and found its serum to become haemolytic for the corpuscles of the sheep, although the animals are so closely allied. They subsequently (VI. 1901, p. 8, repr.) found that a rabbit treated with ox corpuscles developed a serum which was haemolytic for the corpuscles of the goat and sheep, as well as those of the ox, although the haemolysin acted less on the goat corpuscles than on those of the ox. This demonstrated a similarity but not an identity between these corpuscles of Bovidae. Similarly, a rabbit treated with goat corpuscles gave haemolysins for goat and ox corpuscles, acting less on the latter. They explain that the specificity of the immune-body formed in the blood-treated animal is not to be confused with the conception of species held by systematists in botany or zoology. All those elements will be affected by an immune-serum which have receptors corresponding in type to that of the original element injected, and consequently the influence will be most marked where there are most receptors belonging to the common type. This view also applies to the corresponding results obtained with precipitins.

Deutsch (9, VIII. 1900, and later) has suggested the use of artificial haemolysins *in legal medicine*, in the identification of bloods, both fresh and dried. He found that a powerful haemolytic serum dissolved powdered blood completely, the latter being suspended in $9\,^{0}/_{00}$ salt solution. Dried blood to which saline is added brings the haemoglobin of the injured corpuscles into solution, the uninjured corpuscles do not however dissolve even after 24 hours at 37° C. If the dried blood is extracted in normal rabbit serum more haemoglobin goes into solution than with saline, when the proportion added is 1 : 2, whereas the normal serum acts like saline when added in the proportion of 1 : 4. When two samples of the same dry blood are brought into suspension in normal and artificially haemolytic serum respectively, a little phenol or toluol being added, the antiserum brings about complete haemolysis after 24 hours, besides leading to the formation of a precipitum, due to the action of precipitins formed in the blood-treated animal in consequence of the serum which was injected together with the corpuscles. When washed corpuscles alone are injected precipitins are not formed. In view of the specificity of the reactions observed with human blood corpuscles, he considers that the method can be put to use in a practical way. There can, however, be no question but that the precipitins offer many advantages over the haemolysins for such purposes.

In the experiments of Bordet (x. 1898, p. 692), it was found that the serum of a guinea-pig treated with rabbit corpuscles exerted no effect on the corpuscles of the normal guinea-pig, nor upon those of the pigeon. Both normal and treated guinea-pig sera act strongly on the corpuscles of the rat and mouse, the normal serum is however less haemolytic. If 2 c.c. of defibrinated rabbit blood is injected intraperitoneally into the blood-immunified guinea-pig, the rabbit corpuscles are destroyed in 10 minutes. This does not happen in the peritoneum of a normal guinea-pig unless inactivated specific serum (heated to 55° C.) is injected together with the corpuscles. The serum of the treated guinea-pig, injected intravenously into rabbits kills them when 2 c.c. are given. This is in substantial agreement with what has been observed with regard to the specific bacteriolysins, for instance with regard to the cholera germ. Bordet considers the haemolysin (alexine) in the treated guinea-pig serum to be probably identical with the specific bactericidal substance of Pfeiffer.

The study of the relationships amongst animals by means of artificial haemolysins has scarcely been begun, as we see, if we except the work of Ehrlich and Morgenroth above cited, which gives corresponding results to those which have been observed with the precipitins. It remains to be seen whether the haemolysins offer greater advantages in this respect than do the precipitins.

Isolysin and Autolysin.

Ehrlich and Morgenroth (*Ueber Hämolysine*, III. 1900) term the haemolysins formed in the bodies of animals treated with the corpuscles of another species "heterolysins," so as to distinguish them from the "isolysins," which are formed for the corpuscles of the same species. They injected 920 c.c. of goats' blood, obtained from three animals, into a goat. No haemolysis occurred, but after five days or more the treated goat's serum was found to be markedly haemolytic for the blood corpuscles of nine different goats, the tests being conducted *in vitro*.

An "autolysin" is a haemolysin which acts upon the blood corpuscles of the animal which yields the serum. Ehrlich (1901, *Schlussbetracht-ungen*, p. 16, repr.) only succeeded once with Morgenroth in obtaining autolysin, this being formed under the same conditions as the isolysin. He injected the blood of a goat (*a*) into another goat (*b*), and found the serum of goat (*b*) to haemolyse the corpuscles of (*a*). The existence of isolysins and autolysins point therefore to individual differences in animals belonging to the same species.

Besredka (25, x. 1901, p. 788) considers the immune-body in autocy-totoxins to be probably identical with the cytotoxins of another animal when we exclude the complement which is peculiar to each species.

The existence of isolysins and isoagglutinins in disease does not bear upon the immediate subject of this paper. It is however well to mention that Grünbaum (*Brit. Med. Journ.* I. p. 1089, 1900) found the serum of typhoid and scarlatina patients to agglutinate the corpuscles of normal persons, or those affected not with the same but with other diseases. Shattock (*Journ. Pathol. and Bacteriol.* VI. p. 303, 1900) confirmed this in cases of croupous pneumonia, typhoid, erysipelas, acute articular rheumatism. Panichi has observed isoagglutinins in the serum of persons experimentally infected with malaria, after six days, when no parasites could be discovered in their blood. Eisenberg (17, x. 1901) as also Ascoli have found isolysins and isoagglutinins in the blood after crisis from croupous pneumonia, their appearance corresponding with the absorption of the hemorrhagic infiltration. It would therefore seem as if the appearance of these antibodies in the blood depended upon the resorption of blood corpuscles or their constituents. In seven out of eight cases of scarlatina with slight clinical signs of blood destruction, Eisenberg observed these antibodies. He denies that the isolysins and isoagglutinins are specific for the diseases named, but that they simply indicate that corpuscles are being destroyed, or reabsorbed, and consequently that their presence is not of diagnostic importance.

The relation of Cytolysins to Agglutinins.

The evidence in favour of the haemolysins and agglutinins being distinct seems fairly conclusive. It is true that Emmerich, Loew, and Baumgarten consider that agglutination always precedes haemolysis. Baumgarten's view receives some support from the observation of Shibayama, who states that very powerful antisera may haemolyse without agglutinating, Baumgarten claiming that rapid haemolysis may mask the phenomenon of agglutination.

Most observers agree with Ehrlich and Morgenroth in considering the antibodies distinct. These authors, as also Bordet and others, have found the haemolytic action to be prevented by heating haemolytic serum to 55° C., the agglutinins remaining unaffected. Mertens, as we have seen (p. 27), has shown the non-identity of agglutinin and immune-body. In the following pages evidence is given as to the non-identity of agglutinins and precipitins. Jacoby determined that ricin-toxin and agglutinin are different. We see that there is every evidence as to a multitude and diversity of antibodies in sera. Further, whereas lysins pass through a collodium film by osmosis, Gengou claims that agglutinins do not. The bacteriolysins in a serum may be increased by sodium carbonate injection, but this does not affect the amount of agglutinin present. It has been repeatedly noted that no definite relation exists between the agglutinative and bacteriolytic power of typhoid sera.

The studies of Mitchell and Stewart (1897) have shown that the phenomenon of agglutination precedes haemolysis by snake venoms, this being comparable to what we see in Pfeiffer's reaction, where the agglutination of cholera spirilla precedes lysis by the immune serum. In snake venom, Flexner and Noguchi (1902) have also found that they are able to separate the lysin from the agglutinin. Agglutination by venom did not affect haemolysis, but rapid haemolysis limited or prevented agglutination. On adding ricin, which agglutinates blood corpuscles powerfully, and allowing it to act for less than 30 minutes on the corpuscles, it was found that a further addition of venom effected haemolysis within the usual time. On the other hand, when ricin had acted for two or more hours, solution took place on the addition of venom, but the corpuscular stroma remained at the bottom of the tube in the form of a white conglutinated mass. It appears therefore that the stroma undergoes a form of coagulation under the influence of agglutinins, the venom releasing the haemoglobin. No interaction took place between venom and ricin. The haemolysin in venom resisted an exposure of 30 minutes at 75—80° C., being slightly reduced in power after 15 minutes at 100° C. Venom agglutinins exposed for 30 minutes at 75—80° C. were destroyed.

Dubois (25, IX. 1902) considers the agglutinins and immune-bodies of artificially haemolytic sera to be distinct, although they behave similarly in their resistance to heat, putrefaction, light, etc. He states that he was able to separate them by repeatedly injecting rabbits, (*a*) with intact washed fowl's corpuscles, and (*b*) with the same corpuscles previously heated to 115° C. for 15 minutes. The fresh serum of the rabbit receiving treatment (*a*), agglutinated and haemolysed fowl's corpuscles, whereas the second (*b*) did not haemolyse, although it contained some agglutinin. Dubois cites Defalle as having found that certain germs (*B. typhosus, B. mycoides, B. mesentericus*, etc.) when heated to 115° C., and injected into animals, produced an agglutinative serum not containing immune-body, whereas the latter is readily obtained by treating animals with unheated germs.

SECTION V.

AGGLUTININS AND ANTIAGGLUTININS.

It is not my wish to dwell at any length upon the agglutinins and antiagglutinins. They are frequent in haemolytic sera, the serum of one animal generally agglutinating the blood corpuscles of another; thus fowl serum has a powerful agglutinating action upon the corpuscles of the rat and rabbit (as observed by Bordet, x. 1898), this action depending upon the presence of normal agglutinins in the serum of the fowl. Powerful agglutinins are also present in the venoms of serpents. The agglutinins may or may not be specific. Where artificially produced they belong to the latter category, as also when they develope in consequence of disease. If an animal is immunified against the blood of another, it developes agglutinins which act more or less powerfully on the foreign corpuscles. The agglutinins may coexist with other antibodies in a given serum, the complexity of serum in this respect appearing to be unlimited. There may moreover be several agglutinins present at once in a serum. Thus Myers (14, vii. 1900) found that a precipitating antiserum for sheep's globulin, obtained from a rabbit, agglutinated the washed blood corpuscles of the sheep as well as those of the fowl. That the agglutinins which acted on the two kinds of corpuscles were different was shown by adding one kind of corpuscles to the antiserum until it ceased to agglutinate them, it being then found that it was still capable of agglutinating the other kind; the same result following a reversal of the order, in which the corpuscles were subjected to the action of the serum.

The agglutinins have however received chief attention from bacteriologists, their presence in the serum in certain infective diseases, and after recovery, being as is well known a valuable aid in diagnosis. The agglutination test is applicable moreover to a large number of germs against which animals may be immunified. As in the case of the haemo-agglutinins, so with the bacterio-agglutinins, they may be natural or artificial, the latter are more or less markedly specific. Since the discovery of the bacterio-agglutinins, the principles of whose action were worked out by Gruber and Durham, and practically applied by Grünbaum, as also by Widal, they have been used both for the identi-fication of micro-organisms and diagnosis of disease. In the first case, a particular germ, *B. typhosus, Micrococcus melitensis, B. coli*, etc., has been subjected to the action of graded dilutions of a specific antiserum ; in the second case, a well-identified germ has been exposed to the action of the serum from a patient suffering from supposed typhoid fever, and the like. At first it was thought that the reaction of agglu-tination was strictly specific, as in the case of the precipitins, but time has proved that they are but relatively specific. Their mode of action has been the subject of much controversy, the literature relating thereto being well-nigh inexhaustible. My object in mentioning agglutinins at such length is to show that they possess many properties in common with other antibodies, but more especially with those which are the immediate subject of this paper, the precipitins and haemolysins. For the reason that the agglutinins have been most studied in their relation to bacteria I have drawn my material chiefly from this source, for there can be no question but that the haemo-agglutinins and bacterio-agglutinins are essentially similar. They differ as to their value in one respect, the agglutinins are of great practical use in the identification of bacteria, but of little use in the comparative study of blood, zoologically or medico-legally.

Agglutinins of some kind are always present in normal blood, and it would appear that they reside chiefly in the blood, there being less present in the organs. Almost all observers, who have studied the question, have found agglutinins in the lymphoid and blood-forming organs. Gruber (1896), observing that the polynuclear leucocytes took up injured micro-organisms, when cholera and typhoid germs were injected intraperitoneally into animals, concluded that the agglutinins were formed within the leucocytes. Courmont (1897) tested the blood and organs of typhoid cadavers, and almost invariably found the blood to contain most agglutinin. Similar results were obtained by Arloing (1898) with animals infected with *Pneumobacillus bovis*. Fodor and Rigler (1898) found agglutinins made their first appearance in the serum of guinea-pigs rendered immune to typhoid bacilli. Rath (29, IV. 1899), experimenting on rabbits with the same germs, concluded that the spleen (which had been extirpated in some animals), lymph glands and bone-marrow exert no demonstrable effect upon agglutinin-formation. van Emden (1899), experimenting with *B. aerogenes* on rabbits, found the agglutinins chiefly in the lymphoid tissue, less being present in the liver and kidneys. Deutsch (IX. 1899, p. 720) found only traces of typhoid agglutinins in the liver, kidneys, suprarenals, variable quantities in the spleen, bone-marrow and lymph glands, but never as much as in the serum. He considers that they originate in the blood, and this view is supported by a number of observers (Deutsch, 25, VII. 1900). Ruffer and Crendiropoulo (5, IV. 1902), studying the normal agglutinins in the rabbit and guinea-pig, observed (what was already known) that the serum possessed agglutinating power. A solution of red blood corpuscles, previously washed, did not agglutinate, neither did similar solutions of formed elements from immunified animals exert an agglutinating action. They on the other hand bring forward evidence to show that agglutinins may be formed in the leucocytes, for an extract of leucocytes from an immune animal had greater agglutinating power than did the same animal's serum. These observations are certainly suggestive. Agglutinins are transmitted to the foetus in utero. They have also been found in humor aqueus.

The formation of agglutinins in animals subject to immunization, follows the same laws as does that of other antibodies. As Deutsch (p. 720) says, they gradually appear (3—4th day), increase in quantity for a time (10—13 days) and then gradually decrease in amount, there being of course individual differences amongst animals in this respect.

Pick (1902, pp. 30—34) found that when he mixed typhoid (from horse) and cholera (from goat) antisera, they acted independently of one another when they were brought in contact with these germs. This shows, as in Myers's experiment above noted, that there is no interference exerted between the antibodies. The same thing has been observed by others who have immunified animals with two species of micro-organisms. Exactly comparable are the experiments made by Uhlenhuth (11, IX. 1902) with different precipitating antisera, and (in a measure) my earlier experiments (1, VII. 1901), in which I however reversed the order, adding one antiserum to a mixture of bloods in solution.

The influence of salts upon agglutination is in a sense comparable to their action upon the precipitins. Bordet (III. 1899, p. 236), who first studied the question, found that after cholera germs had been agglutinated in the ordinary way and were resuspended in saline solution, he could reagglutinate them, but that they were not reagglutinated if resuspended in water alone. Joos (1901) found that antityphoid serum did not agglutinate *B. typhosus* in the absence of salts. For agglutination to take place he considers salts as necessary as the agglutinin and agglutinatable substance. He believes that salts play an *active* part in the process, combining with both the other substances, a conception which is contrary to Bordet's, that the absence of salts offers only a physical impediment to agglutination. Joos (p. 429) found that but a trace of salt added to washed bacilli which had been impregnated with typhoid agglutinin immediately led to agglutination[1]. These results have been confirmed in the main by Friedberger (IX. 1901), who does not however consider that salts act chemically, for he found agglutination to take place in the presence of grape-sugar, asparagin, etc.

The agreement is fairly general as to the chemical nature of the reaction, both components being used up in the process, as was originally observed by Gruber. In this they are similar to the other antibodies[2]. How stable the union is has not been made clear. Landsteiner (1901) finds that when red corpuscles have been agglutinated with serum or abrin, the agglutinin may be again obtained, after they have been washed, by simply warming them in salt solution. This would scarcely point to a well-established union.

Widal (*Semaine Médicale*, 1897, No. 5, cited by Kraus) found that typhoid bacilli which had been killed by exposure to a temperature of 56° C. were still subject to agglutination; a fact which Kraus (12, VIII. 1897) confirmed both for the typhoid bacillus and that of cholera. The reaction does not therefore depend upon the germs being alive.

Agglutinins are more resistant to heat than are haemolysins or bacteriolysins, withstanding a temperature of 55°C. and over; for instance, Laveran and Mesnil (25, IX. 1901, p. 695) found specific Trypanosoma-agglutinin to be inactivated at 63—65° C. ($\frac{1}{2}$ hour), and Bordet (III. 1899, p. 243) found haemagglutinins (of rabbit treated with fowl's corpuscles) to be destroyed at 70° C. in half an hour. Like the precipitins, they are stable bodies, retaining their agglutinating power at times for many months *in vitro*. Pick (1902, p. 21) cites Widal and Sicard (1897) and Winterberg (1899) as having found the agglutinins to be precipitated with the globulin fraction from serum. He has confirmed this, working with typhoid-agglutinin. The latter was present in the pseudoglobulin fraction, only traces (presumably impurities) being found with euglobulin, when horse serum was examined. On the other hand in immune goat, rabbit, and guinea-pig serum, the agglutinin was almost entirely confined to the euglobulin fraction. Similarly cholera-agglutinin was almost entirely confined to the euglobulin fraction, both in immune horse and goat serum. Here again we find a similarity with other antibodies (see precipitins, antitoxins, etc.).

The significance of the agglutinins is but imperfectly understood. They may persist in the serum of persons who have had typhoid fever for months and even years. In some cases this would appear to be referable to a retention of these

---

[1] See further Bordet (III. 1899).
[2] See the investigation of Eisenberg and Volk (2, v. 1902).

bacilli in some part of the body. Although anti-typhoid sera, to confine myself to these, usually show marked agglutinating power for these germs, when potent, it has at times been observed that such sera were strongly protective and but slightly agglutinative. It has been claimed that such agglutinins may be accepted as an index of immunity, but this seems scarcely justified in view of the fact that strongly agglutinative sera have been obtained from persons dying from typhoid fever. Van Emden (1899, p. 31) has made a similar observation upon rabbits dying of infection with *B. aerogenes*. I am inclined to believe that the agglutinins represent but a group of the many antibodies with which the organism reacts to infection or the like, that they are immune substances directed against some of many injurious factors. It follows from what we know of the individual differences which exist in animals that the reaction may take place more along some lines than upon others, and this would explain the presence of an excess of one immune substance, another, perhaps more vital one, being deficient. The use to the economy of the agglutinins under normal and pathological conditions is certainly obscure as yet.

The occurrence of *Isoagglutinins* together with isolysins in human blood in disease has already been referred to on p. 42.

The existence of *Agglutinoids*, analogous to toxoids and precipitoids, is indicated by the observations of Eisenberg and Volk (6, VIII. 1902), there being apparently a stable combining group, and a labile precipitating group, in agglutinating sera. They found inactivated, as also old antisera, to retard the action of fresh antiserum. Bail (*Arch. f. Hyg.* 1902, cited by Kraus and Pirquet) also believes in the existence of two groups in such antisera. (Compare with precipitins.)

Antiagglutinins have been obtained through the treatment of animals with haemagglutinins. Venom-agglutinins may be neutralized by antivenene (Kanthack), ricin-agglutinins by anti-ricin serum (Ehrlich). Ford (3, VII. 1902, p. 367), taking either the normal agglutinin of rabbits, or the artificial agglutinin developed in rabbits through fowl-blood injection, injected it into fowls, and obtained antiagglutinin. It seems however premature for him to conclude that the immune-agglutinins are only quantitatively different from normal agglutinins. He states that Wassermann has been unable to obtain an antiagglutinin for the agglutinin of *B. pyocyaneus* (formed in pyocyaneus-treated goats), and that Kraus and Eisenberg had been equally unsuccessful in obtaining bacterial antiagglutinins.

Differences between Agglutinins and Precipitins, etc.

Tchistovitch (v, 1899, p. 418) has already drawn attention to the fact that precipitins are distinct from agglutinins, on the ground that anti-tetanus serum agglutinates *B. tetani*, but does not produce a precipitum in culture fluids of this micro-organism. Nolf (v. 1900) found that a serum may contain a precipitin but no agglutinin, and concluded therefrom that they are distinct. Radziewsky (1900, p. 434) also considers the antibodies distinct. Bail (1901) found that after all the precipitable substance had been removed, by adding antiserum to a diluted agglutinating serum, the latter retained its power to agglutinate unimpaired, as tested on *B. typhosus*. Both Eisenberg (5, v. 1902) and Beljajew (1902) note the parallelism between the power of agglutination and precipitation possessed by antisera. The latter observed that precipitins appear much more slowly during immunization than agglutinins, and consequently they must be distinct. In both cases we have to deal with substances which appear to enter into chemical combination in definite proportions; both are stable bodies, and resist heat more than do others. There appears to be a difference though, in this resistance to heat: thus Pick (1902, III. p. 81), working with antityphoid serum, which he heated to 58—60° C., found that the bacterio-precipitins were inactivated; not so the agglutinins, for which reason he considers them distinct antibodies. Kraus and Eisenberg (27, II. 1902) found that the precipitins acting upon their homologous blood did not carry down diphtheria antitoxin, or the typhoid agglutinin they contained. They were unable to obtain diphtheria anti-antitoxin, or typhoid antiagglutinin, by treating rabbits with the respective antitoxic and agglutinative sera, whereas they obtained anti-lactoserum (see under Lactosera), viz. antiprecipitins. The resistance of precipitins to heat, as proved by a number of observers, will be referred to presently. Ford (3, VII. 1902, p. 371) does not consider that the non-identity of these antibodies is proved. See also Wassermann (1903).

PART II.

THE PRECIPITINS.

SECTION I.

METHODS.

Methods of treating rabbits for the production of precipitating antisera.

DIFFERENT methods of injecting blood, serum, etc. have been used by various authors. Tchistovitch and Bordet injected rabbits intraperitoneally, a method followed by Myers, Uhlenhuth, Nuttall, and many others since. The subcutaneous method has been used by Wassermann, Stern, Dieudonné, Zuelzer; the intravenous by Mertens, Leclainche and Vallée, Strube. The subcutaneous method has been claimed to have the advantage of offering less chance of infection, a poor reason if proper aseptic precautions are employed. In my experience it has been difficult to inject quantities of fluid at all comparable to what may be introduced intraperitoneally, sloughing being produced when larger quantities are injected beneath the skin. The intravenous method has recently been advocated by Strube, and Kister and Wolff, on the ground that it is possible thereby to shorten the time of treatment needed by the intraperitoneal method, a smaller amount of blood being required to be injected.

Rabbits when properly treated by any of the above methods will yield effective antisera. The blood injected should be sterile, if necessary filtered as serum through porcelain, as recommended by myself, Uhlenhuth and others. Otherwise freshly defibrinated blood may be used; some claim it yields more powerful antisera. Serum is best for intravenous injections. Pleuritic and peritoneal exudation, hydrocele fluid, albuminous urine, have been used in place of serum by a number of different workers, the impression prevailing, however, that these yield less potent antisera. I have frequently used sera from cadavers, both human and animal, the latter more especially from animals dying at the Zoological Gardens, London.

As a rule no difficulties are encountered, the rabbits bearing the treatment well, a slight loss of weight being noticeable after the first injections. My experience suggests that it is better to *grade the dosage of serum*, as is after all usual in immunization experiments, beginning with a small dose and gradually raising the amount administered, being guided by the animal's body-weight. It is especially necessary when treating an animal with a serum whose possibly toxic effects are unknown to begin with a small dose. Occasionally accidents will happen, well-immunified rabbits dying, perhaps after the last injection, apparently from intoxication; this has however happened rarely in my experience. Of the sera injected, I found rabbits to tolerate larger quantities emanating from the horse than from other animals. It therefore seems strange to me that Rostoski (1902, b, p. 26) should have experienced any difficulty when treating rabbits with horse serum, most of his animals dying, and he concludes, "so scheint es mir sehr schwer bezw. unmöglich zu sein, Kaninchen gegen die specifische Giftwirkung des Pferdeserums zu immunisiren." This is so contrary to the experience of others that I am inclined to attribute it to his simply having used excessive dosage. Even under identical conditions of treatment rabbits of similar weight will yield antisera of different strengths, so that the individual element enters also into the problem.

Attempts made by myself, and others (Whitney, 1902, etc.), to obtain *antiserum after a single injection* of blood, or the like, have failed. Michaëlis (9, x. 1902, p. 734) states that he only once observed a powerful antiserum to have developed in a rabbit bled 14 days after a single injection of 20 c.c. of ox serum, the rabbit's blood having contained no precipitin three days after the injection, and all traces thereof having disappeared three weeks after its presence was observed. Otherwise all workers have subjected their animals to repeated injections of serum, a varying number of days intervening between each.

Minovici (12, vi. 1902) states that ox blood produces precipitin more rapidly than does human blood, an observation which I have not as yet been able to confirm. There is no reason why an animal should not react more rapidly to one blood than to another, but to establish this requires not a few, but many carefully conducted experiments where the conditions are equal.

Uhlenhuth (15, XI. 1900) was able to obtain antiserum in a rabbit *fed* on egg-white, the solution being introduced daily by means of a sound, and the precipitin appearing after the feeding had been continued 24 days. Similarly Michaëlis and Oppenheimer (1902, p. 355) fed rabbits with ox serum in large quantities, and obtained antiserum for ox blood. They attribute the formation of antiserum, through excessive feeding with heterologous albumen, to the escape of some of it from the action of the peptic digestion, for precipitins for milk, etc. are not found in the serum of normally fed man.

In the following table I have summarized the methods of treatment of rabbits adopted by different workers, the data in some cases being rather incomplete.

Antiserum for blood or milk of	Mode of injection	Amount injected at a time in c.c.	Total amount injected in c.c.	Duration of treatment in days	Day when bled after last injection	Author and Reference with date
Man	subcut.	10	50—60	14	6?	Wassermann and Schütze, 18. ii. 01
,,	perit.	5—10		14—21		Stern, 28. ii. 01
,,	,,	10	80—200	16—40		Kister and Wolff, 18. xi. 02
,, *	venous	1	6	8	7	Strube, 12. vi. 02
Ox	perit.	10	50—60	30—40		Uhlenhuth, 7. i. 01
Cow-milk etc.	subcut.	10—50		21	6	Wassermann and Schütze, 1900–1901
,,	perit.	10	30	7	7	Moro, 31. x. 01
,, †	venous	7	56	28		Uhlenhuth, 6. xii. 02
Human albuminous urine‡	,,	20	150—200	90	15	Leclainche and Vallée, 25. i. 01

* Gave reaction with 1 : 1000 blood dilutions.
† Milk sterilized 90 minutes at 65° C., injected every 4 days.
‡ Contained 1—2 g. of albumen per litre.

Uhlenhuth made injections every 5—6 days, Stern every two days, Wassermann and Schütze every 3—4 days, others at intervals which are generally not stated. In Bordet's experiments (III. 1899, p. 240, not cited in the table), the milk was partially sterilised at 65° C. prior to injection intraperitoneally. Wassermann and Schütze sterilized it by means of chloroform.

Without making an attempt at completeness, the above table very clearly shows a great divergence in the mode of treatment, more especially with regard to the duration of treatment and the amounts of substance

Table showing the Influence of Different Modes of Treatment upon the Strength of Precipitating Antisera obtained from Rabbits.

	I	II	III	IV	V	VI	VII	VIII	IX	X	XI
	Antiserum for	Rabbit's weight in g.	Duration of treatment in days	Day when bled after last injection	Number of injections	Total amount injected in c.c.	Doses at each injection in c.c. and days intervening between injections	Precipitum in c.c.	Strength of serum dilution tested	Material injected and remarks	Method of injecting
1	Pig 6. vi. 02	3320—3360	27	9	8	59	5-5-6-7-7-9-10-10 / 4-3-3-4-5-4	·07—·055†	1:100	Used one lot of porcelain-filtered serum 48 hrs old at start, kept on ice throughout	Intraperitoneal
2	Antelope 19. iii. 02	2830—2930	31	4	8	59	5-6-6-8-8-8-10 / 5-4-5-5-4-3	·055	,,	One lot of serum from Cobus unctuosus, d. at Zoo., preserved with chloroform, 4 days old at start	,,
3	Pig 5. xii. 01	3300—2840*	37	5	8	55	4-5-8-7-7-8-8-8 / 4-5-4-3-7-7-7	·045	,,	One lot of porcelain-filtered serum, 24 hrs old at start, kept on ice	,,
4	Ostrich 5. ii. 02	2340—2460	33	6	7	46	4-4-4-6-8-10-10 / 6-7-4-5-6-5	·042	1:40	Chloroform-preserved serum of Struthio molybdophanes (d. Zoo.), 14 days on ice when treatment started	,,
5	Fowl 1. iv. 02	2050—2180	23	8	7	36	3-3-4-4-5-7-8 / 4-5-3-4-4-3	·035	1:100	Serum of two birds, the first 7 days old at start, the second used after three injections, 24 hrs old at start. Chloroform-preserved	,,
6	Man 8. ii. 02	2850—2200	29	7	10	102	4-8-10-10-10-10-10-10-20 / 5-3-3-3-3-3-2-4	·031	1:40	Porcelain-filtered serum from autopsy, sealed in bulbs kept on ice, fresh at start	,,
7	Hyaena 18. x. 02	2630—2430*	18	8	6	30	4-4-6-6-6 / 4-3-4-4-3	·031	1:100	Chloroform-preserved serum, Hyaena striata, d. at Zoo., kept on ice, 52 days old at start	,,
8	Hedgehog 18. x. 02	2550—2450*	18	8	6	31	4-4-4-6-7 / 4-3-4-4-3	·022	,,	Mixed sera of four Erinaceus europaeus, chloroform-preserved, 42 days on ice at start	,,
9	Sheep 10. ii. 02	1700—1500	25	5	9	54	4-4-4-6-6-8-8-8 / 1-2-2-2-2-2-3	·02	,,	Filtered serum, pure, sealed in bulbs, kept in dark for one year at room-temperature	,,
10	Man 18. x. 02	2200—2150	14	8	5	17	1-2-2-6-6 / 3-4-4-3	·018	,,	Pure filtered serum, kept in test-tubes one month on ice at the start. Obtained by venesection	First 3 injections intraven., rest intraperitoneal
11	Mex. Deer 26. v. 02	1950—2080	26	7	8	56	5-5-5-7-7-8-9-10 / 5-3-4-3-4-3-4	·015	,,	Chloroform-preserved serum of Cariacus mexicanus, d. at Zoo, serum 21 days on ice at start	Intraperitoneal
12	Dog 25. iii. 02	2370—2400	29	6	7	47	3-5-7-6-8-8-10 / 4-9-4-5-3-4	·015	,,	First three injections with serum 3 mos. old, rest with serum 24 hrs old when first used. Chloroform-preserved, on ice	,,

No.	Animal / date	Weight (g.)					Dose / days interval	Precipitum		Injection
13	Sheep. A 10. ii. 02	1900—1880	25	5	9	54	4-4-4-6-6-8-8-8 / 1-2-2-2-2-2-3	·015‡	Sterile, filtered serum, sealed, kept 10 months at room-temperature in dark	"
14	Sheep. B	2650—2000	"	"	"	"	"	·014‡	Chloroform-preserved serum of same date as preceding kept at room-temperature in dark	"
15	" C	1770—1550	"	"	"	"	"	·0145‡		"
16	" D	1850—1850	"	"	"	"	"	·0135‡	Filtered serum, dried in scales 10 months, kept in glass-stoppered bottle in dark at room-temperature, diluted 1 : 10 in saline for injection	"
17	" E	1720—1700	"	"	"	"	"	·012‡		"
18	Zebra 14. xi. 02	1700—2050	11	7	6	14	1·5-2-1·5-3-3-3 / 2-2—3-2-2	·012	Chloroform-preserved serum of *Equus grevyi*, d. at Zoo, kept on ice 14 days at start	Intravenous
19	Ox 4. ii. 02	2660—2600	27	6	6	43	3-6-6-8-10-10 / 7-4-5-6-5	·011	Filtered serum, 24 hrs old at start, kept on ice, kept sealed in bulbs	Intraperitoneal
20	Man 23. x. 02	1800—2150*	14	13	5	24	4-4-4-6-6 / 3-4-4-3	·01	Filtered, sterile serum, preserved on ice, one month old at start	"
21	Horse 26. v. 02	2330—2410	26	7	8	56	5-5-5-7-6-8-10-10 / 5-3-4-3-4-3-4	·008	Filtered serum, 48 hrs old at start, kept on ice	"
22	Ourang 14. xi. 02	2600—2620	11	7	6	14	2-2-2-3-2-3 / 2-2-3-2-2	·008	Chloroform-preserved serum of *Simia satyrus*, d. at Zoo. 20 days before starting treatment. Kept on ice	Intravenous
23	Wallaby 26. v. 02	1580—1610	26	7	8	48	4-5-4-6-7-7-8-7 / 5-3-4-3-4-3-4	·007	Filtered, sealed, cold-stored serum of *Onychogale unguifera*, d. at Zoo., 38 days old at start	Intraperitoneal
24	Hog-deer 25. iii. 02	2040—2300	21	6	6	39	4-6-7-6-8-8 / 5-4-3-4	·0063	Chloroform-preserved serum of *Cervus porcinus*, d. at Zoo, 17 days on ice at start	"
25	Seal 18. x. 02	2500—2800*	14	8	5	21	5-4-4-5 / 3-4-4-5	·006	Chloroform-preserved bloody serum of *Phoca vitulina*, d. at Zoo, 65 days on ice	"
26	Cat 26. v. 02	2800—2670	26	7	8	57	5-5-5-7-8-10-10 / 5-3-4-3-4-3-4	·005	Chloroform-preserved serum of a cat bled one week before starting treatment, kept on ice	"

I. The date in the first column refers to when the rabbits were bled.

II. The weights in the second column were taken before the first and last injections except in those marked * where the weight was taken when the rabbits were killed.

III. The duration of treatment in days, stated in column III., covers the time between the first and last injection only.

VII. In the seventh column, the amounts injected on each occasion are stated in the upper line, the number of days intervening between injections in the line below.

In all instances ·1 c.c. of antiserum was added to ·5 c.c. of serum dilution of homologous blood, usually dilutions of 1 : 100 being used. As more precipitum is formed with 1 : 40 dilution than with 1 : 100 dilution, the figures obtained with antisera 4 and 6 are not comparable with the others.

† Two pig bloods were tested here, the blood giving the high figure was 2 months older than the second.
‡ The amount of precipitum given here is the mean obtained from tests made on two sheep sera, both fresh.

Note: The animals yielding antisera 6, 9, 13—17 were treated by Mr Strangeways.

injected. Most workers followed the original method laid down by Tchistovitch and Bordet, injecting what are evidently too large amounts; others have pursued a course of treatment contrary to what is usual in other immunizations, the dose at the start being as large as that at the finish. Very few appear to have kept accurate records of the weights of their animals during treatment, as given in my paper of VII. 1901, although some state that they were guided to some extent by the variations in weight. *It is evident that a more scientific method of treatment has to be worked out*, it being, certainly for practical reasons, desirable to discover the least amount of substance which gives a maximum of precipitin, as also the shortest period of treatment necessary. In my protocols (p. 54) the duration of treatment is stated in terms of the number of days intervening between the first and last injection.

In the preceding table I have summarized the method pursued by me in treating a number of rabbits with different bloods and sera, excluding all the older antisera which were not standardized by my quantitative method, so as to give some conception of the amount of precipitum yielded by antisera obtained in different ways. Owing to other work I have been unable to study this question as I should have desired. The results of Strube, using the intravenous method, had already indicated, what I have also found, that the duration of treatment and dosage required are much smaller than when other modes of treatment are used. It is true that my most powerful antisera have not been obtained by the intravenous method, but this is doubtless due to my not continuing the treatment thus begun longer, there being no necessity for doing so, as the antisera obtained were as strong as was desirable for my purposes, quite as strong in fact as some obtained by the other methods. *I am certainly inclined, therefore, to accept the intravenous method as being the most advantageous*[1].

In the experiments by Graham-Smith and Sanger (1903, p. 260), the anti-human serum, and some of the others, were prepared by intravenous injection, but much smaller quantities than those usually employed were found to suffice[2]. " For example 18 c.c. of human serum injected in doses of 5, 5, 5 and 3 c.c. at intervals of 2, 3 and 4 days produced a powerful anti-human serum. The animal was bled 14 days

---

[1] See further a similar table to mine in the chapter dealing with the experiments of Graham-Smith.

[2] Powerful anti-ox and anti-sheep sera were made by injections of 9·2 and 12 c.c. in doses of 1, 2, 1, 3·2 and 1·5, 2, 2, 3·5, and 3 c.c. respectively. The intervals between the injections were 4, 4, 2 and 5 days in each case, and the animals were bled 7 and 10 days after the last injections.

after the last injection. Continental workers have used quantities ranging to hundreds of c.cs., and have frequently found that their animals stood the operation badly, whereas the above animal, and most of the others we have treated, continued to gain weight, and appeared to be healthy. Some of our control antisera was prepared by the intra-peritoneal method and quantities ranging from 30 to 60 c.c. were used. The animals stood the treatment very well."

The fact that *long-continued treatment of rabbits leads to the dis-appearance of precipitins* from the blood has been already noted by Tchistovitch (1899) in rabbits rendered immune to eel serum. I was able to confirm this (16, XII. 1901) in rabbits treated with ox and sheep serum, the treatment having been too prolonged. As I stated in my paper, just cited, "There is therefore a point in the treatment of animals, when, for purposes of obtaining an antiserum, a maximum of reaction is reached and the animal should be bled; this can be determined by periodic bleedings from the ear-vein."

Mode of Injection.

If the serum to be injected has been preserved by means of chloroform, as is frequently the case, it is desirable to remove as much of this as possible, as the injection may prove fatal from the chloroform contained in the serum. This is best accomplished by pouring the serum into Petri dishes, these being exposed, half covered with their lids, in the thermostat at 37° C. for about half-an-hour, or until no perceptible odour of chloroform emanates from the serum. For intra-peritoneal injections I have used an all-metal 10 c.c. syringe, one of smaller capacity (5 c.c.) being used for intravenous injections.

Intraperitoneal Injections.

The rabbit being held by an assistant, belly up, is lathered over the seat of operation by means of liquid soap containing lysol. A small area about 5 cm. across is then shaved in the lower abdominal region on the left side. After disinfecting the skin with lysol solution, applied with cotton pledgets, a small puncture is made through the skin with a fine scalpel, this facilitating the penetration of the needle of the syringe, the point of which it is better to render somewhat blunt, for the reason that this lessens the chance of penetrating the intestine. Penetrate the abdominal wall, holding the needle at right angles to its surface. Having gently bored the needle through the abdominal wall, through which it passes not infrequently with a slight jerk, immediately proceed to empty the syringe, slightly retracting the point if there appears to be any impediment to the outflow. After the injection has been made, and the needle withdrawn, the parts are dried, and tincture of benzoin is applied, sealing the small opening made by the scalpel (usually but 1·5—2 mm. long). The wound usually heals by first intention, the animal experiencing no discomfort. In this way I have repeatedly injected animals in the same region, all the punctures made during the treatment being included within the area shaved as stated above. It will be seen that the operation is exceedingly simple, and I should not describe it but for the fact that many do not know how to perform it.

Intravenous Injections.

Intravenous injection is practised by introducing a fine-needled syringe into the marginal ear-vein, and injecting smaller quantities than in the preceding method, for the reason, not only that they suffice, but that much smaller doses exert a toxic action when introduced intravenously as against intraperitoneally. Intravenous injections are very easily practised after a little experience. Before introducing the needle the skin of the ear should be shaved and disinfected as in the preceding case.

Subcutaneous injections are not to be recommended from my experience. Absorption of injected material takes place more slowly this way than through the peritoneum, and the mechanical effects of large injections are liable to lead to sloughing.

Marking the Rabbits.

It is absolutely essential that there should be no confusion amongst the animals treated, especially where these are at all numerous. The best way to secure this is to *tattoo* the rabbits inside the ear. This is easily done by dipping a fairly large needle into India-ink and pricking it through the skin, a proceeding which the rabbit does not appear to mind. Various letters and simple signs can be made in this way, and there is then no possibility of confusing the animals.

Weighing the Animals.

It is desirable to weigh the animals during treatment. I have always done this before each injection, stopping treatment as a rule when a notable fall in weight occurred. I have had less loss in weight in the few animals I have treated by the intravenous method. Of the 26 animals noted in the preceding table (p. 54) 11 show a loss of weight, in the figures I have given, whereas the others regained their weight or even greatly increased in weight. Several of those which had lost weight, as recorded, had regained it when killed. Of the 11 that lost in weight, 3 only lost 50—70 g., 5 lost 130—220 g., a loss of no great moment. On the other hand 3 lost heavily, viz. Nos. 3, 6, and 14. Of these, No. 3 was treated with pig serum; at the third injection it exceeded its original weight, by the last it had fallen 400 g., 60 g. more being lost by the time it was killed. A greater loss of weight is noticeable in Nos. 6 and 14 (treated by Mr Strangeways) where the treatment was continued longer, the intervals between injections being shorter, and in No. 6 the dosage very high. This animal however gave a powerful antiserum.

When to kill the treated Animals.

When the animals have received usually 5 to 6 injections, and some days have elapsed, it is well to draw off samples of blood and to test these for precipitins. This is accomplished by shaving the ear, making a small cut into a vein, and allowing the drops of blood which flow out to fall into a Petri dish, which is then covered and set aside, being slightly tilted. After an hour or two the serum will have separated and can be tested. If it prove insufficiently powerful treatment may be continued, whereas if it shows power the animal may be killed, the time for killing being preferably 7 to 12 days after the last injection.

Killing the Animals and Collecting the Antiserum.

The animals may be killed in a variety of ways. Uhlenhuth (25, VII. 1901) chloroforms them, opens the thoracic cavity under aseptic precautions; and, cutting through the beating heart, the blood is allowed to flow into the thoracic cavity, whence it is removed by means of sterile pipettes to suitable vessels. Others isolate the carotid, insert a sterile cannula, and allow the blood to flow through it into suitable sterilized vessels.

The method I have pursued is very much simpler, and usually gives quite as good results, and certainly more blood. Having shaved the neck, and disinfected the skin with lysol solution, the hair on the head and fore-part of the body being moistened with lysol to prevent hair flying about, the animal's head is bent backward, putting the skin of

the neck on the stretch. The animal being held in this position by an
assistant, the operator's left hand is placed upon the back of the rabbit's
neck, which is uppermost, the fingers and thumb encircling it. These,
being drawn upward, put the skin on the stretch transversely to the
long axis of the animal. By pressing downward, the neck arches ven-
trally. A sharp sterilized knife is now allowed to make a clean sweep
through the tense skin to and through the vessels. The skin retracts,
and the blood spurts out into a large sterile dish which is immediately
covered when the main flow has ceased. Drippings are caught in a
similar manner in Petri dishes, the thorax etc. being compressed to
expel as much blood as possible from the body.

The dishes are placed horizontally until a clot has formed; they are
then slightly tilted, and as soon as serum enough has been expressed,
this is pipetted off into sterile test-tubes, and thence transferred to
sealed bulbs of small size. The serum which comes off later is similarly
collected, but it may be necessary to add chloroform thereto or to pass
it through a Chamberland filter to exclude bacterial development[1].
Even when a hair or two falls in, the first serum will generally be
sterile, the clot retaining the microorganisms, the bacteria being moreover in part destroyed through the bactericidal properties of the fresh
serum. After serum has been drawn off once or twice, a further amount
may be gained by slashing the clot with a sterilized knife. This last
serum is bloody, and has to be centrifugalized or placed in the refrigerator to sediment.

Storing Antisera.

The most suitable form of bulb in which to store antisera is represented in figure 3 B (p. 65), being drawn out at both ends. The bulb
is sterilized in the process of drawing out the ends; the latter are cut
off with a diamond. After cooling, a number of bulbs are grasped in
the hand, the ends being sterilized in the flame, after which they are
allowed to cool, resting upon sterilized wire-gauze or otherwise. The
antiserum is now drawn up into the bulbs until these are filled, as little
air as possible being left in the portion where it narrows. By tilting
the tube a little serum flows into the opposite drawn-out portion. The
capillaries are now sealed near to the serum by means of the small
flame, such as serves to relight the large flame of certain Bunsen
burners. A large flame is unsuitable. The serum must not be heated
in the process of sealing. The tubes are now placed vertically in tin

---

[1] The effects of filtration are noted on p. 118.

boxes, containing wool at the bottom, and suitably labelled. The object in placing the tubes vertically and not horizontally, is to ensure the gravitation of any suspended matter into the lower drawn-out end, where it accumulates, and can be readily disposed of by cutting the tube above the layer of deposit, by means of a diamond, when about to use the serum for testing. (Regarding the use of preservatives see p. 64.)

The Choice of Animals for Immunization.

Rabbits appear to be the most suitable animals for the production of antisera. 32 kinds have been produced by me through injecting rabbits with mammalian (non-rodent), avian, reptilian, amphibian, and crustacean bloods. Most observers have used the rabbit in their work, doubtless because of the greater convenience and moderate expense, the guinea-pig not being so suitable, for the reason that it yields less antiserum. Guinea-pigs have been used by but a few observers, including myself. The rabbit is therefore to be selected, apparently in all cases where antisera are required for bloods other than those of rodents. We have sufficient evidence to show that precipitins are not formed in the serum of closely related animals. Thus Bordet and Hamburger obtained no precipitins for rabbit blood by treating guinea-pigs with rabbit serum, and all that Gengou could claim was that the serum of such guinea-pigs produced a slight opalescence in rabbit blood dilutions, never a precipitum. Nolf was unable to produce antisera in pigeons treated with fowl serum, the reason being apparent from my extensive investigation of avian bloods, which, throughout, have been found to react to antifowl and antiostrich sera. In other words, birds appear to be too closely allied for the purpose of inter-avian antiserum-production.

In Tchistovitch's classical experiments precipitins for eel serum were obtained by treating goats, dogs, rabbits and guinea-pigs. Anti-rodent serum (for rabbit) has been produced by Nolf by injecting rabbit blood into the fowl. Hamburger (6, XI. 02, p. 1190) was unable to obtain antiserum for egg-white from a dog treated therewith. Corin has on the other hand used the dog for the production of antisera, according to Strube (1902). Uhlenhuth (18, IX. 02, p. 680) has not had encouraging results with other animals than rabbits. For instance, he treated a lamb and a goat with human blood and ascitic fluid during 2 months, injecting respectively a total of 1·5 and 2 litres thereof into the animals; the antiserum was too weak for practical use. De Lisle claims to have obtained antisera from eels treated with rabbit serum, and Hamburger obtained effective antisera for rabbit blood by treating goats therewith. Wassermann and Schütze have apparently also failed to obtain antihuman serum from the goat.

Judging from the blood reactions obtained, I am inclined to the theory that an antirodent serum will act upon the rodent group, as I have found with other antisera in other groups. The rodents form such a well-defined order among the mammalia that it appears theoretically probable that they will produce antisera for practically any animal outside a rodent. The rabbit is therefore the animal *par excellence* for the purposes of such experiments.

Collecting Bloods and testing them with Antisera.

The bloods or sera tested were collected in two ways, viz. fluid and dry. The fluid sera were relatively few in number because of the practical difficulty of obtaining sera in this way. The fluid sera were obtained by collecting blood in covered vessels where it was allowed to clot, the serum being subsequently removed to clean or sterilized well-stoppered bottles and a small amount of chloroform added for purposes of preservation. Dilutions of fluid sera were made in salt-solution, the dilutions being usually of 1:100 or thereabouts. A number of fluid sera were filtered through sterilised Chamberland and Berkefeld filters and subsequently preserved in sealed tubes. In other cases the dilutions were preserved by the addition of small quantities of chloroform, and it was found that such dilutions gave good reactions after over a year, the bottles having been kept at room temperature in the dark. Although long-stored bloods usually give somewhat less reaction than do fresh ones, the reverse may be the case. It is a remarkable fact, which will be brought out more clearly in describing the results of the quantitative tests, that sera or serum-dilutions preserved for six months or more will at times give more precipitum on the addition of an anti-serum than do the fresh sera. This applies not only to the amount of precipitum obtained by the addition of an homologous anti-serum to a serum dilution, but also to what I have termed the "mammalian reaction," namely one taking place between a powerful non-homologous antiserum and a more distantly related blood. Whereas this increased amount of precipitation obtained with older sera (the increase may amount to 50 %) will naturally greatly affect the quantitative tests, it can be practically ignored in the qualitative tests here described. On what this change in stored blood depends is a matter for future investigation. It appears also to take place in stored antisera at times. It is not due to evaporation in stored dilutions, nor to the prolonged action of the chloroform added for purposes of preservation, nor to bacterial development, for pure sterile sera preserved in sealed tubes have repeatedly given greater reactions. This observation has not as yet been recorded. Although it appears at first sight improbable that a dried serum or blood may (as long as it remains soluble) give an increased reaction after a time, the matter will have to be submitted to experiment.

In a few instances the sera were dried on glass plates at 37° C., the scales being preserved in bottles with slightly vaselined stoppers. In the majority of cases the bloods or sera were collected on strips of filter-paper of fairly uniform size (3 by 5 inches), as noted in the paper by Nuttall and Dinkelspiel (July 1901, p. 378) and in a circular letter which I sent out to those who have kindly aided me in collecting bloods in various parts of the world. The filter-paper is immersed in the serum or defibrinated blood, clots being avoided, after which it is hung up to dry. One end of the strip is left clean, so that the name of the animal (both common and scientific), the date, place of collection or natural habitat, and the name of the collector, can be noted thereon in lead-pencil. The notes can be written clearly if the paper is rested on a hard surface. The strips of paper dry rapidly when suspended in the air, it being convenient to pin them against the edge of a table or shelf or upon the branch of a tree. Collectors were particularly warned not to allow strips of paper saturated with different bloods to come in contact with one another, especially in a moist state. Where it was impossible to wait for the strips to dry, as when out shooting, collectors were requested to place each sample separately in the paraffined paper covers which I supplied, these being of the pattern usually used by photographers. The outfit for collecting which I sent out consisted, then, simply of pure filter-paper of fairly uniform thickness, accompanied by paraffined envelopes and outer envelopes of ordinary stiff paper. On returning from shooting, the moist strips were removed from the paraffined covers and hung up to dry. The drying may be facilitated in our climate by exposing the strips to the sun, but this proceeding does not appear to do in the tropics, for I have found several samples sent from hot countries to have become insoluble. The fact that blood-stains may frequently become insoluble in hot countries was noted in a letter to me from Mr E. H. Hankin, of Agra, India, dated 23 August, 1901. He wrote, "There is a practical difficulty in carrying out the test here in the frequent insolubility of blood stains, which in this climate may be dried at a temperature of 115° Fahr. or even more. Blood stains usually don't give the spectroscope test when extracted. If more dried or older, they successively refuse to give the hæmin and, I believe, the guaiacum tests." It will be seen from my tabulated results that bloods sent to me from hot countries dried on filter-paper not infrequently refused to go into solution in saline. Where I had to deal with such bloods I allowed them to remain some hours in salt solution at 37° C. Some bloods however refused to go into solution even after 24 hours under these conditions. Such bloods naturally gave no reactions. Other bloods from the same source gave full or feeble reactions. This difference in behaviour may depend upon the different temperatures at which they were dried. These observations are of importance from a medico-legal standpoint. I did not find it advisable to use a weak soda-solution (0·1 to 1 %) as recommended by Uhlenhuth (17 Oct. 1901, and by Ziemke, p. 732), for the reason that it frequently yielded solutions which gave rise to a cloudiness on the addition of any antiserum[1], the cloudiness being in some cases quite marked, and sufficient to obscure the mammalian reaction to which I first drew attention, and which will be presently referred to. In the examination of the large number of blood samples investigated it seemed scarcely worth while to try the method recommended by Ziemke (17 Oct. 1901) of extracting with cyanide of potash solution. He does not state the strength of the cyanide solution used as a solvent, but claims that it has even a greater solvent action than the soda solution for bloods which will not go into solution in saline. Ziemke adds a few crystals of tartaric acid to the cyanide-blood solution, shakes the fluid and tests continuously with red and blue litmus paper, until the reaction is almost neutral. The solution is now decanted and filtered. If the solution is to remain clear it must be slightly alkaline. The solution is now diluted to the degree desired and tested with antiserum in the usual way. The method of Ziemke was too complicated for my purposes, although it may be of use medico-legally. Uhlenhuth has also recommended the use of saline chloroform-water for extracting difficultly soluble blood stains. In my investigations I however preferred to pass over the insoluble bloods, that is bloods insoluble in salt solution, and this for the reason that, in the case of chloroform, slight cloudings frequently occurred in solutions where all the chloroform was not driven off, upon the addition of antiserum of any kind. These cloudings due to

[1] See further in the section relating to the medico-legal application of the test.

chloroform or to the presence of soda should not give cloudings which would be mistaken for a full reaction with an homologous antiserum, but as stated they at times obscured the delicate cloudings which in my tests gave indications of blood relationships between different animals, and consequently if used would have vitiated the purposes of my investigation.

This leads me to a brief mention of the method of preserving antisera. In the paper by Nuttall and Dinkelspiel (1, VII. 1901, p. 378), it was stated that neither the addition of chloroform, nor of trikresol to an antiserum prevented the reaction taking place with an homologous blood. It appeared therefrom that the addition of such reagents to antiserum might serve a useful purpose in their preservation. I however soon found that there were disadvantages in adding any preservative; that it was much better to either collect the serum under aseptic conditions, or, failing this, to pass it through Chamberland filters with a view to excluding putrefactive organisms from it[1]. Antisera to which trikresol had been added frequently gave marked cloudings with non-homologous bloods. Antiserum to which chloroform has been added behaves in the same manner unless the chloroform has been removed by evaporation. In evaporating the chloroform one concentrates the antiserum and alters its power. I was led to use trikresol through the statement by Uhlenhuth (25 April, 1901) that he had obtained reactions with an antiserum preserved three months with $0·5\%$ carbolic acid. My experience certainly points to the inadvisability of adding any preservative to an antiserum, for the reason that it may produce pseudo-reactions, and I have no

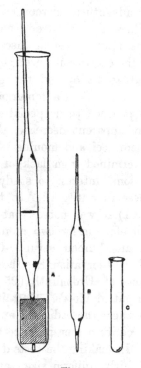

Fig. 3.

A. Test-tube containing square of filter-paper which has been soaked in saline. The remaining blood-dilution has been drawn up into the pipette.

B. Antiserum contained in bulb which has been sealed at both ends, without the serum being heated.

C. Small test-tube such as is used for testing, being figured in the racks in fig. 4.

(The figures are reduced to ½ natural size.)

[1] Note the effects of filtration on serum, p. 118.

doubt but that in certain cases such reactions have led investigators astray. (See further pp. 76—79.)

The tests here recorded were carried out with pure antisera, upon bloods which proved soluble in salt solution. In some cases, through misadventure, microorganisms developed in the sealed tubes of antiserum. Where this occurred the serum was centrifugalised, being cleared in this manner in some cases, not in others. The microorganisms collected in the drawn-out lower portion of the tube (Fig. 3), and when the latter had to be opened the glass was cut above the part including the deposit. It is a remarkable fact that even putrid antisera continued to produce their special reactions, in some cases at any rate, without any apparent decrease. The microorganisms present included bacilli, micrococci and moulds of various kinds, the species not having been determined from lack of time. It appeared to me to be a matter of some interest to study the effects of different germs by cultivating these in antisera[1]. In the paper by Nuttall and Dinkelspiel (11, v. 1901) it was noted that human blood which had undergone stinking putrefaction for two months still gave a full reaction on the addition of anti-human serum, this blood not reacting to other antisera. This observation has been confirmed by the independent investigations of Uhlenhuth (25 April, 1901) and Ziemke (27 June, 1901) on putrid bloods. Certain putrefactive processes do not therefore seem to affect the bodies in antiserum and serum in so far as the amount of precipitum they produce by their interaction is concerned.

In making the tests described below square-cut pieces of filter-paper, of fairly uniform size, saturated with blood and dried, were placed in a definite quantity of normal salt solution. As I have noted elsewhere (July 1901, p. 378) blood and serum which is dried on glass plates only dissolves slowly and frequently gives clouded solutions. On the other hand, the filter-paper method, as I have shown, rapidly yields clear serum solutions, the fibres of the filter-paper retaining the fine particles which would otherwise go into suspension. To clear such clouded solutions Uhlenhuth has resorted to the tedious method of filtering them through Chamberland filters. I have found this to be usually unnecessary. Where I have to deal with a clouded blood, I drop some of it on to filter-paper and place the paper for half-an-hour or so in the thermostat to dry, after which I place it in saline and usually obtain a clear solution without more ado.

Where serum only is present on the filter-paper the solution may be

[1] See the experiments by Graham-Smith and Sanger (1903) given on pp. 119—122.

colourless, whereas when hæmoglobin is present the solution may be more or less coloured. I have attached no particular importance to the amount of colouring matter which has gone into solution, but have throughout judged a blood sample to have been soluble or not by means of the *foam-test*. In earlier papers by Uhlenhuth, as also in those of Ziemke, stress is laid upon a trace of colour in solution as an index of the blood being in solution. In a later paper by Uhlenhuth (18, IX. p. 679, 1902) that author for the first time mentions the foam-test, which I have throughout my work considered essential in this respect. I judge the solubility by blowing air gently through the pipette which is used for transferring the solution into the test-tubes (Fig. 3 A). If air is blown violently through the fluid containing the filter-paper, particles thereof, as also the particulate matter in the blood attached to the paper, are loosened and the solution is rendered cloudy. Solutions of blood or serum of 1 : 1000 and over still foam well, the limit being apparently reached with solutions of about 1 : 5000, when, in several quantitative experiments, it was seen that the minute bubbles formed "held" an instant longer than they did when air was bubbled simply through normal salt solution. I find it possible, moreover, to tell approximately the degree of concentration of a solution by the length of time the bubbles remain at the surface after air has been bubbled through the fluid. Even from a medico-legal point of view the colour test of solubility as noted above may prove fallacious, for stains due to serum, which we know contains the acting substance, might well be overlooked in the absence of the foam-test to which I unqualifiedly attach more importance.

When the filter-paper squares are placed in salt solution, the serum contained in the meshes of the paper usually goes into solution very promptly, often in a few minutes. At times the solution is delayed for half-an-hour or more. Once solution has taken place, it is inadvisable to keep the paper soaking too long as the solid particles may in turn become detached. Solution may be entirely prevented through the paper having become soiled with fat, this having been the case apparently with a few of the samples collected for me from animals. In such cases the immersed paper tended to float, and remained covered by minute silvery bubbles of attached air.

As soon as solution has taken place, as evidenced by the foam-test, the mixed solution is transferred to small test-tubes (Fig. 3 C) placed in suitable racks. The test-tubes I used usually had a capacity of about 1 c.c. and were kept scrupulously clean, and were always well dried.

5—2

The test-tube racks (Fig. 4) were made in the laboratory, and consisted of a flat strip of wood upon which three equidistant blocks of wood were nailed to support a strip of cardboard perforated with holes of suitable size 2 cm. apart. The holes were bored by means of a red-hot iron rod pointed at one end. Each rack contained holes for 20 tubes, the strip below being indented by means of a nail-punch with holes corresponding to those above in the cardboard; these indentations received the convex bottoms of the small tubes and fixed them in place. Each rack was lettered according to the alphabet at one end, and numbers 1 to 20 were written in ink upon a strip of white celluloid which was tacked

Fig. 4.

Showing the test-tube racks on the stand, and a single rack in the foreground. The black riband is shown at the top, forming a background for the top row of tubes. For a further description see the text, p. 70. Every part of the stand and racks, except the celluloid strip bearing the numbers, is blackened.

(The figure is reduced to $\frac{1}{10}$ natural size.)

along one side of the upper surface of the base. When a number of blood samples, my usual limit being 80, were to be tested by different antisera, the racks were placed in rows of suitable length upon the table. Where 80 bloods were being tested, the racks were ordered as follows:

Racks

Row 1.   A B C D

„   2.   E F G H

„   3.   I J K L

„   4.   M N O P

the number of rows corresponding to the number of antisera to be used in testing the series of 80. In rows of 20 bloods the number of antisera used might number 12; in sets of 80, the number of antisera used was usually 4 to 6.

Each blood, whether on filter-paper or in bottles, received a number on arrival. This number is given in the following tables (in the second column) in small figures, and corresponds to the order in which the bloods were tested. *My plan was to test every blood received by means of every antiserum produced.* In this way a "network" was, so to speak, formed through which all the bloods were passed in testing. In some cases this scheme was departed from. For instance there appeared to be no object in continuing to test avian bloods with mammalian antisera after ten or more such antisera had continually given negative results. Moreover, in some cases the supply of antiserum or of blood to be tested ceased, and the series of tests was necessarily interrupted. In other cases again the general results obtained with one antiserum sufficed, so that it appeared only necessary to test a given group with a particular antiserum, this for instance being the case with the tests made with anti-chimpanzee and anti-ourang serum upon the bloods of Primates, etc. The bloods received were not tested in their zoological order, but in the order in which they were received. For example:

696.   *Sus scrofa:* Pig,

697.   *Ateles vellerosus:* Spider Monkey,

698.   *Balaenoptera rostrata:* Rorqual,

699.   *Ornithorhynchus paradoxus:* Platypus,

700.   *Podiceps:* Diver,

were actually tested in succession. Similarly the tests with antisera were made in the order of production of the antisera, which depended entirely upon the time when the bloods for injecting rabbits were received. For reasons of convenience the order of these antisera, as produced, is incidentally given in connection with the tests on the bloods of Pisces. It will be seen that no general picture, other than a subjective one, could be gained of the results until the conclusion of the work, when the tables had been completely rearranged, the bloods being placed in their zoological order and the antisera in theirs.

On receipt, the samples on filter-paper were cut in two after numbering, the one half being placed in order according to system, the other half in order according to number. The systematic collection aided materially in the final arrangement of the tables, and permitted of small series of tests being made upon isolated groups, the bloods being readily found. In both collections the samples were pinned upon clean sheets of paper which prevented their coming in contact with one another. The slips of the collection by numbers were ordered in groups of 20 in paraffined envelopes enclosed within envelopes of stiff paper. All the samples on filter-paper were kept at room temperature.

When a new antiserum had been produced I began by testing down from blood No. 1 to the end of the list.

The test-tubes containing the blood solutions were numbered and ordered in a manner corresponding to the numbers on the racks. The solutions were drawn up by means of the pipette (Fig. 3 A, p. 65) and transferred to tube 1 of racks A, E, I, M (for a series of 80), then the second blood dilution was allowed to flow into tube 2 of the same racks, and so on to the end of the series. This system rendered any confusion practically impossible.

When the tubes had all been filled, the racks were transferred to a stand (Fig. 4) which I constructed at small cost and have found essential in the work. The stand consists of narrow shelves of deal attached to wooden uprights by means of galvanized brackets. Metal T-pieces are attached to both ends of the uprights, the one in each case serving for the attachment of a strip of wood which serves as a base, the other for the attachment (at one extremity of the T), of a brass wire cord which runs down vertically to its attachment, a screw eyelet, screwed into the basal strip of wood. A piece of dull black silk riband sewed to two strips of bent zinc at both ends slides up and down along the wires, wear being prevented by the zinc, within the bend of which the wires run. Small tacks are placed on the posterior upper margin of the shelves to prevent the racks being pushed backward off the shelf. The whole stand, and the racks, are painted black. Each stand carries two racks side by side. Where a series of 80 bloods is being tested, it is necessary to place two stands side by side to accommodate the racks in their proper order.

The racks having been ordered, the solutions are viewed by transmitted light to see if any are slightly cloudy. Where the cloudiness is likely to obscure even a slight reaction the solution must be filtered or discarded. The contents of the sealed tube containing antiserum (Fig. 3 B) are now allowed to flow into a small test-tube such as is used for testing; sediment in the sealed tube being excluded. By means of a small tube of similar shape to that used in filling the solution into the small test-tubes in the racks, the serum is now added in equal quantity to all the solutions in the series. A drop or two suffices (equal to ·05 or even less of a c.c.)[1]. Where the antiserum does not flow down to the bottom of the blood solution it is made to do so by gently tapping the test-tube or by inserting a clean sealed capillary. On reaching the end of the row, a slip bearing the name of the antiserum used, the time, temperature, etc. is attached to the last rack in the row. In this way a different antiserum is added to each row of bloods, until all have received the test serum. The manipulation is exceedingly simple and rapid, drops of antiserum of very uniform size being allowed to fall from the pipette into each tube in the row.

Where a powerful antiserum is added to its homologous blood dilution, the reaction is almost instantaneous, in other cases it takes place more slowly. In the case of a strong antiserum, the reaction takes place as a rule rapidly in related bloods, more slowly in distantly related bloods. The rate at which the reaction takes place may depend also upon the concentration of the blood dilution, the more concentrated dilutions, within limits, reacting earlier than higher dilutions. A weak antiserum will act more slowly than a powerful one. Consequently a " time limit " (Nuttall, 16, XII. '01) may have to be put on each antiserum when testing for a particular blood. In my earlier tests I placed an arbitrary time limit of five to fifteen minutes upon the reactions. Subsequently I lengthened the period of observation, periodically noting the reactions which took place up to several hours, indeed up to 24 hours at room temperature (varied between 12 and 20° C. in winter and summer), the deposits in the tubes being noted at the end of the time mentioned. The black riband noted above, in describing the stand, has proved of considerable value in studying the slight reactions which are observable in more distantly related bloods, or when testing more nearly related bloods by means of weak antisera, where the reactions take place slowly and faintly. Many of the

[1] It is unnecessary to operate with larger quantities as other authors have done, as this only entails a waste of antiserum.

cloudings or feeble reactions observed, for instance in the generalized mammalian reaction already noted, have certainly been overlooked by other observers because of the *mode of illumination employed*. For the same reason it is of advantage to have the stands and racks painted black. By moving the riband up and down, the obliquity of the rays of light falling across it from the window and illuminating the test-tubes can be regulated to suit each row of racks, and render slight reactions quite evident. I can but recommend the use of this simple apparatus to those engaged in similar work.

SOURCES OF ERROR IN PRECIPITIN TESTS.

1. *Opalescent Antisera.*

In the absence of more obvious sources of error, such as I shall mention below, the use of what may be described as "opalescent antisera" constitutes one of the gravest in using the test. A little familiarity in the process of testing, and the making of one or two control tests on heterologous sera, will however suffice to exclude the error which might otherwise be made by using the antisera about to be described. Uhlenhuth (11—18, IX. 1902, p. 680), Miessner, and Rostoski (1902, *b.* p. 29) have observed these. It was due to the fact that my anti-bear serum (referred to in an earlier paper, 5, IV. 1902) belonged to this category, that I never published any tests made with it. I found that it *clouded every blood dilution, mammalian and otherwise*, to which I added it. Although it gave a great reaction with its homologous serum, it gave such marked reactions with all other sera, that I immediately discarded it, without being able to explain why it behaved so peculiarly. Since then I have fortunately rarely encountered such antisera. Uhlenhuth has been the first to sound a warning against their use, medico-legally. As he says, a slight opalescence is usually perceptible when any serum or antiserum is added to blood dilutions, the tube being viewed by strong transmitted light, but the clouding here referred to is totally different and much more marked, and takes place even in salt solution. I tried to clear my anti-bear serum by filtration through porcelain, but was unsuccessful. Uhlenhuth has had the same result. We are still in the dark as to the cause of this peculiar milkiness or opalescence of certain antisera. Uhlenhuth thought it might depend upon the stage of digestion at which the animal was killed, but this seems to me unlikely for the reason that I have killed my animals at no fixed time with regard to meals, and on the whole have encountered the condition but rarely. Moreover in a rabbit recently treated by Dr Graham-Smith, the opalescence was seen to continue for days, samples of antiserum being taken from the ear-vein so as to test its strength[1]. It would therefore appear desirable to take samples of blood from the rabbits' ear-veins to see if the serum is clear before proceeding to kill the animals. Uhlenhuth states that he has encountered the condition much more frequently in intravenously treated animals, but this has not been our experience here. Uhlenhuth suggests that some of the results of Kister and Wolff, as also of Strube, on non-homologous bloods with different antisera, may receive an explanation from their having used opalescent antisera, their results being contrary to those of about 40 other authors. I am certainly inclined to Uhlenhuth's view on this point.

Rostoski (1902 *b.* p. 29) treated rabbits with different constituents of horse serum, and seems to have encountered the condition rather frequently. This is unfortunate as it necessarily detracts from the value of his results. One rabbit (his No. II.) was treated with the crystallized serum albumen of the horse, yielding an antiserum which was "stark milchig getrübt," and he goes on to say, "doch habe ich gelegentlich auch schon andere milchig getrübte Sera beobachtet, die nicht eine besondere starke präcipitirende Kraft besassen." Linossier (25, III. 1902) states that he has found anti-human serum to precipitate not only human, but also ox, horse, dog, sheep, guinea pig, and fowl (!) sera, the reaction, he adds, being however incomparably greater with human blood. He does not state with what concentrations he worked, but says the difficulty can be surmounted by the use of higher dilutions[2]. This is certainly not the case with opalescent antisera. Kister and Wolff (18, XI. 1902) experimenting with anti-horse serum, which had "surprised" them because of the non-specific character of the reactions it gave, tried to see if there were any law with regard to its action on different bloods. These authors expressly state (p. 412) that they only used clear sera, but perhaps they overlooked opalescence which may not be very marked, and which in my experience might deceive one into accepting such antisera. In any case adding it to salt solution would have cleared the matter up. I have referred to their results elsewhere (under tests with anti-horse serum), and it will be seen that they are not in accord with those of the majority of observers, as they obtained quite marked reactions with ox and sheep serum.

[1] Dr Graham-Smith has found several rabbits yielding opalescent antisera to be affected with Cysticerci. This may explain the condition.

[2] See p. 74.

2. *Overpowerful and Weak Antisera.*

Powerful antisera have been of considerable use to me in my work, and have led to my being able to establish more remote relationships, especially amongst the mammalia, as I have shown in speaking of the "mammalian reaction." They may however be a source of error in medico-legal work, as has been pointed out by Uhlenhuth. When an antiserum is so powerful as to produce a reaction in non-related bloods, its action should be weakened by dilution with salt solution as suggested by Uhlenhuth, Kister and Wolff, and Ewing. This lessens the amount of antibody they contain, and by suitable dilution their strength may be graded to suit requirements. Weak antisera may also be a source of error through their reaction being delayed and bacterial development (see below, § 7) occurring in the mixture of antiserum and blood dilution.

3. *Suitable Dilution of Bloods tested.*

In view of the fact that concentrated dilutions of different bloods may give reactions with non-homologous antisera, it is desirable to dilute bloods as far as possible when testing them. This remark applies more especially to medico-legal tests. Tests in which concentrated sera are brought together and allowed to interact, as in the experiments of Friedenthal and others[1], are liable to lead to false conclusions, especially when reactions with non-homologous bloods are to be excluded[2]. This is also true of tests carried out with strong solutions.

Linossier and Lemoine (25, I. '02), working with anti-human serum, having, they state, great power, found it necessary to add it in the proportion of 25 : 1 of human blood in dilution, to remove all the precipitin; in other words, 25 volumes of antiserum were neutralized by one of blood. The proportion of antiserum added to a given blood for this purpose will naturally vary according to its strength. As is stated on p. 89, an excess of blood leads to precipitum solution. This has also been observed by L. Camus, and Linossier and Lemoine (21, III. '02), who dwell upon the importance of not using an excess of serum in solutions to which antiserum is added. Eisenberg (v. '02, p. 293) finds that dilution *per se* affects the interaction of equivalent proportions of precipitin and precipitable substance. Whereas equivalent proportions lead to precipitation in small quantities of fluid, he states that no

---

[1] See also under Antiprecipitins, p. 149, and Normal Precipitins, p. 150.

[2] See p. 73, Linossier.

precipitation occurs when the same proportions act upon each other in greater dilution, although the bodies combine. A precipitum which had been resuspended in a very large volume of fluid did not become redeposited, whereas it did become redeposited in a smaller volume of fluid. He states further (p. 294) that antiserum added to different concentrations of albumen gave a greater reaction with higher dilutions than with more concentrated ones. As has already been noted above, he also finds that an excess of precipitable substance checks the reaction, he thinks mechanically, for if the precipitum is collected and resuspended in saline and in concentrated albuminous solutions respectively, the precipitum is rapidly redeposited in the first case, not at all in the second. He appears to be somewhat confused as to the interpretation of this result. It is scarcely due to viscidity alone, an excess of albuminous substance causes re-solution of the precipitated particles. (See Antiprecipitins, p. 149).

Rostoski (1902 b, p. 53) found that more precipitation occurred in 0·5% and 1% than in 6% serum albumin solutions, under otherwise similar conditions. This was still clearer in corresponding globulin solutions. Strong dilutions of serum pseudoglobulin hindered precipitation more markedly than serum albumin dilutions, and pseudoglobulin hindered the precipitation of serum albumin. He concludes that the globulin exerts an antiprecipitating action, besides having an effect through concentration. He compares the antiprecipitin to antitrypsin and antirennet in blood (Fuld and Spiro, 1900—01), and to the normal antihaemolysin of pseudoglobulin (Pick, 1901).

Rostoski (p. 58) also found higher dilutions to favour precipitation of egg albumin by its antiserum, for precipitation took place in 1·5% solutions, whereas it was impeded to some extent in 3·5% solutions.

In medico-legal work it will be desirable to progressively dilute a suspected blood sample, and to reach a conclusion upon the highest dilution (within limits) which reacts to a given antiserum. In routine work, as I have stated, I have worked with dilutions of usually 1 : 100 to 1 : 200. As the dilution increases, the reaction narrows down more and more, the reactions with the highest dilutions being practically specific. Of course, in medico-legal work, the possibility of blood stains being due not to one blood, but to two or more, has to be taken into consideration. On one or two occasions I have found an avian blood react to an anti-mammalian serum. For instance, the blood of a swan (reported in my paper of VII. 1901), gave a very slight clouding with anti-sheep serum. In such quite exceptional cases I have attributed the result to mammalian bloods having come in contact with avian when the samples were collected, the damp filter-paper strips having been placed side by side, with the result that the one contaminated the other. (See tests with blood mixtures, p. 140.)

Reverting to what I have said with regard to the grade of dilutions tested, I would add that Aschoff (1902, p. 192), who cites an earlier paper of mine in this connection, pertinently adds, " Man muss stets den Verdünnungsgrad der Sera und die Menge des zugesetzten Präcipitins berücksichtigen, um die specifische Wirkung richtig zu beurtheilen."

4. *Effect of Preservatives on Antisera and Bloods.*

In my paper of VII. 1901 I stated that the addition of chloroform to an antiserum did not impede its giving a reaction. Its use for purposes of preservation has since been recommended by Uhlenhuth, Stockis, Biondi, Robin, Rostoski, and Strube. Ziemke (17, x. 1901) also used chloroform, but seems to find antisera thus preserved to lose in potency, adding, however, that they also did so when stored dry. He goes too far however when he adds, " Gegenwärtig ist keine Conservir-ungsmethode im Stande, uns das Serum in frischem Zustande zu ersetzen," for as I have shown elsewhere (p. 123) antisera may at times be preserved for a considerable length of time. As I noted in my paper of 16, XII. 1901 (and the observation still holds), none of these authors " state the fact which I have observed, namely, that it is necessary, before using it, to drive off the chloroform by placing the test-tube in the thermostat. A small quantity of chloroform in a test-serum will frequently cause it to produce a considerable clouding, which may be a source of error, as the clouding is produced in almost any serum to which the test-serum is added." It is therefore necessary to guard against such a possibility of error due to the preservative being present except perhaps in minimal amount.

For the same reason I have discarded the use of trikresol, and would advise caution in the use of carbolic acid, as recommended by Uhlenhuth, Strube and others.

In further support of the above statement I will refer to the experiments made in our Laboratory by Graham-Smith and Sanger (1903, p. 285) upon the effects of various volatile antiseptics, as well as such agents as formalin, mercuric perchloride, and copper sulphate, etc.

"As an example of the important volatile antiseptics chloroform may be taken. In solutions containing much of this reagent on the addition of serum a white cloud, and later a deposit, occur. More dilute solutions give rise to slight cloudings. The results of experiments with a series of such volatile antiseptics are given below. When only present in small quantities in preserved sera the possible error due to their presence can be eliminated by placing them in the incubator for half-an-hour to evaporate off the reagent. When present to a greater extent it was found that the supernatant serum above the deposit caused by them still retained its specific properties. This is in accord with what Nuttall has found.

		Corrosive sublimate	Copper sulphate	Formalin	Thymol	Chloroform	Alcohol	Benzol	Toluol	Xylol	Ether
		1—25									
1—10	30 mins.	C	C	+		×	C	×	*	*	*
	24 hrs.	D	C	D		×	D	×	*	*	*
1—100		×	C	*	*	*	·	*	*	*	·
		d	D	*	*	·	·	*	*	*	*
1—1000		*	C	·	·	·	·	·	·	·	·
		*	D	·	·	·	·	·	·	·	·
1—10,000		*	+	·	·	·	·	·	·	·	·
		·	+	·	·	·	·	·	·	·	·
1—100,000		·	*	·	·	·	·	·	·	·	·
		·	d	·	·	·	·	·	·	·	·

See Footnote.

Corrosive sublimate and ferrous and copper sulphates were found to produce very marked effects. They apparently destroy the serum in contact with them and except when present in very small quantities it was found impossible to carry out the test. The effects of dilutions of corrosive sublimate and copper sulphate are given below.

In this and other tables by these authors the following symbols have been used :

C = coagulation.
+ = marked cloud—full reaction.
+ = less marked cloud.
× = medium cloud.
* = slight cloud.
*? = very slight cloud.

D = large deposit after 24 hours.
D = smaller ,, ,, ,,
d = smaller ,, ,, ,,
tr = trace of deposit.
· = no result.
— = no reaction.

Dilutions	Corrosive sublimate and anti-ox serum		Copper sulphate and anti-ox serum	
	1 hour	24 hours	1 hour	24 hours
1—25	immediate coagulation	large deposit	immediate coagulation	large deposit and cloud
1—100	dense cloud	,, ,,	,, ,,	,, ,,
1—500	cloud	cloud	coagulation and cloud	deposit & cloud
1—1000	,,	slight cloud	,, ,,	,, ,,
1—10,000	slight cloud	nil	marked cloud	,, ,,
1—100,000	nil	nil	slight cloud	small deposit

Silver nitrate causes an opaque white cloud on dilution with salt solution up to 1 in 10,000. Dilutions below this do not affect serum when added to them.

Formalin in 1 in 10 dilutions causes marked clouding, which increases till the whole contents of the tube are opaque white. Dilutions below 1 in 100 do not cause sufficient clouding to interfere with the specific reaction. Solutions of thymol of 1 in 100 cause slight cloudings, but lower dilutions do not apparently affect sera.

Lysol and lysoform both cause great turbidity when added to salt solution even in low dilutions, and moreover even in very low clear dilutions the addition of serum causes clouding. No method has been devised for getting rid of these effects; consequently the presence of these substances except in very small quantities would render the test of doubtful value.

The effects of the reagents, which for the sake of convenience we have grouped under the heading of antiseptics, are very marked except in the case of the volatile class. Some of the latter when added in full strength to liquid sera produce heavy deposits, but the supernatant fluid retains its properties. Formalin and corrosive sublimate in strong solutions, as well as the sulphates of copper and iron and nitrate of silver in much weaker dilutions, completely destroy the precipitum-forming property. Lysol, lysoform, and similar antiseptics, owing to their property of forming cloudings with salt solution, render the application of the test of doubtful value in their presence."

In order to determine quantitatively[1] the effects of the prolonged action of antiseptics on fluid sera, the following experiments have been carried out by Graham-Smith (29, VII. '03, p. 358).

"Antiseptics in the proportions given below were added to fluid ox and sheep sera, and allowed to act in sealed bulbs for 4 months. None completely checked bacterial growth. After this period dilutions of 1—21

---

[1] For description of my quantitative method, see Section VII.

in salt solution were made, and all were allowed to stand in open dishes for 2 hours in order that the volatile antiseptics should evaporate off. The results were compared with serum kept under the same conditions but without the addition of any antiseptic. The following table shows that in nearly all cases the precipitum-forming power was slightly reduced, but in a few completely destroyed.

The effects of the presence of these substances in fluids to be tested have been given above.

		Ox serum		Sheep serum	
		Precipitum	Percentage	Precipitum	Percentage
Normal ox and sheep sera		·0356	100	·0140	100
Chloroform	1—1000	·0338	95	·0103	73
	1—500	·0328	89		
	1—100	·0187	55	·0093	65
Xylol	1—500	·0300	84		
	1—100	·0300	84	·0140	100
	1—25	·0281	79		
Benzol	1—1000	·0281	79		
	1—500	·0187	55		
	1—100	·0225	63	·0112	80
Toluol	1—100			·0140	100
Ether	1—500	·0347	97		
	1—100	·0244	68	·0084	60
Formalin	1—10,000	·0262	73		
	1—1000	·	0		
	1—500	·	0		
Alcohol	1—1000			·0075	53
	1—100			·0103	73
Lysol	1—500	·0187	55		
	1—100	·0169	47		
	1—25	·	0		
Lysoform	1—1000	·	0		
	1—100	·	0		
Chinosol	1—500	·0262	73	·0112	80 "
	1—200	·0262	73		

The results obtained by Graham-Smith and Sanger with carbolic acid and chinosol are considered on page 82, those obtained with lime in Section IX.

5. *The acid or alkaline reaction of the medium.*

Normal blood dilutions possess an alkaline reaction. Tchistovitch (v. 1899) found that precipitation only took place when the reaction was alkaline; when it was neutral, a slight opalescence was observable; no reaction took place when the reaction was acid. Linossier and Lemoine (21, III. '02) found that sulphuric acid in the proportion of ·49 : 1000 lessened reaction, that it was reduced to a minimum when

2·45 : 1000 was present, reaction being entirely prevented by acid 4·9 : 1000. Sodium carbonate, added in the proportion ·66 : 1000 had no influence, stronger dilutions thereof checked reaction in proportion to their strength. Nevertheless precipitation may occur in dilutions containing 5·3 : 1000, and 10 : 1000 does not quite prevent reaction. In other words, the reaction takes place best in neutral or slightly alkaline fluids, an excess of acid being more inhibitive than an excess of alkali.

Rostoski (1902, *a* and *b*, p. 52) comes to diametrically opposite results. He states that precipitation is impeded by an alkaline, aided by an acid reaction. He acidified with acetic acid or an acid salt (Mononatrium phosphate) or very small amounts of HCl, as little acid is required to prevent reaction. Reaction, he states, takes place to a lesser degree in neutral than in acid fluids. Slightly alkaline solutions (degree of alkalinity 5·0) also react well, but as the alkalinity is increased precipitation is impeded. It is probable that the discordant results of Rostoski are in part attributable to his having worked with opalescent antisera (see page 73).

It is not impossible therefore that the changes in the alkalinity of the blood in health, but more especially in disease (see von Rigler, 1901) may to a certain extent affect the degree of reaction obtained with precipitins. We propose to pursue the question by means of quantitative tests, as purely qualitative tests would scarcely show any marked difference.

I have referred elsewhere to the abnormally high figures obtained with anti-human serum upon the (concentrated?) bloods of diseased monkeys. In this connection it is of interest to note an observation by Graham-Smith and Sanger (1903, p. 285) published since the above was written.

" Our experiments on this subject are only three in number but suggest that important differences may be found in diseased blood by means of this test. The following observations were made on sera from tuberculous cattle. The first required 2 c.c. of decinormal caustic soda per 100 c.c. of serum to give a pink tint with phenolphthalein, and the others 1·25 c.c. and ·8 c.c. respectively. Also the former required 4·25 c.c. of decinormal caustic soda per 100 c.c. to produce a condition in which the serum was liquid when hot and solid when cold, and the latter 2 c.c. and 1·2 c.c. respectively. As a mean of three estimations in each case these sera produced ·0375 c.c., ·0328 c.c. and ·0244 c.c. of precipitum.

Sera	$\frac{NaOH}{10}$ per 100 c.c. to give pink with phenolphthalein	$\frac{NaOH}{10}$ required per 1000 c.c. to make serum liquid when hot, solid when cold	Precipitum
1.	2·0 c.c.	4·25 c.c.	·0375 c.c.
2.	1·25 ,,	2·0 ,,	·0328 ,,
3.	·8 ,,	1·2 ,,	·0244 ,,

Strangeways has made numerous observations (unpublished) on the differences in precipitum-forming power of the sera in disease."

We have found that an acid-reacting blood dilution will because of the acidity alone produce precipitation (see below and Section IX.). The observation has considerable practical importance medico-legally, for the reason that blood stains may occur on acid fabrics, or acid-reacting leather, the acidity in the substratum impeding the test by the anti-serum in the blood solution made by steeping the stained article in saline solution. The subject has been carefully gone into by Graham-Smith and Sanger (p. 281), who report as follows:

" Observations on the reaction to litmus of extracts of coarse cloths and leather, as well as the possibility of the treatment of blood-stains in forensic practice with chemical reagents, made it desirable to investigate the action of such reagents on blood.

Acids.

Several experiments were made with dilutions of both organic and inorganic acids in distilled water and salt solution. Dilutions from 1 in 10 to 1 in 100,000 were tested by dropping in serum and noting the effects up to 2 hours and after standing for 24 hours. In these observations one drop of antiserum was added to about ·5 c.c. of the dilution since this has been the quantity uniformly used in qualitative work.

The addition of larger quantities produced slightly different results, probably owing to the alkalinity of the serum itself, and moreover, perhaps for the same reason, the effects of different sera were noticed to vary slightly. This remark applies to all the following experiments of a similar nature.

With the inorganic acids a noteworthy phenomenon was observed. Strong solutions (1 in 10) in salt solution caused coagulation of the serum and destruction of the precipitating substance, whereas weak solutions (1 in 100) produced no result. Dilution between 1—500 and 1 in 10,000 caused more or less clouding, in the latter case taking place half-way up the tube. These cloudings probably resulted from the

N. 6

precipitation of the albumen by the dilute acid and were observable within a few minutes. With greater dilutions nothing occurred. It was also found that neutralisation previously with sodium carbonate prevented these cloudings and in some cases even dissolved them after they had been formed.

One example is given below in detail.

Dilutions of Nitric Acid in normal salt solution	Anti-human serum		Anti-human serum after neutralisation	
	1 hour	24 hours	1 hour	24 hours
1—10	coagulation	coagulated mass	—	—
1—100	very faint cloud	cloud	—	
1—1000	marked cloud	cloud	—	—
1—10,000	slight cloud half-way up tube	—	—	—
1—100,000	—	—	—	—

The precise degrees of clouding caused by the various dilutions differ with different acids.

Some organic acids behaved differently, acetic, oxalic, and tartaric causing little clouding in dilutions of 1 in 10, but marked cloudings in 1 in 100 and in 1—1000.

Details of the action of tartaric acid :—

Dilutions of Tartaric Acid in salt solution	Anti-human serum		Anti-human serum after neutralisation	
	1 hour	24 hours	1 hour	24 hours
1—10	very slight cloud	slight cloud	—	—
1—100	„ „ „	medium cloud	—	—
1—1000	medium cloud	marked cloud	—	—
1—10,000	—	slight cloud half-way up tube	—	—
1—100,000	—	—	—	—

The accompanying table shows the chief actions of acid dilutions in salt solution on serum. These effects are more marked when the dilutions are made in distilled water.

		Sulphuric	Nitric	Hydrochloric	Acetic	Oxalic	Tartaric	Picric	Carbolic	Citric	Salicylic (sat. sol.)	Chinosol
1—10	30 mins.	C	C	+	.	*	*	.	C	.	+	.
	24 hrs.	D	D	D	.	*	*	.	D	.	D	.
1—100		.	*	.	*	*	*	C	*	*	*	*
		.	*	.	×	*	*	D	d	*	.	*
1—1000		*	×	×	*	×	×	+	.	×	.	*
		*	×	×	×	×	×	d	.	*	.	*
1—10,000		*	*	*	.	*	.	*	.	*	.	.
1—100,000		.	.	.	.	.	.	.	.	.	.	.

Strong alkalis and salts.

Experiments with the more powerful alkalis showed that in strong solutions cloudings were also produced in them on the addition of serum. Ziemke (17, VIII. 1901) has recommended the use of ·1 %, caustic soda in distilled water for extracting blood-stains under certain conditions. Our observations show that in such dilutions cloudings are apt to occur on the addition of any antiserum, and render it thus an unsuitable agent for the process[1]. These cloudings are better marked in dilutions in distilled water than with those in salt solution.

The effects of dilutions of caustic soda are given in detail below :—

Solutions of caustic soda in salt solution	Anti-human serum	
	1 hour	24 hours
1—10	slight cloud	slight cloud
1—100	medium cloud	medium cloud
1—1000	slight cloud	—
1—10,000	—	—
1—100,000	—	—

The following table shows the actions of dilutions of alkalis and salts on serum :—

		Caustic soda	Caustic potash	Sodium carbonate	Ammonia	Ammonium sulphate	Ammonium tartrate	Sodium & potassium tartrate	Sodium acetate	Potassium cyanide	Sodium citrate	Magnesium sulphate	Potassium nitrite	Potassium chloride	Borax
										1—20				1—20	
1—10	30 mins.	*	*	·	·	*	*	*	*	*	·	·	·	·	
	24 hrs.	*	*	·	·	*	*	·	*	*	·	·	·	·	
1—100		×	×	·	·	·	·	·	*	*	·	·	·	·	*
		×	×	·	·	·	·	·	·	*	·	·	·	·	·
1—1000		*	*	·	·	·	·	·	·	·	·	·	·	·	·
1—10,000		·	·	·	·	·	·	·	·	·	·	·	·	·	·
1—100,000		·	·	·	·	·	·	·	·	·	·	·	·	·	·

Owing to the absence of any bad results from the addition of serum to sodium carbonate dilutions, we chose this reagent as being the most suitable for neutralising the effects of acids.

We next made some experiments to ascertain to what extent the acidity or alkalinity of the medium influenced the specific reaction. For this purpose both quantitative and qualitative experiments were conducted. The solutions in each case were made up in the following

[1] Confirmation of my statement on p. 64.

way. Three series of 11 tubes were prepared, each containing ·5 c.c. of a 1 in 21 dilution of human serum in salt solution. To the first tube were added 5 drops of a solution of acid, to numbers 2, 3, 4 and 5 were added 4, 2, 3 and 1 drops of acid respectively. The sixth tube was not treated in any way. Numbers 7 to 11 received 1 to 5 drops of alkali respectively.

In the first series very small drops of a 1 in 100 dilution of hydrochloric acid was used, in the second series large drops of the same solution, and in the third series large drops of 1 in 10 solution of the same acid. Large drops of corresponding dilutions of sodium carbonate were used in series two and three.

In the quantitative experiments ·1 c.c. of anti-human serum was run into each tube, and in the qualitative one drop.

No. of tube		Drops	Series I. Small drops	%	Series II. Large drops	%	Series III.		%
1.		5	·0309	(61)	·0009	(2)		nil	(0 %)
2.	1 in 100	4	·0319	(63)	·0018	(4)	1 in 10	,,	,,
3.	Hydrochloric	3	·0431	(85)	·0140	(27)	Hydrochloric	,,	,,
4.	Acid	2	·0431	(85)	·0187	(37)	Acid	,,	,,
5.		1	·0478	(93)	·0422	(83)		,,	,,
6.	Normal dilution of serum		·0507	(100)	·0510	(100)		·0516	(100 %)
7.					·0422	(83)		·0169	(31 ,,)
8.	1 in 100				·0441	(86)	1 in 10	·0150	(29 ,,)
9.	Sodium				·0422	(83)	Sodium	tr	?
10.	Carbonate				·0469	?	Carbonate	nil	(0 %)
11.					·0591	?		,,	,,

In series II the acidity and alkalinity varied from about 1—1000 to 1—5000 and in series III from about 1—100 to 1—500.

These experiments show that the presence of even small quantities of acid or alkali rapidly reduce the quantity of precipitum formed (see Plate, fig. 5). The apparent exceptions of numbers 10 and 11 of series II are due to the fact that the precipita produced were more flocculent and occupied more space than the more compact precipita elsewhere obtained. They also indicate that the presence of small quantities of acid or alkali do not alter the specificity of the reaction, for controls with anti-sheep serum were all negative.

Qualitative experiments undertaken on the same lines showed that with 1 in 100 solutions of acid and alkali, cloudings first occurred in the normal tube, next in the alkaline series, the times of their appearance increasing from No. 6 to 11. The last two showed faint traces of the specific reaction and also general opacity. On the acid side cloudings due to the acid rapidly appeared, but later specific cloudings were superadded in Nos. 3 to 5. Control tubes tested with anti-ox serum showed general opacity in the last of the alkaline series and slight clouds in the acid series. Similar experiments with 1 in 10 solutions showed cloudings in the normal serum and first three specimens of the alkaline series only.

In the light of these observations it becomes necessary to test the reaction to litmus of all solutions which are to be examined and, if found decidedly acid or alkaline, to neutralize them.

It must also be remembered that the addition of strong acid or alkali to fluid or dried blood completely destroys it."

The effects of the prolonged action of Acids and Alkalis on Fluid Sera.

These experiments were continued by Graham-Smith (1903, p. 359) as follows:

" Ox, and sheep, sera with acids and alkalis added in the proportions given below were kept under conditions similar to those just mentioned.

		Ox serum		Sheep serum	
		Precipitum	Percentage	Precipitum	Percentage
Normal ox and sheep sera		·0356	100	·0140	100
Hydrochloric acid	1—1000			·0093	65
	1—500	trace	?		
	1—100	·	0		
Nitric acid	1—1000			·0103	73
	1—500	·0046	13		
	1—100	·	0	·	0
Sulphuric acid	1—1000			·0112	80
	1—100			·	0
Acetic acid	1—1000			·0112	80
	1—500	·0244	68		
	1—100	·0300	84	·0112	80
	1—10	trace	?		
Oxalic acid	1—1000	·0225	63	·0140	100
	1—500	·0244	68		
	1—10			·0028	20
Carbolic acid	1—1000			·0131	93
	1—100			·0112	80
Citric acid	1—1000			·0131	93
	1—100			·	0
Caustic potash	1—1000			·0122	87
	1—500	·0150	42		
	1—100	·	0	·	0
Caustic soda	1—1000			·0122	87
	1—100			·	0
Sodium carbonate	1—1000			·0103	73
	1—500	·0244	68		
	1—100			·0103	73
	1—10	·0244	68		
Ammonia	1—1000			·0140	100
	1—500	·0309	84		
	1—100	·0206	57	·0131	93
	1—10	·0065	18		

After dilution, the solutions were neutralized, and then tested quantitatively. Control antisera were also used in all cases, but gave no reactions.

The effects of the presence of unneutralized acids and alkalis have been given previously (p. 82).

The preceding table (p. 85) shows that, except in very small quantities, the prolonged action of inorganic acids completely destroys the precipitable substance, but that organic acids do not exert so deleterious an influence. Strong alkalis act in the same way as inorganic acids."

6. *Precautions with regard to Quantitative Tests by my Method.*

It is obvious that if either antiserum or the serum to be tested undergo concentration by evaporation the figures obtained will be fallacious. The possibility of serum becoming concentrated *in corpore,* in consequence of disease, has been referred to on page 80. It is naturally essential that there shall be no matter in suspension, or bacterial multiplication occurring in either the test serum or blood dilution.

7. *Bacterial Multiplication.*

I have referred to an objection to weak and slowly-acting antisera being the opportunity which is given to bacteria to develope, when tubes are left standing as long as 24 hours, especially when they are placed at higher temperatures as has been done in experiments of some investigators. In my experiments I have on several occasions had my later readings vitiated by bacterial development, especially during summer, the readings being made after 24 hours and the tubes kept at room temperature. The object of thus prolonging the experiment was to note the deposits formed in the solutions overnight. I sought to guard against this by noting the deposits earlier or later, depending upon the temperature of the room in which the experiments were made; in summer the deposits were noted after 12—16 hours, in winter after 24 or more hours.

Where bacteria develope, provided that the readings are made sufficiently early, there will be little chance of confusing the *clouding* produced by microorganisms, with what I wished to note, namely the *deposit* at the bottom of the tube. A little experience renders it easy to as a rule recognize early bacterial cloudings, for they generally arise from the lower strata of the fluid, gradually float upward, the bacteria seeking oxygen. The bacterial cloud at times remains stationary, or moves gradually upward in the fluid, so that clear fluid may be seen above and below. The behaviour of various bacteria under somewhat similar conditions, has been the subject of an interesting investigation by Beyerinck (Ueber Athmungsfiguren beweglicher Bakterien. *Centralbl. f. Bakteriol.* 1893, XIV. pp. 827—845), who refers to the "Bakterienniveau" as the point in the fluid where the bacterial cloud may hover at times for days.

See further under medico-legal applications of the precipitin method, Section IX., also p. 63, regarding insoluble bloods.

SECTION II.

THE NATURE OF PRECIPITIN REACTIONS.

THAT the precipitins are used up in the process of precipitation, was already observed by Myers (14, VII. 1900) who took it as evidence of the action being chemical in nature. Michaëlis and Oppenheimer (1902, p. 357) consider that the precipitins cannot be regarded as a special form of coagulative ferment for the reason that they are used up quantitatively in considerable amount. Whereas, it is true, that ferments are also ultimately used up, this occurs but slowly and in a very slight degree as compared to the amount of substratum affected. Precipitins are used up in a manner analogous to what takes place when toxin is added to antitoxin, or acid to base.

Eisenberg (5, v. 1902) following Ehrlich's absorption method, determined the amounts of the mixed interacting substances, before and after reaction had taken place. In this way he was able to find that *both substances combine quantitatively*, for measureable amounts of both had disappeared from the mixture. The results of Müller (1902) and of Leblanc (1901) with lactoprecipitins and haemoglobinprecipitins gave similar results, pointing to the fact that *both substances are present in the precipitum.* Eisenberg therefore adds that Halban and Landsteiner (1902) are scarcely justified in denying the chemical nature of the reaction. These authors claim in proof of this that salt-free serum albumin heated to 100° C. partly loses its precipitability, although it contains as much albumin as before. It may well be said in rejoinder that this was no longer native albumen. I have stated elsewhere that the precipitins act independently of the amount of albumen, the latter must therefore possess certain specific properties if they are to combine with the precipitins.

In this connection it is of interest to cite some very careful analyses and tests made here by Mr F. G. Hopkins, to whom I am exceedingly indebted for the loan of his notes, the experiments not having as yet been published. The animals were treated by me with substances supplied by Mr Hopkins.

Observation I. Two rabbits were treated with intraperitoneal injections of crystalline egg albumen, receiving 5 injections, the total volume injected representing 45 c.c. The animals were bled 11 days after the last injection (22, VII. 1901).

The two antisera were mixed. The mixture remained clear on standing.

35 c.c. of antiserum were fully precipitated by crystalline egg albumen solution, the amount of which was not determined.

The dried precipitum weighed only ·0246 g.

Observation II. Three rabbits similarly treated to the above, except that one received 35 c.c., and that the animals were bled 7 days after the last injection (27, VII. 1901).

The three antisera remained clear when mixed.

50 c.c. of antiserum were fully precipitated and the dried precipitum weighed ·0325 g.

Observation III. Two rabbits were treated with ox-serum, receiving 5 injections of a total amount of 45 c.c. intraperitoneally. Bled 23, v. 1901.

The two antisera remained clear when mixed.

When the antiserum was added to undiluted ox-serum, drop by drop, *the instantaneous precipitum produced was redissolved on stirring.* Ten c.c. of ox-serum requiring 5—6 drops of antiserum before permanent clouding was obtained. Evidently then *a precipitum is soluble in excess of serum.* (See Antiprecipitins, p. 149.)

The ox-serum was next added to the antiserum. When 6·5 c.c. of ox-serum had been added to 30 c.c. of antiserum no further precipitation occurred, and there was no further clouding subsequently when more antiserum was added to the clear supernatant fluid. The total precipitum obtained from the above mixture of 30 c.c. antiserum and 6·5 c.c. ox-serum, weighed when dried at 110° C. ·553 g. Prior to drying, it should be added, the precipitum first obtained with the aid of the centrifuge had been thoroughly washed with 5 % NaCl solution, being resuspended and recentrifugalized three times in saline, this being followed by repeated washings with water, the washing being similarly conducted, until the washings gave no trace of biuret or xanthoproteic reaction. The tests conducted upon this precipitum will be considered presently.

Observation IV. A rabbit treated in the usual way with sheep-serum, yielded an antiserum of which 25 c.c. took 4·5 c.c. of sheep-serum to fully precipitate it. The washed precipitum weighed, when dry, ·158 g.

Observation V. Two rabbits were treated with horse-serum, receiving 6 and 5 injections each (55 and 45 c.c. total). Bled 3 days after last injection. The mixed sera remained clear, and 40 c.c. thereof took about 5 c.c. of horse-serum to fully precipitate it. The washed precipitum weighed when dry ·366 g.

Strube (12, VII. 1902) states that his results contradict those of Biondi. Strube, like others, has found a quantitative relation to exist between the interacting substances, for on adding 5 c.c. of a blood solution to different proportions of its homologous antiserum (1:10 to 1:5000) he obtained, as I have done, decreasing quantities of precipitum. In very dilute solutions, as Biondi is also stated to have observed, large quantities of precipitum may apparently be formed, this being attributed by Strube to the looser character of the precipitum making it appear more than it actually represents. He also notes, that by repeatedly adding an antiserum to blood dilution, and filtering each time after precipitation takes place, fresh precipitations occur every time antiserum is added, the amounts of antiserum of course being but fractions of the total amount required to induce complete precipitation. This also indicates very clearly *that the reaction is quantitative and not due to enzyme action.*

Inactivated Antisera.

Müller (18, II. 1902) found a lactoserum which had been inactivated at 70° C. to have acquired the power of neutralizing precipitins, in the sense that it *prevented precipitation* when active serum was added to inactivated, the latter having been previously mixed with a given milk. Lactoserum from which the precipitin had been removed by the addition of casein is incapable of neutralizing active lactoserum. This indicates that *the neutralizing substances are not present in fresh lactoserum,* but that they are formed in consequence of heating. Inactivated *normal* rabbit serum possesses no neutralizing power. The neutralizing substances are not affected by the presence of lime salts, and may be precipitated from inactivated lactoserum by means of dilute acetic acid, for they are absent in the neutralized filtrate. Müller finds that *precipitins do not combine with the neutralizing substance* in the presence of (acetic) acid, on the other hand, under similar conditions, the precipitins combine with casein, precipitation occurring although reaction is retarded. *Inactivated lactoserum dissolved the precipitum* after some hours, whereas inactivated normal rabbit serum did not. Lactoserum robbed of its precipitin by the addition of milk, acquires no neutralizing properties even when heated to 75° C. Müller therefore concludes that *the neutralizing substances are derivatives of precipitins, originating from these when they are heated.*

Whereas normal inactivated rabbit serum possesses no anti-rennet action, inactivated lactoserum does prevent coagulation through rennet ferment just as it prevents precipitation through fresh lactoserum. He concludes therefore that *the neutralizing action depends upon the precipitable* (casein) *and neutralizing substances combining*, thus protecting the casein from the action of the coagulating agents.

Müller considers the neutralizing substances "precipitoids," analogous to the "agglutinoids" recently described by Eisenberg and Volk (see p. 49). According to Ehrlich's theory, the precipitoids would be receptors whose "zymophoric" group has been destroyed by heat, but whose "haptophoric" group is retained (see p. 12).

Two views may be held with regard to the nature of the neutralizing or combining body in inactivated serum, according to Eisenberg (v. 1902, p. 301). Either, (a) the combining body preexists in unheated serum, its action being obscured by that of the precipitin, or (b) it originates from another substance in precipitin through the action of heat. The latter view, as we have seen, is held by Müller, and apparently also by Eisenberg. Supposition (a) appears unlikely to Eisenberg, for the reason that if it is correct an excess of precipitatable substance should have no effect in preventing the reaction of the bacterioprecipitin with which Eisenberg experimented. He found, on the contrary, that an excess thereof increased such action[1]. Moreover *active serum robbed of its precipitin through the addition of precipitable substance no longer antagonizes fresh precipitin*. Neither does deprecipitinated antiserum, as Müller showed, when heated, prevent the action of precipitins.

---

[1] See "Observation III" p. 89 where I have stated that an excess may prevent precipitation by haematosera. This statement of Eisenberg's may therefore have to be modified. He was working, however, with bacterioprecipitins which possibly behave differently.

Eisenberg therefore concludes that precipitins contain two bodies, (*a*) which is labile and produces the visible reaction of precipitation, (*b*) which is stable and represents the haptophoric or combining group of the precipitin, the part which possesses special affinity for the precipitable substance in the homologous serum. On theoretical grounds Ehrlich (*Croon. Lect.*, 1900) assumed the precipitins to possess such a constitution, defining them as receptors of the second order (see p. 12), and we find that Müller and Eisenberg reach the same conclusion in consequence of their experiments.

Eisenberg (p. 297) found inactivated antiserum to prevent precipitation by fresh antiserum both when it was added to precipitable substance before, or when added simultaneously with, fresh antiserum. In such a reaction there are three components, *a*. precipitable substances, *b*. heated antiserum, *c*. fresh antiserum. On adding a constant amount of a mixture of *a*. and *b*. to a varying quantity of *c*., it will be seen that the antagonism is overcome as the amount of *c*. added is increased. Eisenberg notes that small amounts of *b*. are more antagonistic than large amounts, which is not easy to understand, and requires confirmation. He does not consider the antagonism due to the action of *b*. on *c*., for *the inactivated substance antagonizes best when added to the precipitable substance before the active antiserum.* In other words, it does not antagonize as completely when added together with fresh precipitin. From the fact that precipitation may be prevented when *b*. and *c*. are added simultaneously, it seems clear that *b*. has greater affinity for *a*. than has *c*. It is only when *c*. is in excess that *b*. exerts a limited influence, this being due to the preponderating amount of *c*. giving it a greater chance of being the first to enter into combination with the precipitable substance. Where a reaction is brought about by adding an excess of *c*., we have possibly to deal with a " Massenwirkung " such as has been observed by Dönitz, who found that an excess of antitoxin apparently removed diphtheria toxin which had already become anchored to body cells.

Similar observations with regard to the bacterioprecipitins have been made by Kraus and von Pirquet (5, VII. 1902, p. 68). They found that the bacterioprecipitins for typhoid and cholera were inactivated by heat (58° C.), and that such inactivated sera when added to fresh culture filtrates prevented precipitation upon the addition of active bacterioprecipitin. A fact which the authors state has been independently observed by Pick. Inactivated precipitin was allowed to act for 10 hours at 37° C. before adding fresh antiserum. They do not think that the "antiprecipitin[1]" in inactivated serum acts upon the precipitins as Pick supposed, for mixed active and inactive serum allowed to stand for 10 hours at 37° C. gave a precipitum on being added to culture filtrate. They conclude that heated serum retains its combining, but loses its precipitating power, for the reason that it consists of (stable) combining and (labile) precipitating groups. In other words, they agree with Müller and Eisenberg, as stated above.

It was further found by Kraus and von Pirquet (p. 71) that an excess of *old* cholera antiserum prevented precipitation in culture filtrates, as did the inactivated fresh antiserum. *The combining groups would therefore appear to outlast the precipitating groups in such antisera when stored.* It also appears evident that the combining groups, or, as we shall term them precipitoids, acquire a greater affinity for the precipitable substance than the remaining intact precipitin. This is completely in accord with what has been observed with agglutinins and agglutinoids by Eisenberg and Volk (see p. 49). The breaking down of precipitin to precipitoid is comparable to the changes undergone by toxin into toxoid as observed by Ehrlich.

Michaëlis (25, IX. 1902) inactivated antiserum for ox-serum globulin by exposure to a temperature of 68° C. for 15 minutes. He does not think the precipitin is converted into precipitoid, for the reason that when the heated antiserum is added to precipitable substance it retards but does not prevent the action of fresh antiserum. He considers the preventive action as non-specific, but depending upon physical causes, viz. the addition of neutral colloidal substance, for any other dilution of albuminous substance will exert the same effect. This is contradictory to the views previously expressed, where the authors agreed that the action was specific, for inactivated normal rabbit serum added under similar conditions to inactivated antiserum (also from rabbits) did not prevent precipitation. Michaëlis (p. 459) moreover claims that heated antiserum markedly reinforces the action of a minimal quantity of fresh antiserum, as much as if one had added the corresponding amount of fresh antiserum. These results require confirmation and suggest, it seems to me, that he may have worked with incompletely inactivated antisera. Michaëlis believes that precipitation is not due to a single substance, the "precipitin," but to a complex similar to that possessed by the haemolysins. With precipitins we have the conjoint action of two substances, *both absent* in normal sera,

---

[1] Evidently a misnomer, precipitoid is a more suitable term.

in contrast to what is the case with the haemolysins. In the haemato-sera then there are two substances, the thermolabile being predominant, the thermostable being small in amount.

Referring to the neutralizing action of inactivated precipitating antisera as observed by Müller, Eisenberg, Kraus and von Pirquet, above described, I would add that an analogous phenomenon has also been observed with regard to specific bacteriolytic sera by Neisser and Wechsberg, Lipstein, and Walker (see Bacteriolysins, p. 21), the inacti-vated bacteriolytic serum, containing an immune body, which antagonizes the action of fresh antiserum, when it is added to the latter in excess.

The Non-reactivatability of Heated Precipitating Antisera.

In the paper by Myers (14, VII. 1900) that author stated that he had succeeded in obtaining an antiserum for Witte's pepton, that he had been able to inactivate it by heat (see p. 112) and to reactivate it with normal rabbit serum. His observations in this regard have not been confirmed, although he was the first to refer to inactivated pre-cipitin as precipitoid, drawing an analogy between the change observed in precipitin with that which takes place when toxin is converted into toxoid. If Myers' observations were correct, then precipitins would constitute receptors of the third order (see p. 12) according to Ehrlich.

No observer has been able to reactivate precipitating antisera since Myers. Attempts have been made by Eisenberg (v. 1902, p. 302), Kraus and von Pirquet (5, VII. 1902, p. 68), Michaëlis (9, x. 1902, p. 734), and myself. Since the addition of a complement to inactivated antisera does not reactivate them, we must conclude in the light of Ehrlich's theory that *the precipitins constitute receptors of the second order* (see fig. 2, p. 12), as is assumed to be the case with toxins and agglutinins. In Eisenberg's experiments (p. 297) the antiserum was inactivated by an exposure of 1 hour to 72° C., reactivation being attempted both with homologous and heterologous sera, and also by adding quantities of active antiserum, too small to produce by them-selves any reaction.

On Immune Bodies in Precipitating Antisera.

Until 1902 immune bodies had only been observed and studied in sera which act on formed elements, bacteria or different cells. Gengou (25, x. 1902, p. 739) has sought for them in precipitating antisera (from rabbits) for milk, egg-albumen, horse fibrinogen, and also in the antiserum from guinea-pigs treated with heated rabbit and dog serum. He found that all these antisera contained immune body, which for instance in lactoserum attached itself to milk, and thus allowed the precipitin to act, just as is the case with the anti-microbic and other sera above referred to. He tested this as follows:

He prepared

LS. Fresh lactoserum heated to 56° C. for 30 minutes.

NS. Normal rabbit serum heated to 56° C. for 30 minutes.

A. Normal rabbit serum 24 hours old, containing alexin.

C. Washed fowl's corpuscles suspended in saline after having been subjected to action of heated immune-serum of a rabbit which had received injections of fowl's corpuscles.

L. Cows' milk.

NaCl. Normal salt solution.

He mixed these in different ways as follows (I have omitted the quantities intentionally):

Test I. mixed L
LS
A
result : precipitation ; fluid added to *C* don't haemolyse, acts as if no alexin were present ; alexin must be attached to milk as it was not anchored to LS in Test III., nor by milk alone in Test II.

II. mixed L
NS
A
result : no precipitation, *C* haemolised as in Tests III. and IV., alexin being free.

III. mixed NaCl
LS
A

IV. mixed NaCl
NS
A
result : haemolised *C* (added 24 hrs. later) in 30 minutes, completely in 1½ hrs. In this and Test IV. the milk being absent, the alexin remained free to act on the sensitized corpuscles.

V. mixed L
LS
NaCl

VI. mixed L
NS
NaCl
result : no haemolysis of *C*, there being no alexin present.

Gengou concludes from the above ingenious experiment that lacto-serum contains immune body ("sensibilisatrice") besides the precipitin.

On the Nature of Precipitins and Precipitable Substances.

The power of precipitation possessed by antisera does not appear to depend upon any alteration in the usual physical properties of a serum, for Beljajew (1902) has found no difference in the specific gravity, nor in the alkalinity between these and normal sera. Whitney (1902) claims that a powerful antiserum clotted more slowly than a weak one, a statement which requires confirmation.

The nature of precipitins has been the subject of considerable investigation, as will appear from what follows. Nolf (v. 1900, p. 300) treated rabbits with (*a*) washed blood corpuscles, and (*b*) the serum of the fowl and dog, and found precipitins only in the blood of the rabbits treated with serum, or with "plasma[1]." He next found that he could produce precipitins in animals treated with globulin solutions, those treated with albumin solution giving a negative result. Artificial globulin solutions gave as good results in this respect as did normal sera. Subsequently Myers (14. VII. 1900) produced antisera for sheep and ox bloods by injecting their serum globulins into rabbits, and Stockis (v. 1901) obtained antisera by globulin injections. Leblanc (31. v. 1901, p. 359) injected different components of ox blood (serum albumin, pseudoglobulin, haemoglobin) into rabbits and found the precipitins in the pseudoglobulin fraction of the immune serum. According to Leblanc the precipitins are pseudoglobulins, or bodies precipitated together with these. Eisenberg (v. 1902, p. 308) on the other hand found the precipitins in the euglobulin fraction of the serum of rabbits treated with horse serum. Strube (12. VI. 1902) cites the results of Corin, Modici, and Biondi as pointing to the precipitins being bound to the globulins. It appears doubtful however whether the globulins themselves exert the precipitating action (Biondi) if we rely on analogous studies upon the antitoxins, in which Dieudonné (*Arb. a. d. kaiserl. Gesundheitsamte*, XIII) found in the case of diphtheria antitoxin that precipitation with CO_2 yielded antitoxin-free globulins, whereas precipitation with magnesium sulphate, as practised by the above observers for the precipitins, caused antitoxins to be carried down with globulins mechanically. Biondi considers the precipitins possess a fermentative action for the reason that he found different amounts of antiserum added to equal amounts of serum to yield no markedly different quantities of precipitum. We have however seen that there is every evidence in

[1] Ide (1902), and Dubois (1902, p. 692) state they have since obtained precipitins by injecting haemoglobin solutions and intact blood corpuscles respectively.

favour of the action not being fermentative, and I do not doubt but that Biondi did not work with a sufficiently large series of dilutions, for if he had he would have convinced himself to the contrary, as can be readily done[1]. Oppenheimer and Michaëlis (18, VII. 1902) consider the components in serum which lead to precipitation to be constituents of albumin, not merely attached thereto. In a later paper Michaëlis (9, X. 1902, p. 734) states that he found precipitins in that fraction of serum which is precipitated by half saturation thereof with ammonium sulphate, namely, in the globulin fraction. He attempted to further isolate the precipitin by repeating the fractional precipitation with ammonium sulphate and found that most of the precipitin came down with the first fraction (0 and 30 $\%$ saturation), the second fraction (precipitated by 30 and 50 $\%$ saturation) only containing traces of precipitin. The precipitable substance behaved similarly. He finds that anti-ox globulins do not precipitate serum-albumin. Corin (1902) considers the active principle which leads to the formation of precipitin to be paraglobulin, and that therefore it would be best to treat animals with paraglobulin solutions to obtain antisera, such as he prepared for the dog. He finds the precipitin bound up with the paraglobulin, and this can be separated and dried, the dried powder being brought into clear watery solution and used directly for testing.

According to Obermayer and Pick (1902) the very rapid appearance (within 15 minutes) of precipitins in the circulation of animals which have received an intraperitoneal injection of precipitating antiserum, indicates that the precipitin is non-albuminous. They found the precipitins to be contained in the euglobulin fraction, both in rabbit and horse immune serum, obtained from the goat and rabbit respectively. In the hope of determining which of the constituents of egg-albumin are concerned in causing the formation of precipitin *in corpore*, they treated rabbits with subcutaneous injections of egg-albumin and tested the action of the antiserum thus obtained upon egg-white and its constituents. They state that normal rabbit serum produces an appreciable precipitum in egg-albumin solutions[2], this being due especially to the action of dysglobulin on the serum, the action of other constituents being much less marked. They consider that the egg-white precipitins do not constitute an integral part of the albuminous sub-

---

[1] An extensive series of tests made by Mr Strangeways at my suggestion during the past year, which have not been hitherto published, distinctly show Biondi to have been wrong.

[2] Strength not stated; see normal precipitins, p. 150.

stance, but that they adhere in different proportions to its constituents. In support of this assumption they state that trypsin-treated albumin gives a reaction with precipitins when all the albumin has been broken up as indicated by the biuret reaction. They conclude that the precipitin-formation, due to immunization with albumins of egg-white, is independent of albumins, and that it depends upon a substance which is difficult to separate from albumins in the process of chemical cleaning, consequently it cannot be a specific process due to the action of albumins of egg-white. The immunifying substance, which they wrongly style "precipitogen[1]," as well as the precipitin they conclude are not albuminous, and consequently the biological test is of no use in the determination of albumins. They found anti-egg serum to resist the action of hot 0.5% nitric acid. These observers and others have found the precipitins in different antisera to be precipitated by alcohol.

The rapidity of the reaction between precipitin and precipitable substance certainly points to the existence of great affinity between the interacting substances.

In relation to the *precipitable substance*, Halban and Landsteiner (25, III. 1902, p. 475) have found that serum-albumin, rendered salt-free by dialysis and subsequently heated to 100° C., reacted but slightly to precipitins, though when boiled 15 minutes it contained scarcely less albumin (as understood by chemists and physiologists) than before. It would therefore appear that precipitation constitutes a reaction with definite chemical groups (within the albuminous molecule?) like the agglutination reaction, etc. Linossier and Lemoine (18, IV. 1902) and Leclainche and Vallée (25, I. 1901) find that the action of antiserum is not comparable to that of other reagents such as heat or nitric acid. This is proved by the fact that the precipitin does not always give reactions proportional to the amount of albumin present. Antisera are more active towards globulins than nitric acid, less so towards serum-albumin. With urine which contains much of the latter substance an antiserum may give no reaction.

Eisenberg (v. 1902, p. 306) notes that heated albumin is more alkaline than native, and that alkalized albumin is not precipitable. He found (p. 307) albumin to be modified by contact with concentrated solution of urea, added in the proportion of 4 : 1 to albumin solution, and

---

[1] A misnomer, as pointed out by Michaëlis and Oppenheimer (1902, p. 341) for the reason that it suggests a relation between precipitin and (" precipitogen ") precipitable substance such as exists, for instance, between pepsin and pepsinogen, in other words that the " precipitogen " is a forerunner of precipitin, which it is not.

also by weak formalin, both of which are known to coagulate albumin, the solutions thus treated being no longer precipitable. Nevertheless the albumin retained its affinity for precipitin. Eisenberg and Volk have made analogous observations upon agglutinins. Michaëlis and Oppenheimer (1902, p. 347) refer to the earlier observations on toxins as possibly throwing some light upon the nature of the precipitable substance. The toxins were at first considered albuminous, and for this reason called toxalbumins. Further investigation showed that they might be freed from albumin. Thus Brieger (1895) and others prepared tetanus toxin which gave no biuret reaction, Jacoby (1900-1) did the same with ricin, and Hausmann (1902) with abrin. Nevertheless Michaëlis and Oppenheimer consider that both in the case of toxin and precipitable substance these substances may be simply torn away from the large albuminous molecule to which they naturally belong.

The Precipitum.

According to Tchistovitch (v. 1899) the precipitum is soluble in dilute acids and alkalis; insoluble in water, solutions of neutral salts and alkaline carbonates. Nolf (v. 1900) considers the precipitum a globulin, giving as evidence thereof that a precipitin added to globulin and albumin solutions separately or mixed, yields a precipitum. Tchisto-vitch states a precipitum is only formed in alkaline solutions, although there may be some opalescence in neutral solutions, whereas acid solutions remain clear. The precipitum (from eel and anti-eel serum) proved non-toxic to rabbits when administered intravenously, although the clear supernatant fluid proved toxic when it contained unneutral-ized ichthiotoxin[1].

Bordet (III. 1899) found different precipita soluble in dilute alkali. Leclainche and Vallée (25, I. 1901) state that washed precipitum gives all the albumin reactions. Leblanc (31, v. 1901, p. 362) states that a repeatedly washed precipitum, obtained by adding anti-ox globulin (from rabbit) to ox globulin solution, had a *pink* tint, due to haemoglobin which had doubtless entered into union with the precipitin. Nitrogen estimations made before and after precipitation lead to the same conclusion regarding the occurrence of a union. Moro (31, x. 1901) found that a lacto-precipitum was almost entirely dissolved in warmed salt solution. Eisenberg (v. 1902, pp. 307-8) has also found lacto-precipitum to be soluble at higher temperatures, being different in

---

[1] Ichthiotoxin, see p. 39.

this respect to the sero-precipita studied by other authors, and it also differed in being soluble in concentrated urea solutions, and in saturated solution of magnesium chloride. The egg-white precipitum, on the other hand, he found to be soluble in dilute acids and alkalis[1], whereas he was unable to confirm the observation of Myers (1900) to the effect that this precipitum is soluble in 2 % saline. It proved insoluble in solutions of alkaline carbonates, and various solutions, even up to saturation point, of sodium chloride. Acid solutions of precipitum were reprecipitated[1] on being neutralized, behaving like an acid albumin. When washed in saline and heated, the precipitum coagulates, and is then insoluble in weak acids, behaving therefore like coagulated albumins which have undergone clotting through corresponding ferments and have then been coagulated by heat.

The precipitum, obtained by adding anti-ox serum (from rabbit) to ox serum as described under "Observation III." (p. 89), examined by Mr Hopkins last year, was found to be soluble in very dilute NaHO, much less easily soluble in Na_2CO_3. It was precipitated from alkaline solution on neutralization, but was soluble in excess of acid, even acetic. The precipitum contained *abundant phosphorus* in organic combination. It gave the biuret, xanthoproteic, and glyoxylic reactions, and was found to contain "loosely bound" sulphur.

Macroscopic and Microscopic Appearances during Reaction.

When antiserum is added to blood dilution, it sinks to the bottom of the tube. If it does not mix with the dilution, but flows down the walls of the vessel, the reaction taking place almost instantaneously at the zone of contact, assuming that the antiserum is powerful, and reacting with its homologous blood dilution. The reaction consists in the formation of a milky layer at the point where antiserum and blood dilution come in contact. The milkiness extends gradually upward, until the whole fluid is clouded. Where the fluids have been partially mixed this generalized clouding occurs more rapidly, being perfectly uniform when the bloods are thoroughly mixed by shaking. Where the fluids have been mixed by shaking the diffuse clouding undergoes a change; after 10 to 20 minutes, or later, very fine granules of precipitum begin to appear, and the upper layers of fluid begin to clear, due to the sedimentation of the particles of precipitum. The fine particles soon become aggregated into coarser ones, and these into flocculi, which

---

[1] Also observed by Michaëlis (9, x. 1902, p. 734).

gradually sinking to the bottom of the tube give rise to more or less deposit, having, when pure serum is used, a whitish appearance. Particles may adhere to the walls of the tube, from which they are readily detached by rotating it. With blood dilutions of, say 1 : 40 to 1 : 200 and over, the deposit formed is usually sharply defined, where more concentrated dilutions are used the deposit may form an irregular mass at the bottom of the tube. The increased viscosity of such dilutions may retard sedimentation. The supernatant fluid remains clear. In some few cases it may remain slightly clouded, more especially in non-homologous blood dilutions. In rare instances where non-homologous, but related bloods are tested, a cloud may form, without, however, leading to a deposit.

Similar appearances are noted when anti-human serum is added to albuminous urine, as described by Leclainche and Vallée (25, I. '01), who made their tests by bringing equal volumes of antiserum and urine in contact with each other in the manner described above. In testing with lactosera the opacity of milk diluted to 1 : 40, as recommended by Wassermann and Schütze, prevents one following the finer details of the reaction. In this case the antiserum may be added in the proportion of 1 : 1 or 1 : 5 as recommended by the last-named author. Here the casein is precipitated by the antiserum.

The reaction may be followed microscopically, as recommended by Tarchetti (1901), and Modica, who used the "hanging-drop" method, familiar in bacteriological work. Grünbaum (18, I. 1902) used this method also, comparing the appearances to what is seen when bacteria are agglutinated, and observed that the particles appeared and were more rapidly aggregated when anti-human serum was added to human than to simian blood-dilution and *vice versâ*. Robin (20, XII. '02) also, apparently unaware of Tarchetti's method, has found it useful, for he reports that he could observe the formation of granules within 10 to 15 minutes microscopically, when a reaction visible to the naked eye was only observable after two hours. He considers that the aggregation of granules is possibly brought about by a process analogous to that of bacterial agglutination.

The Supernatant Fluid.

If antiserum is added in insufficient amount to a blood dilution the supernatant fluid, after removal of the first deposit, still contains precipitable substance, as may easily be proved by adding more antiserum, and *vice versâ*.

Linossier and Lemoine, Eisenberg (5, v. '02, p. 291), and Ascoli (26, VIII. '02, p. 1410), agree in finding that the supernatant fluid may contain an excess of both interacting bodies in solution after precipitation has taken place. This is proved by adding either the one or the other component to the clear supernatant fluid, which leads in either case to fresh precipitation.

The Influence of Salts upon the Reaction.

That concentrated solutions of certain salts impede analogous reactions, such as those with bacteriolysins (Lingelsheim) and agglutinins (Eisenberg and Volk) has already been observed. Linossier and Lemoine (21, III. '02) taking a 1 : 20 serum dilution and adding increasing amounts of NaCl thereto, found that even the presence of 1% salt impeded the precipitin reaction, and that 5% salt prevented precipitation even after a mixture had stood as long as 24 hours. They found the reaction to take place in the presence of small amounts (1%) of ammonium sulphate, magnesium sulphate, and sodium fluoride, as they state was previously noted by Arthus. Eisenberg's (v. '02, p. 307) results are in flat contradiction; he states that concentrated NaCl solutions (up to close on 18%, at which point albumins are thrown down) had no influence on the reaction. The reaction was prevented by 0·25 normal magnesium sulphate, slowed by 0·5 magnesium chloride, prevented by a 2-normal solution of the latter salt. Moreover, Rostoski (1902, b. p. 42) observed no noticeable difference in the precipitations which took place with haematosera in the presence of 10% NaCl, and 2% ammonium sulphate and ammonium chloride. On the other hand, Müller (18, II. '02) found the action of lactosera depends upon the presence of lime salts, although these may be replaced by barium salts. Lactoserum only occasionally precipitated boiled milk after the addition of lime salts. Michaëlis (9, x. '02, p. 734) removed all trace of lime salts both from haematoserum aud its homologous blood by means of oxalate of potassium and found the contrary, namely, that precipitation occurred as well as before. Rostoski (1902, b. p. 42) on the other hand, also experimenting with haematosera, obtained the same result as Müller. He states that precipitation does not take place in the absence of salts, small quantities thereof are sufficient, large amounts of NaCl do not impede precipitation. It was evident therefore that these experiments required to be repeated.

Graham-Smith and Sanger (1903, p. 266) undertook to solve this

problem at my suggestion. Their experiments agree more closely with those of Eisenberg and Rostoski. They report thereon as follows:—
" We have quantitatively estimated the influence of salt in the following way[1]. Tubes containing 1 in 21 dilutions of human serum with gradually increasing percentages of salt were arranged in a rack, and to each ·1 c.c. of anti-human serum was added. We found that the precipita in the tubes containing the most salt were more flocculent, and owing to the increased specific gravity of the medium took longer to settle (see Plate, fig. 2).

Results of measurements showed a slight decrease to 7 °/₀ and later an increase, probably due to the fact that the more flocculent precipitum, though really less in amount, occupies a greater volume.

Results of increasing quantities of salt in human serum dilutions.

Percentage of salt	Precipitum c.c.	Percentage of precipitum as compared with ·6 °/₀ salt solution	Percentage of salt	Precipitum c.c.	Percentage of precipitum as compared with ·6 °/₀ salt solution
·6 °/₀	·0643	100 °/₀	8 °/₀	·0693	107·7 °/₀
1 ,,	·0571	88·8 ,,	9 ,,	·0673	104·6 ,,
2 ,,	·0638	99·2 ,,	10 ,,	·0770	119·7 ,,
3 ,,	·0554	86·1 ,,	12 ,,	·0686	106·6 ,,
4 ,,	·0639	99·2 ,,	14 ,,	·0737	114·6 ,,
5 ,,	·0618	96·1 ,,	16 ,,	·0730	113·5 ,,
6 ,,	·0635	98·7 ,,	18 ,,	·0821	127·0 ,,
7 ,,	·0630	97·9 ,,	Saturated solution	·0854	132·8 ,,

Experiments with sheep and anti-sheep sera, which formed more compact precipita, show the diminution in volume plainly. In this experiment the tubes were all centrifugalized for the same length of time in order to diminish the error due to the increasing specific gravity of the solutions.

Similar experiments to above with sheep serum.

Salt	Precipitum c.c.	Percentage of precipitum
1 °/₀	·0199	100 °/₀
2 ,,	·0203	102 ,,
4 ,,	·0201	101 ,,
10 ,,	·0140	70·3 ,,
Saturated solution	·0122	61·3 ,,

Qualitative estimations showed that when the quantity of salt was increased above 5 °/₀ the antiserum did not sink to the bottom, and

---

[1] The quantitative method of testing is described in Section VII. which follows.

that clouding occurred at the top of the tubes, and also took longer in forming."

Graham-Smith and Sanger (p. 266) moreover studied the behaviour of aqueous blood dilutions to antisera, and report thereon as follows :—

"It has been noticed by many observers that solutions of fluid, or dried, sera in distilled water become cloudy, and that after 24 hours a precipitate occurs. In ·5 c.c. of a 1 in 21 dilution of human serum in distilled water this precipitate amounts to about ·001 c.c. We have, however, found that including this precipitate ·1 c.c. of human antiserum produces a smaller quantity of precipitum with blood diluted with distilled water than with the same specimen diluted with normal salt solution. The mean of three experiments in each case gave ·0384 c.c. of precipitum in salt solution dilutions and ·0328 c.c. in watery dilutions."

Regarding the Claim that Precipitins permit of Distinguishing different Albumins of the same Animal.

The investigations of Nolf (v. 1900) indicated that precipitins are formed for serum-globulin and not for serum-albumin and solutions of blood corpuscles.   Leclainche and Vallée (25, i. 1901) however state that antiserum for serum-albumin is very active for its homologous substance, but almost indifferent towards globulins contained in the albuminous urine with which they experimented. The urine from three cases of interstitial nephritis gave marked reactions, whereas the urine from a case of parenchymatous nephritis, which contained much globulin, gave but slight precipitation. Albuminous horse and cow urine gave no reaction. They add that human pleuritic exudate gave a reaction, but not human blood serum when antiserum for human albuminous urine was added thereto.   Mertens (14, iii. 1901) treated rabbits with human serum and found the antiserum to precipitate albumin in urine, concluding therefrom that the latter must be derived from the blood. The urine reacted both to this antiserum and to one obtained by treating the rabbits with human albuminous urine.   Dieudonné (2, iv. 1901) found that peritoneal exudate and blood of man reacted similarly to anti-human serum. Zuelzer (4, iv. 1901) treated rabbits with urine containing 1—9 % albumin, and obtained antisera as did Mertens, to whose generalization regarding the origin of albumin in the urine from the blood he objects, although he considers that his results warrant the conclusion that at least one albuminous body in urine is derived from the blood[1]. Ide (1, IV.) and his pupil Leblanc (31, V. 1901, p. 355) state that rabbits treated with chemically pure albumin derived from cows' milk formed antisera which possessed a different character, depending upon which albumin of the same species had been used for the treatment of the animals. They go so far as to consider that there are specific antibodies not only for each cell-species but also for as many albumins as the cell may contain. According to Ide the agglutinins represent easily precipitated albuminous bodies, the antitoxins being less precipitable. Leblanc (*loc. cit.*) treated rabbits with ox serum, pseudoglobulin, and albumin. He states that the antisera produced a reaction with ox, but not with sheep, horse, pig, guinea-pig and pigeon sera. Anti-pseudoglobulin for ox, precipitated ox pseudoglobulin solutions but not those of euglobulin and serum albumin. Antiserum for serum-albumin precipitated its homologous substance but not pseudoglobulin. Anti-ox serum had no effect on lactalbumin. Anti-ox haemoglobin precipitated the haemoglobin of the ox. In other words, the different antisera possessed a high degree of specificity for the different albumins of the same species of animal.

Nuttall (V. and VII. 1901) showed that anti-human serum caused reactions with human pleuritic exudate, the fluid from blisters, and to a slight extent with nasal, and lachrymal secretions[2] and a faint clouding even with normal urine. Anti-human serum was produced in rabbits not only by serum but also by old pleuritic exudate of man, preserved as long as 5—6 months with chloroform. Schütze (22, XI. 1901, p. 492) treated rabbits with human muscle albumin producing an antiserum which he states precipitated homologous substance, but not human albuminous urine. Levene (21, XII. 1901) immunified rabbits with milk and found the antiserum to precipitate milk, casein, milk albumin, and ox serum, a result contrary to Leblanc's. This same antiserum failed to act on fowl egg-white, egg-albumin, egg-globulin, fowl serum and sheep haemoglobulin. Linossier and Lemoine (25, I. 1902) found they could not distinguish albumin in urine by precipitins, as claimed by Leclainche and Vallée (*loc. cit.*). Halban and Landsteiner (25, III. 1902, p. 475) treated rabbits with injections of ox spermatozoa, obtaining (spermotoxic) antiserum, which precipitated not only saline extract of spermatozoa but also ox serum dilutions. Linossier and Lemoine (28, III. 1902) found haematosera to also act on other albuminous fluids

---

[1] See a short paper by Aschoff (6, IX. 1902) on this subject.

[2] Confirmed by Biondi (1902, p. 21), who also obtained reactions with milk, vaginal secretion, etc.

from the same animal, viz., saliva, albuminous urine, spermatic secretion, milk serum, muscle-extract. Moreover, blood serum was found to be precipitated by lactoserum, as well as by antiserum for albuminous urine. Repeating the experiment of Nolf (cited above, see p. 104) they obtained a contrary result. Antiserum produced by globulin injections acted not only on globulin but also on solutions of serum-albumin. Moreover they found antiserum for serum-albumin to precipitate serum-albumin, and also globulin. To make sure that this was not due to impurity of the serum-albumin, which might contain globulin, they prepared as pure a substance as was possible (see method in original, p. 370), but obtained an identical result. They conclude that anti-globulin acts most on globulin solutions, less on those of albumin. The antiserum for albumin on the other hand actually produced more reaction with globulin solutions than with its homologous substance. They conclude that for the present the precipitins do not permit us to demonstrate chemical differences between the different albumins of one animal. Strube (12, VI. 1902) injected rabbits intraperitoneally with human spermatozoa and human testicular extract (from cadaver), obtaining in both cases a weak antiserum which acted on spermatozoa solutions. Anti-human haematoserum had however the same effect, quantitatively and qualitatively, on spermatozoa solutions, and *vice versâ*. Meyer and Aschoff (7, VII. 1902) found that injections of blood, spermatozoa, and tracheal epithelium led to the formation of antisera which coagulated milk solutions (1 : 40).

Obermayer and Pick (1902) studied the different constituents of egg-white, finding the albumin to contain: a globulin, crystallizable albumin, a non-crystallizable albumin-constituent (conalbumin), and ovimucoid, also other bodies upon which they will report later. The egg-globulin of other authors they claim to have separated into four different constituents: ovimucin, dysglobulin (both insoluble in water), euglobulin and pseudoglobulin (both soluble in water). These bodies, purified by washing, were injected severally into rabbits and the various antisera tested upon the different constituents of egg-white. Repeatedly crystallized egg-albumin (method of Hoffmeister) did not lead to the formation of antiserum. The result of the tests was that all the constituents of egg-white gave a precipitum, proving that the antisera were not specific for the different constituents. They were surprised to find that a substance might not lead to the formation of a precipitin which acted upon that substance, but to a precipitin which acted on some other body in egg-albumin.

Hamburger (cited by Aschoff, 1902, p. 193) obtained precipitins for casein and albumin of cows' milk, stating that he was able to differentiate the one from the other, although both antisera precipitated ox serum, whereas ox haematoserum did not affect milk.

Michaëlis and Oppenheimer (1902, p. 343), working with anti-ox serum, euglobulin and pseudoglobulin, found these antisera to act upon globulins but not upon albumin. Both acted more upon pseudoglobulin than upon euglobulin, the action of anti-euglobulin being markedly stronger with euglobulin than that of anti-pseudoglobulin. Anti-ox haematoserum acted on ox serum-globulin, pseudoglobulin, somewhat less on euglobulin, slightly or not at all on serum-albumin. They consider (p. 345) these antisera not "specific" for each albumin of the same animal, although they certainly react with some more than with others. They conclude therefore that the specific "combining groups" are not possessed by one form of albuminous molecule, but that they are common to chemically related albumins. The precipitin reactions are therefore of no use for the qualitative chemical separation of the different albumins of the same animal.

Rostoski (1902, *a.*) treated rabbits with horse serum, as also with horse serum-globulin, euglobulin, pseudoglobulin and serum-albumin, but found that all the antisera reacted with the different serum constituents above named. As I have stated elsewhere, this might be ascribed to his using milky antisera (see p. 73) which unfortunately robs his results of their value. A rabbit treated with Bence-Jones albumins gave an antiserum which acted upon these, but also upon human serum, serum-albumin and globulin, but not with the serum or serum derivatives of other animals. Like the preceding authors he concludes that precipitins afford no aid in distinguishing the albumins of the same animal. In a subsequent paper Rostoski (1902, *b.*, p. 21) sums up the results of Leblanc and of Hamburger and points out their obvious contradictions. These are evident from what I have stated above.

Umber (14, VII. 1902) on the whole confirms the results of Obermayer and Pick, Michaëlis, and Rostoski. He treated rabbits with egg-albumin and globulin solutions and tested egg-white and the homologous substances with the antisera produced. Separating fibrinogen, globulin and albumin from the antisera, he dissolved them in saline and tested them upon the egg-white constituents named. Anti-globulin and anti-albumin serum precipitated dilutions of egg-white, of globulins, but not of crystallized egg-albumin. The separated constituents of the antisera were next tested, and it was found that a solution of the fibrinogen fraction, and still more of the globulin fraction, precipitated the same constituents as the preceding, but no precipitin was contained in the albumin fraction. The precipitin was therefore precipitated with globulin, but it is not certain whether it was simply carried down, or whether it is a globulin itself. Umber also concludes that the precipitins do not afford a means of differentiating the albumins of the same animal.

Oppenheimer and Michaëlis (18, VII. 1902) treated rabbits with ox serum-albumin and found their antiserum to precipitate serum-albumin, having less effect upon pseudoglobulin, none on euglobulin, nor on horse serum-globulin. Globulin-treated rabbits formed precipitin for globulin alone. Landsteiner and Calvo (18, VII. 1902, p. 782) treated rabbits with different components of horse serum, the fractions containing *a*, fibrinoglobulin and euglobulin, *b*, pseudoglobulin, *c*, albumin. The antisera for fraction *a* precipitated dilutions of all three fractions, the intensity of reaction being in the order *a*, *b*, *c*, being weak in the last. The antisera for fraction *b* acted similarly, but more on dilutions of *b* than of *a*. The antisera for fraction *c* had no effect in a first experiment, but did have in a second, precipitating both solutions of *c* and of globulins where the fraction had been reprecipitated. The behaviour of globulins to precipitins was inconstant. The authors conclude that the precipitable substances in serum represent several bodies with different reactions as regards precipitation. The precipitable substances do not act in accordance with known albuminous bodies in serum, in other words, there is no reason to identify the precipitable substance with globulin. Ide (27, VII. 1902, p. 266) injected ox haemoglobin into rabbits, and in confirmation of his pupil Leblanc found the antiserum to precipitate ox haemoglobin solutions, the antiserum also haemolyzing ox corpuscles. Ascoli (26, VIII. 1902) treated rabbits with fibrino-, eu-, and pseudoglobulin, serum-albumin, and normal serum. Following a recommendation of Arthus he used 3 % sodium fluoride solution with which to prevent bacterial development in his serum dilutions, which had to be kept at 37° C. during his experiments. He reaches the conclusion that there are qualitative differences in eu- and pseudoglobulin, and serum-albumin solutions made evident by the biological test, and that consequently an antiserum produced through injection of normal serum contains different precipitins, each of which seizes upon different components in normal serum when it acts thereon.

Michaëlis (9, x. 1902, p. 735) injected horse and ox serum-albumin into rabbits, and found the antisera to precipitate both serum-albumin and globulin. Contrary to Nolf he found, possibly owing to his having used different methods, that serum-albumin injections lead to the formation of precipitins which acted on serum-albumin. Gengou (25, x. 1902, p. 746), experimenting in another manner (see *Immune Bodies in precipitating Antisera*, p. 94) and using antisera obtained from rabbits treated with dog serum, found that anti-dog serum acting on dog euglobulin and serum-albumin led to precipitation and fixation of complement. Lactoserum, acting on cow casein and lactoglobulin did likewise, whereas lactalbumin was not precipitated and complement remained free. This experiment, often repeated, he considers contradicts the statement of Hamburger (see p. 107). Schütze (6, xi. 1902, p. 805) found haematosera to also precipitate dilutions of the homologous spermatozoa of different animals, the spermatozoa tested being those of the horse, ox, sheep, pig, and man. Uhlenhuth (1901, p. 501) had previously noted that haematosera also clouded spermatozoa solutions of the same animal. Ziemke (15, ix. 1902) appears to have obtained results which do not point to the specificity of the reaction.

Falloise (25, xi. 1902, p. 836) refers to the discrepancies in the results of other workers as being possibly due to their different method of treating their animals. Working with the sera of the ox and horse, Falloise separated the globulin and albumin after the methods of Leblanc and Nolf, whose results, as I have noted above, are not in accord. The results obtained with anti-globulin and anti-albumin for ox and horse confirmed each other, for Falloise found that, *A*, anti-horse globulin precipitated normal horse serum and globulin solution, albumin solutions being very slightly affected. It exerted a very slight action on ox serum, and globulin, none on ox albumin. (Anti-ox serum behaved in a corresponding manner upon ox and horse serum etc.) *B*, Anti-horse serum-albumin produced slight reactions and deposit in horse serum, in albumin and globulin solutions. With ox serum and globulin solutions there was only a very slight trace of deposit in 24 hrs., no effect on ox serum albumin solutions. (Anti-ox serum-albumin behaved in a corresponding manner towards its homologous serum, etc.)

Falloise would explain the failure of albumin to produce precipitin in larger amount to the presence of but little globulin therein, attributing the precipitin formation entirely to the action of globulin which it contains in consequence of our imperfect method of purification. He agrees therefore with Nolf that serum-albumins do not lead to the formation of precipitins.

Liepmann (18, XII. 1902) treated rabbits in the usual way with placenta emulsion, obtaining antiserum which caused precipitation in the presence of bits of placenta, and possessed slight haemolytic power. Pieces of organs, including pieces of the uterus, and also blood, gave no precipitation even after $\frac{1}{4}$ to 1 hour. This author's technique is certainly open to criticism. Liepmann (29, I. '03, p. 81) further reports that his antiserum for human placenta gives reactions with foetal serum (obtained from the cord), reacting more with placental substance, but giving no reaction with blood serum of a man and non-gravid woman. He proposes to test the serum of gravid women, to see if by means of the test he is able to demonstrate the presence of placental substance (Veit) in their circulation. Nötel (13, III. 1902) following the method suggested by Uhlenhuth for the identification of meats (see Section IX.) states that an antiserum for muscle-albumin gave a greater reaction with muscle extract than did the corresponding haematoserum. The statement requires confirmation, no other experiments of a comparative character having as yet apparently been made.

Uhlenhuth (6, XII. '02) obtained an antiserum for the yolk of egg, which precipitated yolk of egg dilutions, but did not act upon egg-white dilutions, except when these were fairly concentrated, and even then but slight action was observable after a considerable time. This antiserum also clouded avian serum slightly. He concludes therefore that the albuminous constituents of egg-white and yolk are different. Egg-white, as we know, contains albumin and globulin, egg-yolk containing vitelline, lecithine, and nuclein. He recommends this antiserum for the detection of egg-yolk in the examination of foods.

Graham-Smith and Sanger (1903, p. 268) working in our laboratory with my quantitative method (*q. v.*) obtained the following differences in the amount of precipitum when anti-human serum was added to different body fluids in 1 : 21 dilutions. Anti-human serum No. I. was much more powerful than No. II:, which moreover had undergone putrefaction.

Material	Anti-human No. I.	Anti-human No. II.	Percentages from means of these two	Anti-ox	Normal rabbit
1. Fresh human serum (2 days)	·0291 c.c.	·0197 c.c.	100 %	—	—
2. Old ,, ,, (8 months)	·0272 ,,	·0187 ,,	93·8 ,,	—	—
3. Placental serum (8 months)	·0150 ,,	·0112 ,,	54·5 ,,	—	—
4. Pleuritic exudate (2 weeks)	·0065 ,,	·0084 ,,	30·3 ,,	—	—
5. Hydrocele fluid (9 months)	·0046 ,,	·0037 ,,	16·7 ,,	—	—
6. Fluid from ovarian cyst (9 months)	·0018 ,,	trace	6·1 ,,	—	—
7. Amniotic fluid (9 months)	·0009? ,,	trace	3·0 ,, ?	—	—

Effect of Tryptic and Peptic Digestion on Antibodies, especially the Precipitins.

The resistance to tryptic digestion of the precipitin and agglutinin of abrin, after the destruction of the albumen as indicated by chemical reactions, has been observed by Hausmann[1], an observation which had previously been made upon ricin by Jacoby. Landsteiner and Calvo (18, VII. 1902, p. 786) found the precipitable substance in horse serum-globulin solutions to give somewhat less precipitum after tryptic digestion than before. They however succeeded in obtaining a precipitating antiserum from rabbits treated with ox serum which had previously undergone tryptic digestion, after having been coagulated. According to Obermayer and Pick, this resistance to tryptic digestion contraindicates the active substance being of an albuminous nature. Rostoski (1902, *b.* p. 60) found precipitins to resist tryptic digestion, and he adds that Ringer (1902) has found the globulin molecule to also resist. We have referred to the opinions regarding the connection between the globulins and precipitins. Michaëlis and Oppenheimer (1902, p. 34), on the other hand, state that blood serum is only digested with difficulty by trypsin, and that large quantities of it are required to exert an action, this action being exerted slowly. They find that a serum reacts to precipitins as long as it remains coagulable, but that it does not react when trypsin has been in contact for a sufficient length of time. Serum subjected sufficiently to the action of trypsin is incapable of causing the formation of precipitins in animals treated with such serum. The question evidently requires further investigation.

Destruction through peptic digestion. Leblanc (31, v. 1901, p. 361) found precipitin, as also precipitable substances in sera, to be destroyed by peptic digestion, as Dziergowski (1899) had previously done in the case of diphtheria antitoxin. Obermayer and Pick (1902)[2] found the precipitin-generating power of a serum destroyed by peptic digestion, although large amounts of albumoses and peptones were present in the treated solution. This agrees with what is stated below with regard to the very doubtful properties of immunization possessed by peptones. According to Jacoby (*loc. cit.*) peptic digestion destroys the precipitable substance in egg-white, ricin, and in abrin. Michaëlis and Oppenheimer (1902, p. 34) also found that immunizing properties

---

[1] Hofmeister's *Beiträge*, II. p. 134, cited by Landsteiner and Calvo, p. 783.
[2] See also Oppenheimer and Michaëlis (18, VII. 1902).

were lost in a serum subjected to peptic digestion, and the precipitable substance in a serum, which during the first stages of digestion reacts slightly, after a time gives no reaction to precipitins. They found a stage in the action of pepsin when the serum remained coagulable but not precipitable. Rostoski (1902, *b*. p. 60) found the precipitins to be destroyed by peptic digestion. Michaëlis (9, x. 1902, p. 735) found that pepsin and normal HCl, taken separately, exerted no action, but when they acted in conjunction for 1 hour, all the precipitin or precipitable substance in a serum had disappeared. The evidence therefore with regard to the action of peptic digestion is unanimous.

Regarding the supposed Precipitins for Peptones.

Contrary to Tchistovitch (v. 1899), whose experiments will be referred to presently, Myers (14, VII. 1900) claimed to have produced precipitins for peptone. He treated rabbits with solutions of Witte's peptone and obtained what appeared to be an antiserum which precipitated homologous peptone solutions. A curious statement of Myers is that the antiserum lost some of its precipitating power after being heated to 56° C., this being contrary to what has been observed by others regarding the resistance of precipitins to heat (see p. 114). He moreover stated that the heated serum could be markedly reactivated through the addition thereto of fresh normal rabbit serum, although the latter alone had no such effect. Other observers, however, are unanimous in finding that precipitins cannot be reactivated in this manner (see p. 94). Myers found that heating antisera for ox and sheep globulin did not have the same effect as upon the "antipeptone" serum. The only author who appears to have confirmed Myers' observation is Schütze (von Leyden's *Festschrift*, 1902, cited by this author in his paper of 6, XI. 1902, p. 804) who claims to have been able to distinguish different peptones, for instance, those obtained from the muscles of man and ox. By treating a rabbit with human peptone (from muscle) he states that he obtained an antiserum which gave a reaction with human peptone-containing urine, derived from a patient suffering from carcinoma of the peritoneum, no reaction being given with normal rabbit serum, nor with anti-ox peptone. The latter, however, gave a reaction with the stomach-washings of the same patient above referred to after a meal of beef.

As stated above, Tchistovitch was unable to obtain precipitins for peptone, after injecting 10 % solutions in doses of 5 c.c. repeatedly into rabbits. Buchner and Geret (16, VII. 1901) stimulated by the results of Myers, treated rabbits with peptone prepared according to Kühne. Although the peptone possessed a considerable toxicity, they succeeded in immunifying the rabbits, obtaining an "antiserum" which caused a precipitum in peptone solutions, the precipitum being composed of crystalline bodies or "globulites." On further investigation (8, VIII. 1901) they were surprised to find that these globulites consisted of barium sulphate, the barium being derived from the peptone they had used. When barium was excluded no reactions occurred. Briefly, they obtained no antibody for peptone. Michaëlis (9, x. 1902, p. 736) repeated Myers' experiment, using Merck's egg-peptone, and "peptonum siccum Riedel," both barium-free. He also obtained no antibodies for these. On the other hand, I have noted elsewhere (under Phyto-precipitins, Section VI.) that Kowarski has claimed to obtain precipitins for plant albumose. Michaëlis found but few of his rabbits to survive the toxic effects of the peptones he injected. He concludes that egg-peptone no longer possesses the side-chains which are present in the egg-white molecule, giving rise in the latter case to the formation of precipitins. The side-chains are destroyed through peptic digestion. Obermayer and Pick (1902) also repeated Myers' experiment, treating rabbits with Witte's peptone, finding that it produced little or no precipitin when injected. The effects of peptic digestion upon pre-cipitins and precipitable substances has been considered on p. 111.

It would appear from the foregoing, that Myers and Schütze are mistaken with regard to the nature of the "precipitins" they found.

I might add here that Klein (17, VII. 1902) treated rabbits intraperitoneally with injections of starch, glycogen (from fowl and rabbit), grape-sugar, gum, and gelatin but did not observe the formation of precipitins.

The Influence of Temperature upon the Reaction.

The influence of temperature upon the precipitin reaction is well-marked. Myers (14, VII. 1900) stated that it took place more rapidly at 37° than at room temperature. This was subsequently confirmed by Wassermann and Schütze (18, II. 1901), Michaëlis (9, x. 1902, p. 734), and Stockis (v. 1901), the latter stating that 40—42° C. are most favourable. Biondi (1902, p. 16) found the temperature to exert a distinct influence. Linnossier and Lemoine (1902) found reactions to occur at temperatures ranging from 0 to 58° C., the amount of precipitum not being materially altered, although deposits formed badly at 0°. Reactions appeared to occur most rapidly at 35°. Working with bacterioprecipitins Kraus (1897) kept his tubes at 37°, and Nicolle (1898, p. 162) and Radziewsky (1900, p. 434) found this the best temperature for the reaction. Kister and Wolff (18, xi. 1902) state however that there is no special difference in the reaction at room temperature and at 37° C., this being contrary to the experience of all other workers.

Observations made in this laboratory by Mr Strangeways have finally disposed of the question. Blood dilutions and antisera were placed at 37°, at about 12°, and in the ice-chest at about 5° C. prior to being mixed. After having attained the temperature of their surroundings, the antisera were added to the blood solutions, and the time noted when reaction took place, and it was found that a low temperature markedly retarded the reaction, although it had no influence upon the amount of precipitum as measured by my volumetric method.

The statement therefore of Kister and Wolff is wrong. It can doubtless be explained in a measure by the fact that in routine work with powerful antisera, the reaction *begins* almost immediately even at low temperatures.

The effect of Heat upon Precipitating Antisera.

The following table contains the results of experiments by different observers with regard to the effects of various temperatures upon antisera obtained from rabbits. The antisera were tested, after heating, upon their homologous bloods with which they had previously given reactions. Some authors state the time of exposure, others not.

Linossier and Lemoine (21, iii. 1902) state that their anti-horse serum, which had acted on horse blood in the dilution 3:100 before heating, was exposed for 48 hours at a temperature of 60° C. At the end of that time a coagulum had formed, and on separating the clear fluid therefrom it was found to cloud a 10:100 horse blood dilution. Diluted five times in saline, the antiserum showed a still greater resistance to heat, for after 48 hours at 60° it showed no loss of strength. After four days at 60° it still precipitated a 10:100 solution.

It will be seen that *the Bacterioprecipitins are much more readily destroyed by heat* than are the others, a fact which has been brought forward as a reason for considering them antibodies of another nature.

Kind of Antiserum	Temperature centigrade to which it was exposed		Remarks	Authority
	Destroyed at	Resisted		
Haematoserum	70°	—	—	Tchistovitch, 1899
,,	70°	65°	though weakened in power	Bordet, 1899
,,	—	60°	no effect apparent	Rostoski, 1902 (*b*)
,,	—	60°	½ hr., still effective	Obermayer and Pick, 1902
,,	65° in 24 hrs.	60°	48 hrs., weakened	Linossier and Le- moine, 1902
,,	72°	—	—	Eisenberg, 1902
,,	68°, 2 hrs. *	52°	no effect	Michaëlis, 1902
Anti-egg albumin	—	60°	1 hr., scarcely affected	Uhlenhuth, 1900
Anti-urine†	—	58°	2 hrs., still effective	Leclainche and Vallée, 1901
Lactoserum	—	56°	" resisted "	Moro, 1901
Bacterioserum	58°	—	—	Kraus and v. Pir- quet, 1902
,,	58—60°	—	destroyed in ½ to ¾ hr.	Pick, III. 1902

* " Almost quite destroyed."　　† Rabbit treated with human albuminous urine.

It will be seen from the above that all observers agree in finding that a temperature of 60° C. does not destroy the efficacy of haematosera. In the absence of detailed observations regarding the effects of temperature, Dr Graham-Smith (29, VII. '03, p. 354) carried out some experiments at my suggestion by means of my quantitative method, his results being as follows:

" In order to determine quantitatively the effects of heating on the precipitum-forming property, specimens of antisera were heated in small sealed capillary tubes attached to the side of a thermometer in a water-bath.

Specimens of anti-ox serum were heated for 5 minutes each, and of anti-sheep serum for 1·5 minutes, at the temperatures given in the following table. Subsequently ·1 c.c. of each sample was added to ·5 c.c. of a 1 : 21 dilution of its homologous blood, and the resulting precipitum measured quantitatively.

After the process of heating, no visible change was noticed in the anti-ox serum till a temperature of 65° C. was reached, when the fluid became slightly opalescent. At 70° C. this opalescence was very marked, and at 75° C. the serum became gray, opaque, and solid. In the case of the anti-sheep serum slight opalescence was noticed at 66° C., which became more pronounced at 68° C. The following table shows that a marked reduction in the precipitum-forming power coincided with the visible change.

When the slightly opalescent antiserum was added to a serum dilution a slight cloudiness appeared throughout the fluid. The more markedly opalescent serum differentiated itself as it settled to the bottom of the tube as a very definite cloud. After shaking the tube the fluid appeared cloudy throughout, but remained in this condition, no precipitum settling to the bottom.

The precipitum settled most quickly in the unheated specimens, and the rate of formation of precipitum decreased as the temperature, to which the antiserum had been exposed, increased.

Up to 60° C. no change in the precipitum-forming power was found in either the anti-ox or anti-sheep sera, and both gave no trace of precipitum when heated beyond 67° C. Between 60° C. and 67° C. the quantity produced in each case was diminished. The figures given are the mean of two estimations in each case."

Temp.	Anti-ox (heated for 5 minutes)			Anti-sheep (heated for 1·5 minutes)		
	Precipitum	Percentage	Remarks	Precipitum	Percentage	Remarks
37° C.	·0234	100		·0075	100	
40	·0234	,,				
45	·0234	,,				
50	·0234	,,				
55	·0234	,,		·0075	100	
56				·0075	,,	
57				·0075	,,	
58				·0075	,,	
59				·0075	,,	
60	·0234	100		·0075	,,	
61				·0056	74	
62				·0056	,,	
63				·0065	83	
64				·0037	49	
65	·0187	79	Slight opalescence	trace	?	
66				,,	?	Slight opalescence
67	·0103	42	,, ,,	·	0	,, ,,
68				·	0	Marked ,,
69				·	0	,, ,,
70	·	0	Marked opalescence	·	0	,, ,,
75	·	0	Opaque, solid			

The effect of Heat upon the Precipitable Substance.

Substance heated	Temperature centigrade to which it was exposed		Remarks	Authority
	Destroyed at	Resisted		
Eel serum	80°	58°	but gave less reaction	Tchistovitch, 1899
Fowl egg-white	—	56°	½ hr., not appreciably affected	Myers, 1900
Ox and Sheep serum globulin sols.	—	,,	,, ,,	,,
Fowl serum	—	70°	½ hr.	Bordet, 1899
Human albuminous urine	—	58°	2 hrs.	Leclainche and Vallée, 1901
Milk	100°, ½ hr.	—	no reaction	Schütze
,,	—	100°	½ hr., still reacted	Müller
Egg-white dil.	78°, 1—1½ hr.	—	no reaction	Eisenberg, 1902
Serum dil. 1:100	100°, 5 min.	55°	½ hr., no effect	Nuttall, 1901
Serum dil. 1:10	65°, 24 hrs.	60°	4 days, no effect	Linossier and Lemoine, 1902

Linossier and Lemoine (21, III. 1902) state that a serum dilution containing so little albumin that it will not coagulate on boiling, has to be boiled several minutes to destroy its precipitable substance.

The few observations noted in the preceding table made it appear that the precipitable substances in normal serum possess about the same resisting power as the precipitins. The results with milk are in flat contradiction, and probably due to error on the one or the other side (see under Lactosera). Whereas some observers exposed undiluted serum, others exposed diluted serum to the different temperatures, consequently the results cannot be compared.

The following experiments were carried out at my suggestion by Dr Graham-Smith (29, VII. '03, p. 355).

"The heating of specimens of undiluted ox serum (1 c.c. for 3 minutes), was carried out in the same manner as described for antisera. Subsequently 1 : 21 dilutions in salt solution were made, and tested with anti-ox serum.

No visible change in the serum was noticed till a temperature of 56° C. was reached, when the serum became slightly opalescent. This opalescence increased between 63—67° C., and was still further marked at 68° C. All these specimens gave slightly cloudy solutions. At 70° C. the serum became very opaque, and at 75° C. white and solid.

The quantity of precipitum formed remained constant up to 50° C., but from 55° C. to 62° C. a marked diminution was noticed. At 63° C. a further reduction occurred, and at higher temperatures the formation of precipitum ceased. All solutions gave a good foam-test.

The figures given below are the mean of two estimations in each case.

Normal undiluted ox serum heated for 3 minutes.

Temp.	Precipitum	Percentage	Remarks
Unheated	·0262	100	
40° C.	·0262	„	
45	·0262	„	
50	·0262	„	
55	·0225	85	
56	·0225	„	Slight opalescence
57	·0215	„	„　　„
58	·0215	82	„　　„
59	·0206	74	„　　„
60	·0187	71	„　　„
61	·0187	„	„　　„
62	·0187	„	„　　„
63	·0122	46	Increased opalescence
64	.	0	„　　„
65	.	0	„　　„
66	.	0	„　　„
67	.	0	„　　„
68	.	0	Marked　„
69	.	0	„　　„
70	.	0	Opaque
75	.	0	„　and solid

These experiments, as far as they go, appear to indicate that an antiserum can be exposed to a greater degree of heat than its corresponding serum without injury, and that the precipitum-producing property is completely destroyed in the latter at a lower temperature."

The effects of filtration of Normal Sera through "stone" filters.

" It has been already indicated that the substance of "stone" filters when allowed to act on serum exerts some influence on the serum exposed to it. In order to further test this point ox serum was filtered through a new Berkefeld filter, and through a new clean Chamberland filter. After a certain quantity of serum had filtered through it was removed, and specimens from it diluted and tested. It was found that in the former case the precipitum-forming power was at first diminished, but returned to the normal after 110 c.c. had been filtered. No change was noticed during the passage of a further 300 c.c. through the filter.

In the latter case the precipitum-forming property diminished rapidly and fairly uniformly as the filter became choked." (Graham-Smith, 1903, p. 357.)

Ox serum.

Quantity filtered in c.c.	New Berkefeld filter		New clean Chamberland filter	
	Precipitum	Percentage	Precipitum	Percentage
Unfiltered	·0281	100	·0281	100
10	·0187	66		
20	·0210	74		
30	·0229	82	·0272	97
40	·0225	80		
50	·0229	82	·0272	97
60	·0229	82		
70	·0215	76		
80	·0225	80		
90	·0244	86		
100			·0245	87
110	·0225	80		
125	·0281	100		
140	·0272	96		
150			·0158	56
165	·0281	100		
200	·0281	100	·0158	56
250	·0281	100	·0114	40
350			·0114	40
400	·0281	100		

The Precipitins and Precipitable Substances resist Putrefaction and Desiccation.

I have already noted elsewhere that putrid antisera might retain unimpaired precipitating power for their homologous bloods. I have found this to be the case repeatedly, having found a variety of microorganisms, moulds[1], Bacilli and Cocci, present in such sera, the latter at times emitting a very putrid odour. Similarly, putrid bloods have been found to react to their homologous antisera both by Uhlenhuth, Nuttall (1901), and Biondi (1902), the latter incidentally mentioning that he obtained positive reactions with human blood ingested by fleas, bugs, and mosquitos. In my paper of 1, VII. 1901, I stated that human blood which had undergone putrefaction for two

---

[1] Confirmed by Biondi (1902, p. 17).

months still gave excellent reactions with its antiserum but not with other antisera, a fact independently observed by Uhlenhuth (25, VI. 1901).

In the same paper I stated that I had preserved antiserum 42 days dried upon filter paper, a fact which has been also noted by Corin, and Stockis (v. 1901) as cited by Ziemke (17, x. 1901, p. 732). These authors have kept dried antisera (globulin) for two months and found it still active. Ziemke states that he kept dried antiserum for three months but found that it had then lost in power. Biondi (1902) has also preserved dried antisera for several months.

The fact that dried bloods give reactions after the lapse of a considerable time, months or even years, has been fully established by Uhlenhuth, and confirmed by others. It is true nevertheless that the serum becomes insoluble after a time, the length of which appears to vary considerably. It is doubtless for this reason that old dried bloods not infrequently give very feeble reactions.

Ziemke (1901) obtained negative results with blood stains 25 years old. Biondi (1902) obtained reactions with human blood stains which had been dried 10—15, but not with those dried 20, years; Modica (1901) on the other hand claims to have obtained a reaction with blood dried 25 years. Uhlenhuth (5, VI. and 25, VII. 1901) obtained reactions within 1 minute with bloods dried 6 to 12 years. I found (1901) that blood dried and kept at 37° C. in the dark for 42 days and blood hung for 6 months exposed to the air in the laboratory still gave reactions. Dried blood exposed for 30 minutes to a temperature of 100° C. remained unaffected. Ferrai (1901) and also Biondi (1902, p. 30) have found dried blood acted in the way that I have stated. They found however that reactions did not take place after it had been exposed to 130° C. for 1 hour, to 150° for 10 minutes, or to 160° for 5—10 minutes.

Graham-Smith and Sanger (1903, p. 274) have also studied the influence of putrefaction on sera and antisera.

"Following a suggestion of Dr Nuttall's, in order to determine the influence of specific bacteria on serum, 1 in 21 dilutions in salt solution of ox and horse serum were inoculated with a series of organisms. Undiluted human pleuritic exudate was similarly treated. All were incubated for 5 days at 37° C. and then left at room temperature for 36, 50, and 40 days respectively; but the horse serum was allowed to undergo natural putrefaction also for the last 10 days. With the exception of the putrefactive bacteria none gave rise to very considerable growth, and in nearly all cases by the time of examination the organisms had sunk to the bottom, leaving the supernatant fluid clear. When necessary the fluids were filtered through filter-paper. All were slightly alkaline or neutral in reaction.

	Human pleuritic exudate (1—11)			Ox serum (1 : 21)			Horse serum (1 : 21) (contaminated)		
	Anti-human c.c.	%	Control anti-ox	Anti-ox c.c.	%	Control anti-human	Anti-horse c.c.	%	Control normal rabbit
Control. No organisms	·0234	100	—	·0173	100	—	·0572	100	—
Putrefactive organism } No. I.	·0280	119·6	—	·0140	80·9	—	·0713	124·4	—
„ No. II.	·0280	119·6	—	·0112	64·7	—	·0525	91·7	—
„ No. III.	·0280	119·6	—	—	—	—	—	—	—
Streptococcus	—	—	—	·0163	94·2	—	—	—	—
Putrefactive organism } No. IV.	·0215	91·8	—	·0163	94·2	—	·0666	116·4	—
B. anthracis	·0206	87·8	—	·0150	86·7	—	·0591	103·3	—
Hofmann's bacillus	·0187	87·8	—	—	—	—	—	—	—
B. subtilis	·0187	80·0	—	—	—	—	—	—	—
B. typhosus	·0187	80	—	·0140	80·9	—	·0657	114·8	—
B. diphtheriae	·0187	80	—	—	—	—	·0670	117·1	—
Putrefactive organism } No. V.	·0187	80	—	·0084	48·7	—	·0582	101·1	—
Staphylococcus albus	·0187	80	—	·0169	97·6	—	·0754	111·5	—
B. coli	—	—	—	·0131	75·7	—	—	—	—
V. of cholera	—	—	—	·0112	64·7	—	·0670	117·1	—

In considering the above table in detail it is seen that the effects of various organisms on ox and human serum agree fairly closely with a few exceptions. The most striking are the putrefactive organisms I, II, and V. These differences may be due to the fact that growth in nearly all cases was less marked in the undiluted human, than in the diluted ox serum, the latter moreover was a year old and had been preserved in sealed tubes after filtration through porcelain. The effects on horse serum of the action of specific organisms combined with general putrefaction for 10 days agree with those of putrefactive organisms I, II, and III, on ox serum, in that the capacity for forming precipitum is increased.

It appears then from the few quantitative experiments we have made that the results of bacterial growth on sera differ, some reducing the quantity of precipitum produced and others raising it, neither action being however very marked. Such slight changes as do occur do not alter the specific character of the reaction.

Experiments were also made on human and other sera which had undergone natural putrefaction. Most of the materials had been kept for some time and consequently show the combined results of age and putrefaction.

Material			Anti-human No. I.	Anti-human No. II.	Per-centage	Anti-ox	Normal rabbit
1. Fresh human serum	(2 days)		·0291 c.c.	·0197 c.c.	100	—	—
2. Old ,, ,,	(9 months)		·0272 ,,	·0187 ,,	93·8	—	—
3. Putrid ,, ,,	(5 months)		·0262 ,,	·0169 ,,	88·1	—	—
4. Putrid ,, ,,	(8 months)		·0150 ,,	·0140 ,,	59·4	—	—
5. Putrid ,, ,,	(9 months)		·0131 ,,	·0150 ,,	57·4	—	—
6. Putrid placental ,,	(9 months)		·0150 ,,	·0112 ,,	53·7	—	—

			Anti-ox			Anti-human	
7. Ox serum	(1 year old) mean of 8 exps.		·0233 c.c.		100	—	
8. Ox serum, putrid (,, ,,) mean of 3 ,,			·0233 ,,		100	—	

The above table shows that in some cases advanced natural putrefaction seems to exert little influence, for although the precipitum is decreased considerably in Nos. 4, 5, and 6, yet this is not the case in Nos. 3 and 8. All the specimens had been putrefying for the time given in each case. Though time may have influenced Nos. 4, 5, and 6, it is more probable that organisms whose growth deleteriously affected the serum were present.

Finally, from the few experiments we have done we are of the opinion that putrefaction to almost any extent does not affect the specific precipitum-forming body.

Since blood dried in small quantities does not undergo putrefaction to any appreciable extent this factor may be neglected in ordinary medico-legal work.

Experiments quoted on p. 124 with contaminated anti-human and anti-fowl's albumin sera likewise demonstrated that putrefaction in sealed tubes does not affect the antibody in them, as has also been found by Nuttall.

An experiment conducted on the same blood dilution with a normal and a contaminated sample of the same antiserum gave as a mean of four estimations in each case ·0433 c.c. and ·0436 c.c. of precipitum respectively.

Moreover putrid (filtered) sera when injected produce, as several of us have found, powerful and specific antisera, and Strangeways has shown that the power of antisera made with similar doses of fresh and putrid filtered sera is nearly identical."

The Stability of Haematosera and Sera sealed in vitro.

In my paper of 21, XI. 1901, I stated that some of my antisera had given good reactions after being sealed in a pure state for seven months *in vitro*. Wassermann and Schütze (18, II. 1901) had only kept antisera on ice up to two weeks. They state that fresh antisera give greater reactions. I have some antisera which are effective after 14 months of storage. I have preferred to keep them on ice, being under the impression that they remain potent longer at low temperatures. In the majority of cases antisera deteriorate markedly after 3—4 months. My observations in this respect have been confirmed by Uhlenhuth (25, IV. 1901), Rostoski (1902, *b*, p. 17)[1], and Linossier and Lemoine (21, III. 1902) who have kept antisera for three months. Uhlenhuth however added 0·5 % carbolic acid to them, which I consider disadvantageous, for the reason that carbolized antisera tend to cloud blood solutions to which they are added, irrespective of their being homologous. Strube (12, VI. 1902) preserved antisera, both pure and with 0·2 % carbolic acid for three months. Moro (31, X. 1901) found sealed antisera to give reactions after several months, a fact also observed by Biondi (1902, p. 17). Robin (20, XII. 1902) found antiserum preserved four weeks to give a reaction in two hours whereas it gave a reaction in 30 minutes at first[2].

The precipitable substance appears to be even more stable. In my paper of v. 1901 I stated that I had successfully immunified rabbits with old antidiphtherial horse serum preserved two years and seven months in the laboratory at room temperature. This serum was exposed throughout that time to diffuse light and room temperature, preservation being secured through trikresol. The serum has given reactions with antihorse serum after more than four years. Similarly I immunified rabbits with human pleuritic exudate preserved for six months with chloroform at room temperature, and this has also given reactions after being preserved for over two years. Further observations with regard to the durability of sealed fluid sera in this respect have since been made in this laboratory by Graham-Smith and Sanger (1903, p. 273) who report as follows:

" A few quantitative experiments quoted below made on fluid sera, preserved by sealing in glass bulbs, indicate that such sera lose their

---

[1] Preserved with chloroform.
[2] Robin's antiserum was evidently weak at the start.

strength to some extent, though differences exist in the rate at which this occurs.

	Anti-human No. I.	Anti-human No. II.	Per-centage	Anti-ox	Normal rabbit
Human serum (1 week)	·0291 c.c.	·0197 c.c.	100 %	—	—
„ „ (9 months)	·0272 „	·0187 „	93 „	—	—

	Anti-ox			Anti-human	
Ox serum (mean of 8 exps., p. 264) sealed 1 year	·0233 c.c.		100 %	—	
„ „ (mean of 3 exps.) sealed 2 years	·0239 „		102 „	—	

	Anti-fowl's egg, No. I.	Anti-fowl's egg, No. II.			
Fowl's egg albumin (2 days) ...	·0254 c.c.	·0162 c.c.	100 %	—	
„ „ (9 months) ...	·0160 „	·0112 „	67 „	—	
„ „ (14 „) ...	·0225 „	·0144 „	88 „	—	

Antidiphtherial horse serum four years and six months old preserved with trikresol was found to produce a good but somewhat flocculent specific precipitum amounting to ·0572 c.c.

In the above experiments anti-human serum No. I. was only a few days old, whereas No. II. was three and a half months old, and was moreover contaminated by bacterial growths. The first anti-fowl's egg serum was quite fresh and the second three months old.

All sera of the same kind do not give with the same antiserum identical precipita, nor even the sera of the same individual at different times in some cases, consequently an accurate determination of the influence of age is not possible. Our experiments however seem to point to a slight decrease in strength as the result of age, the human serum and fowl's albumin experiments showing a decrease of precipitum of 7 % and 12 % after 9 and 14 months respectively. The fowl's albumin kept for 9 months shows a decrease of 33 %. It is, however, by no means easy to get accurate dilutions of egg albumin, and the relative weakness of the specimen may be due to this cause.

The two experiments just quoted also indicate that antisera lose some of their power, but not to the extent that some observers have stated. Some undoubtedly preserve their power of producing specific reactions after the lapse of 12 months. Others lose this property more rapidly, whilst some, as Nuttall has also found, become untrustworthy after a time, giving cloudings with all sera.

In considering the general results of these tables it appears that in the case of dried bloods time *per se* does not destroy their capacity for reacting with their own antisera. Judging from the control experiments with recently dried bloods we should think that the period between the addition of the antiserum and the formation of the cloud was increased, and the magnitude of the cloud diminished.

Fluid sera appear to deteriorate at any rate in some cases by keeping. It has been occasionally observed, however, in qualitative tests that old sera appear to react better than fresh ones."

SECTION III.

OBSERVATIONS UPON THE PRECIPITIN AND PRECIPITABLE SUBSTANCE IN CORPORE.

WE have seen that precipitins appear in the serum of suitable animals after a longer or shorter treatment with non-homologous albuminous substances, serum, milk, bacterial filtrates etc., administered by intraperitoneal, subcutaneous, or intravenous injections or by excessive feeding. The increase in the amount of precipitin is gradual, as is the case with other immune substances, a fact that can be readily determined by occasionally bleeding the animal and testing its serum upon the blood with which it has been treated. Obermayer and Pick (1902) noted occasional differences in this respect in animals treated with different blood components, some of which produced no effect at first but great effects in the later stages of treatment, whereas in other cases the increase was gradual. In accordance with what has been observed with regard ·to the antitoxin of diphtheria in the horse and the goat, by Salomonsen and Madsen (IV. 1897) (see p. 9) and others, the amount of precipitin present in the serum of the animal during immunization falls after each injection of the precipitin-producing substance, the fall being succeeded in due course by a rise. Curves made by roughly estimating at frequent intervals the amount of precipitin present show that during successful treatment precipitin is gradually formed within the animal's body. No measurements of the amount of precipitin during the growth of immunization have as yet been made, which would correspond to those made upon antitoxin, but it is safe to say that a corresponding *undulation* would be observed. Numerous observers, besides myself, have noted that it is best to *wait* for a minimum of five days, usually a week or more, before bleeding an animal, after the last injection it receives. The object of this is to obtain the maximum amount of precipitin, by allowing a sufficient time for an increase of precipitin after the fall which succeeded the last injection. The maximum of the precipitin content usually appears to be reached about eight days after the final injection.

Following upon a period which may last two weeks or longer, the precipitin. content gradually begins to fall, in animals which have ceased to receive treatment. Thus Strube (12, vi. 1902), who studied eight animals in this respect, found that they furnished for about a month sufficient precipitin to produce a reaction in a blood dilution of 1:1000, but that after eight weeks the precipitin had disappeared, and that then it was possible to again treat the animals with the same blood, and again obtain precipitin. Rostoski (1902, *b*, p. 39) found scarcely a trace of precipitin in the sera of animals 13 weeks after the last injection. The precipitins do not therefore appear to persist as long in the body as do, for instance, the agglutinins.

In animals subjected to long-continued treatment, the precipitins may be seen to gradually disappear, as was noted by Tchistovitch (v. 1899). I have not found a similar observation recorded in publications by other writers, although I was able to confirm it a year ago (Nuttall, 16, xii. 1901, p. 407) in some rabbits which I treated with human serum, in the false hope of increasing the strength of the antiserum they possessed. As I wrote at the time, "There is therefore a point in the treatment of animals, for purposes of obtaining an antiserum, when a maximum of power is reached, and the animal should be bled."

That the precipitins are present in other body-fluids besides the serum is indicated by an observation of Eisenberg's (v. 1902, p. 308) who twice found precipitins in the humor aqueus of rabbits treated with fowl-egg injections. As far back as 1888 I noted the existence of normal bacteriolytic substances in the aqueous humour, and I might add that other antibodies, e.g. typhoid agglutinins, have also been found in this situation by Levy and Giesler. Eisenberg, and Moro, (31, x. 1901 in milk-treated rabbits), were unable to find any precipitins in the urine of their immune animals. It would however be a matter of interest to see if they appear in the urine of rabbits suffering from albuminuria following upon injection with foreign albumins, even though there may be but a moderate amount of precipitin in their blood. I propose to investigate this.

The precipitins are transmitted to the offspring in utero, as was first shown by Mertens (14, iii. 1901) who examined the serum of one out of three newborn rabbits for precipitins, the mother having been treated during pregnancy. This observation was subsequently confirmed by Moro (31, x. 1901) with lactosera, and Biondi (1902, p. 15) with haematosera.

The function of the precipitins in corpore is not as yet cleared up. Presumably they serve to protect the body against the injurious effects of corresponding foreign albumins, and more probably, as is suggested by Michaëlis and Oppenheimer (1902, p. 363) to neutralize the specifically foreign character of the albumin introduced, thereby making it forthwith assimilable. This may explain the physiological significance of the phenomenon, and I might add, is in substantial agreement with Ehrlich's theory as to the function of antibodies in general.

The regeneration of precipitins in the body after large bleedings would appear improbable from the investigations of Rostoski. It will be remembered that a regeneration of diphtheria antitoxin has been observed in animals subjected to large and repeated bleedings (see p. 10). Rostoski (1902, b. p. 35), in a similar manner, subjected his rabbits to one large bleeding (drawing 55 to 64 c.c. from animals weighing 2000 to 2300 g.), and tested their serum for precipitins nine days or so later. He observed a marked decrease in the precipitins and concludes that they are not regenerated. It appears to me that these experiments should be repeated with the aid of my quantitative method. Rostoski notes, however, that the relatively rapid disappearance of precipitins from the bloods, after treatment has ceased, indicates that precipitin production soon ceases, owing possibly to the elimination of the foreign substance which has stimulated their production. He nevertheless considers it possible that the anaemia following large bleedings may also affect the activity of the sources of precipitin formation.

The systemic reaction of treated animals is slight, if we except the slight loss of weight which follows the first injections in properly treated animals, by which I understand animals treated with graded doses which do not injure or destroy them by intoxication. It is easy to observe that a leucocytosis follows the injection of foreign substance, which is usually most marked in immune animals (note immunity to bacteria). As stated by Michaëlis and Oppenheimer (1902, p. 356), rabbits treated by intraperitoneal injection of serum (they used ox serum) develope a leucocytosis due to multi-nuclear granular, and mono-nuclear non-granular leucocytes. On examining the peritoneum of animals receiving an injection of serum for the first time, nothing abnormal is observed after absorption has taken place. On the other hand, Michaëlis and Oppenheimer note that immunified animals present a different appearance if examined two to three days after an injection. Here solid masses of albumin are found, either free or attached to the peritoneal surface, the masses, when viewed in section, showing an outer zone, due to leucocytic infiltration (chiefly microphages, with macrophages at periphery). The parietal and intestinal peritoneum is covered by small tubercle-like nodules made up of nests of leucocytes[1]. Intravenous injections, they found, produced a leucocytosis, due almost exclusively to mono-nuclear elements, the leucocytosis being more marked in immunified animals.

Considerable interest attaches to the observation that *a precipitin and precipitable substance may coexist in the serum of immunified animals*. It was purely by accident that Mr Hopkins (Reader in Physiological Chemistry, Cambridge) and I had occasion to observe this in 1901, in connection with rabbits which had been treated with crystallized horse albumin, and I herewith append the protocol of the experiment.

Three rabbits bled 29, VII. 1901.

Treatment: Intraperitoneal injection of crystallized horse albumin. First injection 22, VI., last 20, VII. 1901, all bled nine days after last injection. Total amount injected 50, 50 and 53 c.c. of solution (strength undetermined) respectively, in graded doses of 5 c.c., rising to 10 c.c. for last three injections, the day intervals between injections being 5, 4, 5, 5, 9. The weights of the rabbits were (in g.):

	At 1st injection	At last injection	When killed
I.	1690	1750	1580
II.	1940	1980	1600
III.	1910	1780	1520

Rabbit I. bled from the ear vein four days after injection 5, showed presence of precipitin for the solution with which treatment had been administered.

It is noticeable that the weights of all the animals had fallen very considerably during the days preceding death, this being somewhat unusual and the reason not clear. The animals were bled to death and their sera collected in the usual way. When we came to test their precipitating power, we found Sera I. and III. to give a large precipitation, whereas *Serum II. gave no reaction* at all. Before this fact was discovered, in order that a test *en gros* might be made, Serum II.

---

[1] The authors found that injection of albumoses, either subcutaneously or intraperitoneally, only produced sterile abscesses at the point of inoculation, or strong peritoneal adhesions. See further on the effects of blood injections in Metchnikoff, *l'Immunité*, Paris, 1901.

N.   9

and one of the others were poured together, when to our astonishment
a massive precipitation occurred.   Fortunately small separate samples
of these sera had been preserved which enabled us to make further tests
to see if we could find a reason for this remarkable behaviour of the
antisera.   On mixing the samples we found that

> Sera I. and III. gave no reaction,
> „   II. and I. gave marked reaction,
> „   II. and III. gave powerful reaction.

Consequently it was due to some *peculiarity in Serum II.*, and it
seemed to us that the only reasonable explanation to be found was
*that some of the precipitable substance, viz. crystallized horse albumin,
was actually present in the rabbits' serum.*   The matter was not pursued
farther at the time, but subsequently I directed the attention of
Mr Strangeways in our Laboratory to this observation, and asked
him to mix various antisera for human blood which he had made,
especially those taken from rabbits which, through prolonged treatment,
had shown a decrease in the amount of precipitin.   On adding a
certain anti-human serum to another, he also obtained a precipitation, and this was interpreted in the same manner as before.   In
the latter case, the antiserum which contained precipitin likewise
contained precipitable substance.   It appears somewhat remarkable
that precipitation does not occur in such antisera on standing in sealed
tubes; perhaps the explanation of the reaction which takes place with
a similar antiserum from another animal of the same species will be
found to depend upon an individual difference in the constitution of
the precipitin.   It is however premature to draw further conclusions.
The matter certainly deserves further enquiry.   Owing to stress of
other work, it has been impossible for me hitherto to pursue the
question, but I hope to do so shortly.   I have not published this
observation before for the reason that I hoped to have had more data
before doing so.   The reason that I do so now is that similar observations have been made by others, as follows.
Obermayer and Pick (1902) find that when egg-white is injected
intraperitoneally into a rabbit which has precipitins for this substance
in its serum, both precipitin and egg-white may be present in its
blood, and, although this is the case, no precipitation occurs.   Ascoli
(26, VIII. 1902) also found that precipitin and precipitable substance
may coexist in the serum of immunified animals.   He thinks the
condition is similar to that which some claim to have observed *in vitro*,

namely the coexistence of the interacting bodies in the supernatant fluid after a precipitum has formed. Hamburger (6, xi. 1902, p. 1190) has also found this to be the case; the egg albumin may, however, be found as readily in rabbits which have received their first injection as in those which are immune. The egg albumin appeared in their sera two hours after injection, and disappeared after four days in both immune and non-immune rabbits, there being at no time any apparent difference in the quantity of egg albumin present in these animals. Hamburger (p. 1191) considers that this observation makes it all the more difficult to explain the disappearance of the albuminuria in immune animals, although this may possibly be due to immunity having been acquired by the kidney cells. The probabilities are that the reaction between the precipitin and precipitable substance constitutes but one of several reactions which are taking place in the body, as to the nature of which we are ignorant. As Rostoski, and Michaëlis, and Oppenheimer (1902, p. 363) state, all we know is that precipitation does not appear to occur *in corpore*.

By analogy with what takes place *in vitro* upon the mixture of precipitin and precipitable substance, precipitation should take place in the body of an animal whose serum contains precipitin when precipitable substance enters its circulation. If such a reaction took place with anything like the rapidity with which it does *in vitro*, it is needless to say that it would be fatal to the animal, leading to the formation of thrombi, etc. The remarkable thing is that the animal remains well. Rostoski (1902, *b*, p. 40) thinks that the absence of precipitation *in corpore* may be due to three causes, (*a*) the strongly alkaline reaction of its blood, (*b*) the large amount of albumin contained in its serum (about 7 %), (*c*) and possibly to the presence of antiprecipitins.

Michaëlis and Oppenheimer (1902, p. 363) draw attention to *the possible significance of the enormous leucocytosis* observable in immune animals upon the injection of fresh precipitable substance, and they consider that the leucocytes may possibly take up the precipitum the moment it is formed. This would be scarcely demonstrable. They do not think the leucocytes would take up the foreign albumin as such, but they might take it up after it has been acted upon by the precipitin which circulates in the plasma.

Hamburger (6, xi. 1902, p. 1191) believes that a combination is effected *in corpore*, for the reason that the amount of precipitin decreases after fresh egg-white injections. In support of the statement

9—2

I have made (p. 126) he found that a rabbit whose serum gave a dense precipitum with a 1:500 dilution of egg-white *before* a fresh injection of egg-white, only gave a faint clouding with a 1:200 dilution 24 hours after another injection of egg-white. No deposit takes place in such a serum when it is removed from the body. The solubility of a precipitum in an excess of precipitable serum has been noted on p. 89.

The seat of origin of the precipitins is unknown. Michaëlis and Oppenheimer (1902, p. 356) followed up the clue which seemed to be presented by the leucocytosis, (if one adopts the view of Metchnikoff and his followers,) basing an analogy upon the origin they claim for the cytolitic complements in the breaking up of leucocytes. Michaëlis, however, found precipitins circulating freely in the plasma, a fact which does not lend support to the view that precipitins originate from leucocytes.

Biondi (1902, p. 15) sought to determine in which organs of humanized rabbits the precipitins were formed. Having bled the animals to death, he washed out the vessels with salt solution, cut up various organs (liver, spleen, kidney, lung, brain, lymph glands, bone-marrow, thyroid gland) and extracted these in saline. All the organ-extracts contained precipitin, which he attributes to the possibility of blood having been retained in the washed organs; nevertheless the retroperitoneal lymph glands appeared to contain more precipitin than did the other organs.

Referring to the experiment in which Uhlenhuth succeeded in obtaining precipitins for egg-white in the serum of a rabbit to which egg-white had been administered *per os*, (an experiment which has been confirmed for precipitins which act on blood, see p. 53,) Michaëlis (9, x. 1902, p. 734) notes that Uhlenhuth's antiserum only became rich in precipitin after the rabbit had received many eggs. This points to the probability of the precipitin being formed in consequence of the excess of egg-white introduced, leading to quantities of it *escaping digestion*. In other words, he thinks it is the non-assimilated egg-white which stimulates the formation of precipitin. He pertinently adds that if this were not the case, the serum of human beings, for instance, those using milk as a part of their ordinary diet, would contain precipitins for milk, whereas as a matter of fact he has not found this to be the case.

I might add here that Hamburger (6, xi. 1902, p. 1190) observed the rapid appearance of egg-white in the serum of a dog fed therewith, although no precipitins were formed in this animal. M. Ascoli (1903) has in a similar manner been able to demonstrate the presence of egg-white, and of substances derived from roast fowl in the lymph of dogs fed therewith. He was unable to note a parallelism between the amount of precipitable substance contained in the dog's lymph and blood serum. The serum of human subjects fed on roast beef also contained substances which were precipitated by an homologous anti-serum for ox meat. Where precipitins were contained in the serum of an animal thus fed, the amount of precipitin therein underwent considerable oscillations when a corresponding food was given. The experiments will be reported upon *in extenso* in a future paper.

The presence of albuminuria in rabbits treated with foreign albumin is noted by Hamburger (6, xi. 1902). This is however an old observation, which has been made not only on animals but also on man, by physiologists and others, in connection with the so-called physiological albuminuria as the result of food rich in albumin. Hamburger, however, has made very important observations with respect to the albuminuria which is observed in the course of immunization with egg-white. He found that the albuminuria disappeared during the process of successful immunization. Moreover *the earlier the albuminuria disappeared, the sooner did the animal form a powerful antiserum.*

In a rabbit in which the albuminuria disappeared after the third injection of egg-white, an antiserum was obtained which reacted with egg-white dilution of 1 : 200,000, whereas in the case of another rabbit in which the albuminuria only disappeared after the sixth injection, the antiserum only reacted with an egg-white dilution of 1 : 40,000. The practical bearing of this observation is clear for those engaged in the preparation of precipitating antisera, and without doubt Hamburger's discovery is of general interest, not only from its physiological aspect, but also from the standpoint of immunity. The degree of albuminuria may very well serve as a guide as to the grading of dosage during such immunizations. And I do not doubt but that the method would prove the need of grading the dosage much more carefully than some observers have done. Some indeed have made no attempt at gradation, but have injected 10 c.c. for example of a foreign serum, I might say blindly, every time, during the whole course of treatment.

That the albuminuria in rabbits is not simply due to an escape of egg-white, was proved by Hamburger through the use of two antisera, anti-rabbit and anti-egg sera. These antisera gave reactions with both albumins, when the coarser tests of boiling and nitric acid ceased to act. Finally the traces of albumin demonstrated by means of antisera disappeared, nevertheless egg-white persisted in the serum of the immune rabbit.

The behaviour of precipitins injected into normal animals has been the subject of investigation by Obermayer and Pick (1902). They injected anti-egg serum intraperitoneally into rabbits, and noted its appearance after 15 minutes in the rabbits' circulation, the quantity undergoing a rapid increase, a fact which could be observed by removing samples of blood from the animals' ear veins at stated intervals. Control experiments with subcutaneous injections showed that the precipitins appeared much more slowly in the blood when this mode of administration was employed. The authors conclude that the rapid diffusibility of these substances argues against their being of an albuminous nature.

SECTION IV.

ON THE SPECIFICITY OF PRECIPITINS.

THAT no special attention had been paid to the reactions which may take place in non-homologous bloods upon the addition of powerful antisera, prior to my publications, would seem clear from a comment upon one of my papers by Rostoski (1902, *b*, p. 18), who writes: "Die Angaben von Nuttall, dass das Serum eines mit Menschenblut behandelten Kaninchens auch eine schwache Trübung in Blutlösungen des Pferdes, des Ochsen und des Schafes hervorrufe, wiederspricht allen andern Beobachtungen direkt"! We shall see, however, that my observations have since been confirmed by several investigators.

Ehrlich (22, III. 1900) appears to have been the first to attribute "a rigidly specific" character to the precipitins, basing his statement upon the investigations of Bordet and Myers, and speaking of the precipitins as "specific coagulines—which act only in a specific manner, *i.e.* precipitate only the albuminous body injected." It was natural that he was led to this conclusion in view of the results of the investigations he cited, which, however, were based on but a very limited number of tests upon non-homologous material. Michaëlis (9, X. 1902, p. 733) refers to Wassermann (1900, p. 501), saying he "führte zuerst grundlegend aus, dass die Reaktionen streng specifisch für die injicirte Eiweissart sind," although on the next page (p. 734) he contradicts this by saying that the specificity of the reaction is not "absolutely strict." Wassermann and Schütze (2, VII. 1900), it will be remembered, stated that ox-lactoserum, for instance, did not act on goat milk and *vice versâ*, whereas other observers since, notably Moro (31, X. 1901, see further under Lactosera, p. 157), have found the contrary to be the case, doubtless because they used stronger antisera, or made their observations at a later period, after milk and antiserum had been brought in contact. Uhlenhuth (7, II. 1901) was apparently the first to consider that it might be worth while to study the reaction given by the bloods of *related* species, mentioning that he intended to see if there were any similarity of reaction between the bloods of horse and donkey, man and monkey. In his paper of 25, IV. 1901, he first demonstrated reactions amongst a variety of species of egg-albumins which were brought about by an antiserum for the egg-albumin of one species (see further under Anti-egg Sera, Section VI), a fact already indicated earlier by Myers (14, VII. 1900), who found that the antiserum for fowl egg-white acted upon both fowl, and to a lesser degree upon duck egg-white, and that antisera for ox acted on sheep globulin solutions and *vice versâ*. I was able (11, V. 1901) to confirm this with corresponding antisera for ox and sheep blood, showing that slighter reactions were produced by anti-ox serum on sheep blood dilutions and *vice versâ*. On the other hand Wassermann and Schütze (18, II. 1901) stated that the action of anti-human serum was "streng specifisch," except with the blood of a monkey, a fact independently established by Stern (see tests with anti-human serum, Section VI). Uhlenhuth (25, VI. 1901) found anti-sheep serum to precipitate sheep, goat and ox blood, anti-horse to precipitate horse and donkey blood, anti-human to precipitate human and monkey blood. In my paper of 1, VII. 1901, in view of the results then obtained, I stated that the " precipitins are specific, although they may produce a slight reaction with the sera of allied animals." *In view of the very limited number of bloods examined by most authors, it seemed to me altogether premature, as some had done, to make any broad generalizations.* For this reason, I began early in 1901 to collect as many bloods as I could from all classes of animals. It was as important to do this from the medico-legal as from the zoological standpoint, and in presenting the results given in this book I am safe in saying that they constitute the first scientific demonstration, on general lines, of the specificity or relative specificity of precipitins.

Although I have been unable to consult the original paper by Schirokich (21, VII. 1901), it would appear that he also noted the action of antisera on certain non-homologous bloods. He put a *time limit* upon the reaction for the reason that he observed an opalescence, as he terms it, after 4—5 hours, when he added anti-human serum to dilutions of ox, goat, hog, horse, camel, cat, guinea-pig and rabbit sera, a precipitum being formed after 24 hours. I do not know with what dilutions he worked, but should think that they were concentrated, for it does not seem impossible to me that some of the deposits may have been due to matter suspended in the antiserum itself, which in the interval of time had been able to settle. I have observed such deposits when antisera were not clear, although not of the opalescent kind (see p. 72). That even rabbit blood gave a precipitum indicates possible errors in his technique.

Uhlenhuth (25, VII. 1901) agrees with me in finding *that the zoological relationships between animals are best demonstrated by means of powerful antisera.* He judged from reactions with such antisera, that the ox is not so closely allied to the sheep, as the sheep is to the goat. He found that weak anti-sheep serum produced no reaction in ox blood. In my paper of 21, XI. 1901, I wrote "*The more powerful the antiserum obtained the greater is its sphere of action upon the bloods of related species.* For instance, a weak anti-human serum[1] produced no reaction with the blood of the *Hapalidae*, whereas a powerful antiserum did produce a reaction, and proved what I may be permitted to call the '*blood relationship*' in the absence of a better expression." This generalization was based upon data, previously published, with regard to the general action apparently possessed by anti-ungulate sera upon certain ungulate bloods (see anti-ox and anti-sheep blood tests, etc.), as well as on the bloods of *Canidae* and *Primates*. I also noted that reactions took place "to a lesser extent, in the bloods of allied animals, than in the homologous blood[2]." This was of paramount importance, the statement being based upon the examination, not of one or two bloods, but of over 200. In a subsequent paper (16, XII. 1901, p. 408) speaking of the reactions amongst the Primates, I wrote "If we accept *the degree of blood reaction as an index of the degree of blood-relationship* within the Anthropoidea, then we find that the Old World apes are more closely allied to man than are the New World apes, and this is exactly in accordance with the opinion expressed by Darwin." I cite these earlier papers for the reason that I wish to make it clear that my results have materially contributed to a modification of the views held with regard to the specificity of these reactions.

I have already shown that a reaction may take place with even the distantly related blood of the horse, very slight it is true, upon the addition of anti-human serum. This was confirmed by Grünbaum (18, I. 1902, p. 143) working with another anti-primate serum, that for the

[1] When a powerful antiserum is diluted, corresponding results may be obtained. See pp. 74 and 142.

[2] Exceptions, notably in the case of ungulate blood (*q. v.*) have since been noted by me and are recorded in this book. They may however be due to the effects of disease on the animals yielding the blood tested.

chimpanzee, which he states " gave a slight but distinct turbidity after a few hours with horse blood." In a paper which I published almost simultaneously (20, I. 1902) I stated that when testing bloods it is necessary " to put a *time limit* upon them. This may appear to be a rather arbitrary proceeding. My time limit has usually been 5 minutes at average temperatures in the laboratory. A powerful antiserum will certainly have acted within that time upon its homologous blood-dilution; with powerful fresh antisera the reaction takes place almost instantaneously. On the other hand, *if we allow mixtures of antisera and bloods to stand, a reaction takes place with non-homologous bloods.* The results I have hitherto obtained tend however to prove that anti-mammalian sera only produce these later reactions in mammalian bloods, anti-avian sera similarly in avian sera alone."

Furthermore Linossier and Lemoine (8, III. 1902) state that in view of the unanimity amongst authors regarding the strictly specific character of these reactions, they were surprised to find antisera not so specific as had been supposed. They found *that a precipitin may act on a number of different bloods, although the degree of reaction varies.* As a rule, they state that the precipitum obtained is least voluminous in bloods of distantly related animals to the one whose blood has been used for the production of the antiserum. Their method of testing with concentrated sera is different from mine and is subject to criticism, as will be seen by reference to pages 74 and 89. And this will account for their obtaining a reaction even with fowl serum, as stated below. They added anti-human, anti-horse, and anti-ox serum in the proportion of 10 volumes to 1 of the different sera tested, viz. those of man, dog, ox, sheep, horse, pig, guinea-pig, fowl, and obtained a precipitum in all of these bloods. With the exception of the fowl's blood in which the reaction was minimal, and the guinea-pig's in which it was feeble, all gave well-marked reactions. None of the antisera acted on rabbit serum. They noted what appeared to be a faint clouding in the serum of the eel. Judging from these results, which are anomalous in some respects, it would appear desirable to make a series of tests with a large number of sera under the same conditions. I have not as yet had an opportunity of doing so. Halban and Landsteiner (25, III. 1902, p. 475) state in a footnote, that their haematosera (presumably anti-human) also acted to a very slight degree on substances (serum and milk probably meant) of other not closely related animals. Anti-human serum produced " spurweise Präcipitation," that is, traces of precipitation, with horse, but not with ox serum. Here again my observation was confirmed. Referring to the specificity of the reaction, Linossier and Lemoine (*loc. cit*) remark "Là où on a cru voir une action spécifique, un examen attentif ne permit de voir qu'une action particulièrement intense," which sums up the question in few words.

Passing over my paper of 5, IV. 1902 (see below) in which I described my method for measuring the degrees of reaction, by volumetric estimations of the precipitum formed, I shall mention some more papers referring to the subject of these generalized reactions. Strube (12, VI. 1902) confirms my observation that the stronger an antiserum is the more powerfully does it act upon non-homologous bloods. He thinks that the quantitative differences of reaction may be explained by assuming that different species of blood do not possess an identical constitution (which is a self-evident proposition) but a closely related constitution, with the result that the antiserum for one reacts but to a limited extent upon the other, as in the case of the agglutinins. Secondly, let us assume that the serum is composed of different albumins (and there appears to be evidence of this) which we shall style *a*, *b*, and *c*. Substance *a* is present in other animals, but *b* and *c* are not. With the homologous antiserum *a*, *b*, and *c* are precipitated, whereas in the non-homologous bloods only *a* is precipitated. He offers this in explanation of my "mammalian reaction," and we have no other at present. Strube, I might add, makes no mention of a time limit, saying only that "several hours" may be necessary for reaction to take place. Test sera which take several hours to act would however be scarcely desirable in practice. Whereas Strube found anti-human serum to produce an equal, though *slight*, reaction in the bloods of the ox, sheep and pig, Kister and Wolff (18, XI. 1902) claimed to have noted that human blood is more strongly acted upon by anti-horse and anti-ox than by anti-sheep or anti-pig sera, this observation being evidently due to some experimental error. They also note that differences are not well observed in concentrated solutions, or when a too large proportion of antiserum is added (note in this connection what I have said on page 74). They note, as I have also observed, that clouds may form in concentrated solutions, whereas flocculent precipitates will form in more dilute solutions. Oppenheimer and Michaëlis (18, VII. 1902), working with but a limited number of bloods, conclude that precipitins are specific for a "bestimmte Thierart." They evidently overlooked the generalized reactions.

Ascoli (26, VIII. 1902) has also observed generalized reactions. He treated rabbits with the serum-globulins of man, horse, mule and sheep, and found that the antisera acted on solutions of all these globulins, only to a different degree.   On the other hand he observed " scarcely any quantitative differences " upon adding anti-horse or anti-mule sera to both of the corresponding globulin solutions.   Much less, and quantitatively decreasing in the order named, were the reactions obtained with similar solutions from the ox, sheep, and man, this again being in accord with my results.   That different substances are acted upon in homologous and non-homologous sera (see Strube, quoted above) is clear, for as Ascoli found, if non-homologous serum is added in excess after precipitation has occurred, no more precipitum forms, but on the addition of homologous antiserum one obtains further precipitation.   According to Ascoli, for example, when anti-horse serum produces a large reaction with horse blood and a small one with human blood, there are two possibilities to be considered regarding the differences of reaction.   Either the antiserum contains several (in this case two) precipitins acting on different constituents of each serum, the one being common to both horse and man, and the other present only in the horse ; or the precipitin is simple, there being less precipitable substance in the human serum.   The latter supposition is not borne out by experiment, for even when we add an excess of anti-horse, the human serum continues to give less precipitum.

Uhlenhuth (11–18, IX. 1902, p. 661) dwells upon the fact that the zoological relationship of animals is brought out by an antiserum ; he however does not bring new facts to bear on the question.

Further details regarding the specificity of the precipitin reactions are to be found under Tests with different antisera, in Section VI which follows.   As will be seen, my most generalized reactions amongst the mammalia were obtained by a powerful anti-pig serum.

It is evident from the foregoing that there are many points of similarity between the precipitins and the haemolysins.   Ehrlich has pointed out that the haemolysins of different species may possess receptors which are identical but quantitatively different.

The Selective Action of Precipitins in Blood Mixtures.

In my paper of 1, VII. 1901, p. 384 I described some tests made with a view of determining whether or no a *mixture* of several kinds of blood in solution would prevent a reaction taking place upon the addition of an antiserum which was effective when added to one of these bloods when alone in solution.   It was found that when two to six different bloods were brought together into solution, so that each blood in the mixture was diluted to about 1 : 500 or 1 : 600, the presence of other bloods did not impede a reaction taking place between an antiserum and its homologous blood in the mixture. The antisera only acted however when a suitable blood was present in such a mixture. (For details see original.) Ziemke (17, x. 1901) subsequently repeated my experiment [1] and confirmed it.

The measurement of Degrees of Reaction.

In my paper of 21, xi. 1901, p. 152, read before the Royal Society, I mentioned that I had undertaken to make quantitative measurements of the degrees of reaction obtained with precipitating antisera upon different bloods. Owing to the labour involved in the qualitative, and approximately quantitative tests here recorded, a report upon the results obtained had to be deferred. I however described my method in the following year (5, iv. 1902). I shall not dwell upon the method here, but refer the reader to Section VII which follows. As stated in the last paper referred to, the amount of precipitum, say 0·03 c.c., obtained by adding 0·1 c.c. of an antiserum to 0·5 c.c. of its homologous blood dilution (1 : 100 or 1 : 200) is taken as 100, and the reactions given by non-homologous bloods are stated in percentages of that figure. At the time I reported two sets of tests, which were, however, not to be taken as final, as follows:

Anti-sheep serum		Anti-pig serum	
Sheep	100	Pig	100
Ox	80	Horse	16
Antelope	50	Hog-deer	14
Hog-deer	47	Cat	14
Reindeer	30	Dog	13
Pig	20	Sheep	13
Horse	16	Wallaby	5
Cat	12		
Dog	7		
Wallaby	5		

I only cite these results to show that *there are measurable differences* in the degrees of reaction, as calculated upon the precipitum obtained with different bloods, the precipitum being actually measured volumetrically. The method I described has not as yet been tried by workers outside our laboratory.

---

[1] Ziemke cites my paper but does not state this as well as some other particulars.

All observers, including myself, who have worked with different blood dilutions, have noted the fact that an antiserum will act upon a higher dilution of homologous than of non-homologous blood. In view of the labour involved, but very few figures have been obtained by this method, employed comparatively. Linossier and Lemoine (8, III. 1902) thus sought to express the reactions with various dilutions, as follows:

Antisera added in the proportion of 15 : 100 of dilutions.

Anti-ox serum gave reactions with Ox serum diluted 1 : 5000
,, ,, Horse ,, 1 : 300
,, ,, Man ,, 1 : 50

Reactions of "apparently equal intensity" were obtained when anti-human serum was added to human serum diluted 1 : 1000, and to ox serum diluted 1 : 20. In another similarly conducted test, anti-human serum acted on human serum diluted 1 : 2500, but only on horse serum diluted 1 : 20. As the authors state, these figures have no absolute value, as they naturally will vary according to the antisera used being different. Strube (12, VI. 1902) testing with anti-human serum, obtained a reaction with human blood diluted 1 : 20,000, other bloods reacting in dilutions of 1 : 100. He said the reactions might take place after some hours, as already mentioned above, and he does not mention anything with regard to the *rate* at which different bloods react in a series, an omission made by most authors. I find *the rate at which a reaction takes place is a very fair index of the degree of relationship.* This rate would appear to be due chiefly to the differences in the amount of matter precipitated; where this is slight, a reaction would only be registered when the particles of precipitum form agglomerations which are visible to the naked eye. On the other hand there may be a slower reaction actually taking place, due to a lesser degree of avidity between the combining substances, but to prove this will be a matter of some difficulty. Experiments made by progressively diluting the antiserum which is added to a blood dilution, have furthermore been reported by Ewing (III. 1902, p. 14), following a suggestion made by Uhlenhuth, and Kister and Wolff. He does not state the blood dilution used, but I presume it was 1 : 100. He found

Anti-ox serum diluted 1 : 5 to act on dilutions of "several bloods."
,, ,, ,, 1 : 30 ,, ,, ,, ,, bloods of ox and goat,
,, ,, ,, 1 : 50 ,, ,, ,, ,, ox blood alone.

Similar observations were made with anti-human serum tested upon the bloods of man, and of the baboon, rhesus and Java monkeys. When added to these bloods in solutions of equal strengths, the anti-human serum in its highest dilution, only acted upon human blood dilutions.

The dilution-method has not been adopted in the study of blood-relationships, and these are the only figures obtained by the method which I have come across. The importance of making a series of dilutions of a suspected blood in medico-legal work, is made particularly clear by the figures of Linossier and Lemoine, given above.

It is evident that one observer will record a reaction and another not, when using the dilution-method without reference to time, and undoubtedly faint cloudings will be completely overlooked if the fluids are examined in an unsuitable light (see description of my apparatus, p. 70 and Fig. 4). There is consequently a considerable subjective element in the tests conducted by the dilution-method, which is absent from my method of actual measurement.

Eisenberg (5, v. 1902, p. 290) considers a "unit of precipitable substance" to be that minimal quantity, which, when contained in a given volume of fluid, suffices to produce a specific reaction. Thus, when a given albuminous solution (dilution 1 : 1000) yields a precipitum, then 1 c.c. thereof contains 1000 units of precipitable substance. The "unit of precipitin," is that minimal quantity which just suffices to produce a reaction in any albuminous solution. For example when an antiserum still produces a precipitation in a 1 : 100 albumin dilution, he would say it contained 100 precipitin units per c.c. It seems to me until we know more about the possible existence of normal anti-precipitins in sera, it is somewhat premature to attempt an exact standardization either of the precipitin or precipitable substance. Nevertheless the method suggested may have its use, unless my method of volumetric measurement is preferred. I have standardized my antisera, simply by stating the amount of precipitum produced as may be seen by my table of measurements quoted on p. 145. This is again stated in most cases for the antisera with which my tests were made in the following tables. It will be seen I express the "strength" of my antisera simply in terms of precipitum amounts.

To study the quantitative relations of the interacting bodies, Eisenberg (*loc. cit.* p. 291) adds a constant amount of one substance to a variable amount of the other, and, after reaction has taken place, he determines the absolute as well as the relative absorption of precipitin. As in agglutination experiments, the amount of absolute absorption is defined in terms of the difference between the number of precipitin-units present in the fluid per unit of volume, before and after reaction has taken place. He expresses the proportion of absorbed substance to the amount originally present in the form of a fraction.

The Delicacy of the Precipitin Test.

Whereas the ordinary chemical tests cease to give reactions in blood dilutions of about 1 : 1000, powerful antisera greatly exceed this limit, as the reported results of independent observers have shown. The tests in each case were conducted with antisera added to their homologous blood dilutions of the strengths indicated:

Strube (12, VI. 1902) anti-human serum gave reactions with a dilution of 1 : 20,000
Stern (1901)   ,,   ,,   ,,   1 : 50,000
Uhlenhuth (15, XI. 1900) anti-egg serum   ,,   ,,   1 : 100,000
Ascoli (26, VIII. 1902)   ,,   ,,   ,,   1 : 1,000,000

Some tests conducted at my suggestion by Mr Strangeways, with a view to determining the limit, have shown that reactions may take place even in dilutions of over 1 : 1,000,000, for on adding a constant amount of antiserum to progressive dilutions, differences in the amount of deposit, measurable to the eye, were observable even in these highest dilutions, when the precipitum had been collected in fine capillary tubes.

The Strength of Antisera.

Precipitating antisera have been termed powerful, or weak, in proportion to their action upon various dilutions of their homologous bloods. Most authors express the power of the antiserum by stating the blood dilution upon which it will act. It is evident that the time and the temperature at which reaction takes place, especially in high blood dilutions, must be taken into account, but this has not been done as yet with sufficient care to permit of accurate standardization by this method. My quantitative method would appear therefore to give more accurate results, consequently I have used it throughout the greater part of this investigation. The following figures give an idea of the "power" of some of my antisera, which have all been standardized by stating the amount of precipitum, measured volumetrically, obtained by adding 0·1 c.c. of antiserum to 0·5 c.c. of a 1 : 100 dilution of its homologous blood.

Antiserum for	Amount of precipitum in c.c.		Antiserum for	Amount of precipitum in c.c.	
Fowl's egg	·06		Dog and Mex. Deer	·015	
Ostrich	·042	most	Zebra	·012	
Pig and Antelope	·055	powerful	Ox	·011	moderate
Fowl	·035		Whale	·009	
Hyaena	·031		Ourang and Horse	·008	
Turtle	·03	powerful	Wallaby	·007	
Hedgehog	·022		Seal	·006	weak
Sheep	·02		Llama	·005	
			Reindeer	·004	

Measurements of ·002 to ·001 can still be made. Weaker antisera than that for the Reindeer were not used or standardized.

On Differences in the Reactions of Individual Sera belonging to the same Species of Animal.

In the investigations which I have made I have not studied the question of individual differences in the reaction of normal sera to their homologous precipitins. Together with Mr Strangeways, I have made a number of measurements on normal sheep sera with anti-sheep serum, and have obtained remarkably uniform results by means of my quantitative method. These results will be published later. On the other hand Mr Strangeways has observed a considerable amount of variation especially in human subjects affected with various diseases. It is not impossible that *disease* has in some cases been the cause of my obtaining greater reactions than would be expected according to theory. I shall not at present lay stress upon this point, but shall only, as an example, draw attention to the very great reactions I obtained with two Cercopithecus bloods (see Section VI) when testing with anti-human serum. Linossier and Lemoine (8, III. 1902) were unable to observe any individual differences in two human sera, the one being obtained from a case of pneumonia, the other from a case of uraemia. One of these sera had been used for the treatment of the rabbit which yielded the antiserum.

That there may be individual differences in health is indicated by the investigation of Moro (31, X. 1901) on lactosera (see p. 159), who found that different women's milks reacted especially to the antisera to which they had given rise. His results will however require confirmation. On the other hand we have evidence of differences between *maternal and foetal blood* in this respect[1]. Thus Halban and Landsteiner (25, III. 1902,

---

[1] Halban and Landsteiner (p. 473) cite Krüger (1886) as having found foetal blood to contain more solids but less fibrin than maternal blood. Scherenziss (1888) found the specific gravity of foetal markedly lower than that of maternal blood, although it con-

p. 475) report that anti-human serum gave reactions with higher dilutions of adult serum than of foetal serum. On the other hand, foetal serum gave more clouding in strong dilutions than did adult blood. Evidently an excess of normal serum checks precipitation, and this seems to me to be due possibly to normal antiprecipitins (see p. 149). It would appear that normal maternal serum contains more precipitable substance, and more antiprecipitin, than foetal serum. These differences however by no means indicate that maternal serum contains more globulins, for the precipitin reaction as we have seen (p. 98) does not constitute a quantitative test for albumins.

Uhlenhuth (5, VII. 1902), using antiserum for fowl's egg-white (from rabbit) found it to precipitate the serum of the hen much more than that of the cock, the sera of sexually mature animals being used in both cases for comparison. The reactions of the sera were so markedly different; that he found it easy to distinguish the one from the other.

He therefore refers to a sexual reaction (Geschlechtreaktion), and proposes to see if it is possible to distinguish between the mammalian sexes in a similar manner.

On the Character of the Precipitins in different Species of Animals treated with the same Blood.

It has been accepted for the bacterial antitoxins obtained from different animals that they are similarly constituted, whatever their source. This would appear to be different for the precipitins, judging from results obtained by Ascoli (26, VIII. 1902), who, to begin with, remarks that because two individuals, even of one species, yield sera which react equally to their homologous precipitin, we have no right to conclude that the sera of the two individuals are chemically identical, the reaction being perhaps insufficiently fine to demonstrate differences which actually exist. That he is right in saying this receives confirmation from what has been observed with regard to the isopreci-

tained more salts. Doléris and Quinquand (1893) made similar observations. Halban (1900) found foetal blood corpuscles to be agglutinated by maternal serum, and *vice versâ*, whilst the sera showed differences in haemolytic and bacteriolytic power. Schumacher also found maternal serum to agglutinate B. typhosus better than foetal serum. According to Halban and Landsteiner (*loc. cit.*) maternal serum is more haemolytic than foetal, it agglutinates red blood corpuscles more powerfully, it is more powerfully bacteriolytic (tested on the cholera vibrio), it is more anti-fermentative (antitryptic), more antitoxic (as against the haemagglutinins of abrin and ricin), and finally it yields more precipitum with precipitating antisera.

pitins, which would scarcely be formed unless there were chemical differences between individuals (see p. 148).

Working with rabbits, Ascoli (p. 1411) found it impossible to demonstrate any differences in the properties of the antisera they produced, as tested against egg-white solutions with which they had been immunified. On the other hand, when he immunified both *rabbits and guinea-pigs* with the defibrinated blood of the same dog, he was able to find a difference in the antisera they produced, for on adding the antiserum, say from the rabbit, to dog blood dilutions, until no more precipitation occurred, he found that he obtained a further precipitation on adding antiserum from the guinea-pig. As an additional proof of the antisera from the rabbit and guinea-pig being different qualitatively, he states that

> Anti-dog serum from rabbit gave a precipitum with normal guinea-pig[1] but not with normal rabbit.
>
> Anti-dog serum from guinea-pig gave a precipitum with normal rabbit but not with normal guinea-pig.

Assuming that Ascoli is correct in this observation, then it is patent that a multiplicity of precipitins are formed in the blood of an animal treated with normal serum of another species. And, moreover, different animals form different precipitins when treated with the blood of the same animal. According to Ascoli, the conclusion seems reasonable (p. 1412) that different precipitins attach themselves to different side-chains of the immensely complex " Rieseneiweissmolekül," in the sense of Ehrlich.

---

[1] Concentration not stated.

10—2

SECTION V.

ISOPRECIPITINS.

THE occurrence of isoprecipitins analogous to isohaemolysins discovered by Ehrlich and Morgenroth (see p. 42) appears to have been observed by Schütze (12, XII. 1901, and 6, XI. 1902, p. 804), who obtained them by injecting animals with the serum of other individuals belonging to the same species. Schütze treated a rabbit every two to three days with 5 to 10 c.c. of the serum of other rabbits, administered subcutaneously, until a total of 60 c.c. had been given. The treated rabbit was bled six days after the last injection, and its serum tested for isoprecipitin upon the sera of 32 different rabbits. Of these 32 sera only two reacted, precipitations occurring in 30 minutes at 37° C. The reaction was " strictly specific," for, upon the addition of 2 to 4 c.c. of the treated rabbit's serum to 3 c.c. of normal guinea-pig or human serum, no reaction was observable after two hours under the same conditions of temperature. Granted that the reaction took place as stated, it seems somewhat premature to conclude as to the strictly specific character of the reaction after only testing two other non-homologous bloods, even that of the guinea-pig being but distantly related, not to mention the human blood.

Schütze (p. 804) states that he found it more convenient to obtain isoprecipitins from goats treated with goat's milk. To do this, he injected 40—50 c.c. of the milk every 4—5 days, until a total of 400 c.c. had been administered in the course of a month. Bled eight days after the last injection, the goat yielded a serum which, when added in the proportion of 0·5 to 1 c.c.—5 c.c. of a 1:40 dilution of goat's milk, gave an immediate reaction.

Evidently then, there would appear to be individual differences in the chemical constitution of different individuals, demonstrable by means of isohaemolysins and isoprecipitins.

ANTIPRECIPITINS.

Normal Antiprecipitins. I have noted elsewhere (p. 145) that Halban and Landsteiner (25, III. 1902, p. 475) observed a difference between maternal and foetal serum with regard to their reactions to precipitins. This difference appears to me to depend upon the possible existence of normal antiprecipitins, analogous to the normal antihaemolysins observed by Besredka (1901) (see p. 22). Halban and Landsteiner found that the addition of large quantities of human serum to its antiserum prevented precipitation (but this might be explained in the light of what I have stated elsewhere, viz., that a precipitum is soluble in an excess of its homologous blood), whereas, additions of large quantities of ox or horse serum were less effective, and dog, fowl, and fresh rabbit serum did not prevent precipitation. Similarly, the authors I have cited found that anti-ox serum acted least upon the addition of much ox serum, more markedly upon adding the serum of the horse, most when the sera of the dog, man, rabbit, and fowl were added. That is to say, the latter did not impede the reaction. It is possible, it seems to me, that this appearance may be due to normal antiprecipitin, or precipito-lysin if I may so term it, which is present only in small quantity in normal serum, and therefore only exerts an obvious effect when large amounts of a serum are added to an homologous antiserum. (See Observation III. p. 89.)

Artificial Antiprecipitins have been produced by Kraus and Eisenberg (*Wiener klin. Wochenschr.*, 1901, p. 1191; see also 27, II. 1902, p. 212—213) through treating goats with goat-lactoserum obtained from rabbits which had been treated with goat's milk. They found that antilactoserum prevented the precipitation of milk by lactoserum. In addition, the antilactoserum contained antihaemolysin, which was natural, for the goat lactoserum (obtained from rabbits) was haemolytic for goat's red blood corpuscles.

Schütze (2, XII. 1901) has also succeeded in obtaining antilactosera, which, as stated above, are analogous to the antihaemolysins (see p. 21). Schütze's antiprecipitin retarded or prevented the precipitation of milk by its homologous antiserum.

It remains to be determined if injections with precipitoids, which have been compared to toxoids, will lead to the formation of antipreci-pitins, as toxoids do of antitoxins, as discovered by Ehrlich (1900 *Croonian Lect.*). I propose to see if this is possible.

NORMAL PRECIPITINS.

Noguchi (XI. 1902) and M. Ascoli (1903), report having found precipitins in certain normal sera. I have noted elsewhere that normal sera may contain antibodies, which are usually present in small amounts only. This appears also to be the case with the normal precipitins. Ascoli was stimulated to search for normal precipitins by a statement of Obermayer and Pick (see p. 97), that they had found normal rabbit serum to precipitate egg-white. Ascoli found ox serum to contain precipitins for the sera of man, dog, pig, goat, rabbit, guinea-pig, fowl; dog serum to contain precipitins for the serum of the fowl, and for egg-white; goat serum contained precipitins for fowl and guinea-pig serum. He judged of the occurrence of a reaction by observing if any clouding occurred at the zone of contact between a serum and serum dilution, made in 0·85 % saline. He used small tubes 4 cm. long by 3 mm. wide, and observed the reactions after the tubes had stood 30—60 minutes at 38° C. By diluting the one or the other of the normal sera he was able to note which contained normal precipitin, and in some cases found both sera to contain it. As has been observed for other normal antibodies, the amount of normal precipitin present varies considerably in different animals of the same species.

Ascoli also states that he has found "autoprecipitins" in normal human and animal sera, comparable to the auto-agglutinins found by him and other observers.

The occurrence of the precipitins in normal sera may account for some aberrant results obtained by various authors, but they do not appear to affect my results in the slightest. Although Ascoli does not state with what *concentrations* he worked, any more than do Obermayer and Pick, it amounts to a certainty that but low dilutions, and, in the latter case, even concentrated sera were used. That the use of concentrated sera or dilutions constitutes a grave source of error when making tests with *specific* precipitating antisera has already been pointed out, the source of error being apparently in part the normal precipitins. The stronger an antiserum the greater will be the dilution of homologous blood with which it will react, and it appears now all the more necessary to work with the highest possible dilutions when attempting to identify a blood. This being in accordance with what Wassermann (10, II. 1903) has just found in the case of the agglutinins for bacteria.

The foregoing remarks apply also to the observations of Noguchi

(XI. 02) who tested the interaction of the blood of cold-blooded animals with reference to haemolysis, agglutination, and precipitation He says that "the results do not lend themselves readily to classification, and from them no systematic conclusions can be drawn."

In regard to precipitins he states that his "studies show conclusively that normal precipitins are not uncommon, and may be compared with the occurrence of normal haemolysins and agglutinins. The technique of the experiments is simple: blood or body-cavity sera are obtained after coagulation has taken place, and are rendered clear by filtration. The clear products are mixed in given proportions, and the resulting fluids are compared with controls of the mixed sera."

No conclusions as to the relationship of these animals can be obtained from these experiments.

SECTION VI.

TESTS WITH PRECIPITINS.

I. PHYTOPRECIPITINS.

Bacterioprecipitins.

UNDER the name of Bacterioprecipitins I include the precipitins discovered by Kraus in the serum of animals treated with certain bacteria, which act upon culture-filtrates of the corresponding germs, or upon a solution of the substance of such micro-organisms. It will be remembered that these were the first precipitins to be discovered.

Kraus (30, IV.[1] and 12, VIII. 1897) found that if he added cholera, plague or typhoid antisera to filtrates of the corresponding cultures, that a precipitation took place. The antisera were obtained from animals which had been immunified. The culture-filtrates were prepared by passing the culture-fluid through porcelain filters. A filtrate of crushed germs gave the same result as the filtered culture-fluid, proving that the precipitable substance was present within the bacterial cell. The antisera only acted upon their homologous cultures. Filtrates of *B. diphtheriae* to which antitoxic horse-serum was added, on the other hand, did not give a precipitum. Kraus allowed the reaction to take place at 37° C., observing the precipitum after 24 hours. He states that the latter appeared to be composed of alkali-albuminate and peptone, and notes that it is necessary to add more antiserum in making these tests than is the case when adding agglutinating sera. Corresponding to the precipitable substance in culture-filtrates of *B. typhosus* and the Cholera germ, there must be a substance therein corresponding to the agglutinin-producing substance in these bacteria, for it may be noted that Widal, Levy and Bruns succeeded in rendering

---

[1] Kraus's earlier paper (30, IV. 1897—*K. Gesellschaft der Aerzte in Wien*) is cited by Bordet (1899, p. 228). I have been unable to gain access to the original.

animals immune by means of such culture-filtrates, the immune serum agglutinating specifically.

Nicolle (III. 1898, p. 162) confirmed the observations of Kraus with regard to *B. typhosus, B. pestis,* and *B. diphtheriae.* He in addition carried out corresponding experiments with *B. coli* and *Vibrio massauah,* obtaining most precipitum from old cultures of the corresponding germs. The reactions were studied in the thermostat and noted after 15—20 hours. Nicolle considered the reaction specific, and found that it was not impeded by antiseptics which were added.

According to Tchistovitch (v. 1899, p. 414) Marmorek subsequently obtained similar results with antistreptococcic serum when this was added to *Streptococcus* culture-filtrates.

Radziewsky (1900, p. 434) repeated and confirmed the observations of Kraus and of Nicolle upon *B. coli.* He, however, found that the reaction was not hastened at 37°, but that it took place as well at room temperature. Bail (1901) found that both typhoid immune serum and the peritoneal exudate of a guinea-pig which had received an intraperitoneal injection of *B. typhosus,* gave a precipitum upon being added to a culture-filtrate of this organism. In the latter case the peritoneal exudate doubtless contained typhoid bacilli substance in solution. Both Radziewsky and Bail consider bacterioprecipitins and bacterioagglutinins to be separate bodies, for, the latter states that after all precipitable substance has been removed from a culture-filtrate through repeated addition of precipitin, the deprecipitinated fluid showed unimpaired agglutinating power. Markl (13, VI. 1901, p. 812) has also found that anti-plague serum precipitated culture-filtrates of *B. pestis.* Kraus (18, VII. 1901) states that Wladimiroff (*Petersburg. med. Wochenschr.,* 1900) has applied the precipitation method in the diagnosis of glanders, finding a precipitum formed in a glycerine-free culture-filtrate of *B. mallei* (aged 46 days) upon the addition of the serum of a provedly glandered horse. Neufeld (2, v. 1902, p. 65) found that anti-pneumococcic serum precipitated a solution of *Pneumococci.* Castellani (28, VI. 1902, p. 1828) injected living cultures of *B. typhosus, Staphylococcus aureus,* and *B. coli* (two varieties) into rabbits and obtained antisera which produced a precipitum in old culture-filtrates of the corresponding germs. A rabbit received combined treatment with culture-filtrates of *B. typhosus* and *B. coli,* its serum subsequently precipitating the two filtrates. One of the cultures of *B. coli* was agglutinated, and its filtrate precipitated by typhoid-precipitin[1]. When

---

[1] The existence in a serum of common receptors for certain races of *B. typhosus* and *B. coli* has also been observed with " specific " bacteriolysins.

animals were treated with culture-filtrates, both dialyzed and non-dialyzed, precipitating and agglutinating antisera were also obtained. In confirmation of Kraus (see above), he was unable to obtain precipitins for *B. diphtheriae*.

There is fairly general agreement that the bacterioprecipitins and agglutinins are distinct chemically. I have noted above that both may coexist in a serum, and will refer the reader to p. 50, where the differences between agglutinins and precipitins are more especially discussed. Kraus and von Pirquet (5, VII. 1902) found that bacterial culture-filtrates contain not only a precipitable substance, but also specific anti-agglutinin, the latter being probably identical, they think, with the agglutinatable substance of bacteria. Pick (1902, p. 54) considers the precipitable substances in old and fresh culture-filtrates to be differently constituted, as they combine differently with bacterio-precipitins. The precipitable substance in old typhoid cultures, according to Pick, is not albumose, nor probably peptone, nor nucleo-proteid, and it is not an albuminous body. The reader is referred to Pick's third communication (1902) for details regarding the effects of various chemical agents upon bacterioprecipitins.

According to Pick (1902, p. 81), typhoid bacterioprecipitins are inactivated by an exposure of 30—45 minutes to 58—60° C., but not the agglutinins. The inactivated serum prevents the action of fresh antiserum. He considers (p. 92) that the inactivated serum acts upon the precipitin, not upon the precipitable substance (compare p. 91). As Rostoski (1902, *b*, p. 23) points out, the temperature at which bacterioprecipitins are inactivated is lower than that at which haematoprecipitins etc. are inactivated, and this indicates that they may be of a different nature. He suggests that the name "coagulin" should perhaps be retained for bacterioprecipitins so as to distinguish them from the precipitins which act upon blood and the like. This term, coagulin, is rarely used. It was applied by Ehrlich (1900, *Croonian Lect.*) and is used by Pick and a few others, but in my opinion it should be dropped, as it will only lead to a confusion with the coagulins for blood (see anticoagulins, p. 17) to which the term properly applies.

Yeast-Precipitins.

Schütze (6, XI. 1902, p. 805) sought to discover if it were possible by the use of precipitins to distinguish between different yeasts, which were severally used for the immunization of rabbits. He cites an earlier paper (*Sitzung. der Gesellschaft der Charitearzte*, 12, XII. 1901) wherein he reported success in obtaining yeast-precipitins. He extracted yeast-albumin by rubbing up yeast in a mortar in sterile 25 % soda solution, facilitating the rupture of the cell membranes by the addition of powdered glass and sand. The thick paste at first formed soon became fluid and thin, and a clear fluid was obtained by centrifugalizing off the sand and cellular detritus. The clear fluid was injected in doses of 5—10 c.c. every 3—4 days, during two months, into rabbits, a total of 100 c.c. being administered.

In this way he treated rabbits with yeast-albumin of four kinds: Potato yeast, Baker's yeast, Top and Bottom Beer yeast. When he came to test the antisera he found that all the yeasts reacted to the various antisera, in other words that it was not possible to distinguish the species of yeasts by means of these precipitins.

Precipitins for Albumins of Higher Plants.

Kowarski (4, VII. 1901) injected wheat albumose[1] into rabbits, obtaining precipitins, which acted promptly on wheat albumose in solution, less markedly on rye and barley, but not on corresponding solutions from oats; peas giving a very weak reaction (faint clouding). Normal rabbit serum had no such effect.

Schütze (22, XI. 1901, p. 493) treated rabbits with a vegetable proteid "Roborat," obtaining an antiserum which precipitated roborat solutions, but not muscle albumin, the antiserum for the latter having in its turn no effect on roborat solutions. Castellani (28, VI. 1902, p. 1828) confirmed this observation, finding that anti-roborat did not act on somatose solutions. He also appears to have produced a weak anti-somatose, which had no action on roborat solutions.

Jacoby (1901, cited by Bashford) found that animals which had been rendered immune to *ricin* yielded a serum, which, whilst antitoxic, gave a precipitum on being added to ricin solutions. Bashford (1902, *Journ. of Pathol. and Bacteriol.* VIII. p. 59) states that he has recently made a similar observation in rabbits treated with *crotin*. The rabbits' serum gave a dense precipitum with a solution of crotin, normal rabbit serum not exerting this effect[2].

---

[1] These results are to be received with caution, see p. 111.

[2] Note analogous observations on snake venom, see Section VI.

II. ZOOPRECIPITINS.

I. Lactosera.

The term "lactosera" was applied by Bordet (III. 1899, pp. 240—241), who discovered them, to antisera which precipitated milk casein. Having partially sterilized milk at a temperature of 65° C. he injected it intraperitoneally into rabbits at stated intervals. After a time he bled the rabbits, and found their serum to possess the property of causing precipitation in the milk with which the animals had been treated. He performed the test by placing about 3 c.c. each of the antiserum and of normal serum in tubes, to which he subsequently added 6—15 drops of the milk. Whereas the mixture with normal serum retained its opaque white appearance, the mixture with antiserum soon underwent a change; fine granules appeared, and these rapidly growing in size led to the formation of flocculent masses. Soon the fluid became limpid above, the flocculi sinking to the bottom. This was best observed in milk which had been deprived of an excess of fat by being passed two or three times through filter-paper. When milk rich in fat is used, the flocculi are carried upward by the fat globules. In studying the reaction, it is therefore best to first *filter the milk.* The supernatant fluid remains clear, but may at times again show a cloud upon the addition of fresh antiserum.

Fich (II. 1900, cited by Schütze, 29, I. 1901, p. 7, footnote) soon afterwards found that an emulsion of udder-cells, as well as milk, gave rise to antisera which precipitated milk, when rabbits were treated with these substances. The observation of Bordet has been since amply confirmed by a number of investigators. The first of these to suggest a practical application of the method in the differentiation of milks were Wassermann and Schütze (2, VII. 1900 and 29, I. 1901), who treated rabbits with three different milks, namely those of the cow, goat, and human subject. They claimed that these milks could be distinguished from one another, the corresponding lactoserum acting *only* upon its homologous milk. Their method of treating rabbits will be found described under Methods (p. 53). Schütze (29, I. 1901, p. 5) subsequently described their investigations in detail, and drew attention to the possibility of using the lactosera in the examination for milk adulteration.

Uhlenhuth (6, XII. 1902), in testing with lactosera, adds 5 c.c. of antiserum to 3 c.c. of a 1 : 60 milk dilution in salt solution. In his tests, lactoserum for cow's milk produced a thick flocculent precipitate .in cow milk dilutions within a few minutes, whereas it left human and donkey milk dilutions unaffected.

The specific character of the reaction was dwelt upon by Wassermann and Schütze, who stated that no reaction took place, for instance between the lactoserum for cow's milk and goat's milk, and *vice versâ*. This result may have been due to their using weak lactosera, for the results of other workers have been different with the milks of animals so closely related. Thus Moro (31, x. 1901) found the lactoserum for cow's milk to precipitate goat's milk. This is in perfect agreement with what I found with regard to the antisera for the corresponding bloods, and is also in accordance with the observations of Ehrlich and Morgenroth, who found that injections of the blood corpuscles of the ox into experimental animals led to the formation of an haemolysin which acted not only on the corpuscles of the ox, but also on those of the goat.

The question of specificity was next studied by Gengou (25, x. 1902, p. 749) with lactosera from rabbits treated with cow's milk. He states that the reactions which took place in the milks of the cow, sheep and goat were indistinguishable when he made the following tests. He prepared two tubes containing:

I. Milk (of cow, goat, or sheep, etc.)   II. Milk (as in I.)
    Lactoserum (heated to 56° C.)   Normal rabbit serum (heated to 56° C.)
    Normal rabbit serum, fresh   Normal rabbit serum, fresh

He found that the complement or "alexine" (of the fresh normal rabbit serum) was absorbed in I. and not in II., throughout the series, when a lactoserum was added to its homologous milk. It was in this that the three milks above named were indistinguishable. The immune substance acted to a lesser extent upon human and mare's milk. The lactoserum, he adds (p. 754), did not act upon ox serum (see contrary below). Schütze (6, XI. 1902, p. 804) found that a powerful isoprecipitin (see p. 148) for goat's milk did not produce a reaction in human milk dilution even after a considerable time, though he does not state how long.

Milk boiled for half-an-hour, according to Schütze (*loc. cit.*) still gives a slight reaction upon the addition of antiserum. Moro (31, x. 1901, p. 1076) obtained an antiserum from rabbits treated with milk which had been boiled for half-an-hour or upwards. He found that milk which had been boiled for 30 minutes reacted immediately to its antiserum. Müller (18, II. 1902) dissolved lacto-precipitum in boiling saline, and found that it could be reprecipitated by fresh lactoserum, and also by rennet ferment. He considers that the *precipitum consists of unaltered casein combined with the precipitin.* Finally Fuld (1902) also found lactosera to act upon boiled milk. He considers that the casein is chiefly concerned in the reaction. He found the presence of lime salts essential. Contrary to Moro, he was unable to obtain an antiserum from rabbits treated with boiled milk. This difference, it appears to me, may very well be due to the milk being "boiled" in a different manner by the two workers. The experiment should be repeated, the temperature of the milk within the receptacle being accurately determined. Uhlenhuth (6, XII. 1902) obtained reactions when antiserum was added to milk which had been boiled, and even with milk which had been exposed to a temperature of 114° C. for 30 minutes in an autoclave.

Lactosera do not act on milk only, for Halban and Landsteiner (25, III. 1902, p. 475), following Hamburger, found that human lacto-serum acted also upon human blood serum. They also found that human haematoserum precipitated human milk, just as Schütze (2, XII. 1901, cited in this author's paper of 6, XI. 1902, p. 805) had found antiserum for ox blood to also precipitate cow's milk.

Meyer and Aschoff (7, VII. 1902) found ox lactoserum to be haemolytic for ox blood corpuscles, and that it immobilized ox spermatozoa. The milk which they injected into their rabbits had been previously centrifugalized to rid it of cellular elements which of themselves might have led to the formation of cytolysins. They also found lactoserum to precipitate ox serum. Injections of blood, spermatozoa, and tracheal epithelium, led to the formation of antisera which also precipitated 1 : 40 milk dilutions. They found that lactosera (which were also haemolytic) could be neutralized by homologous blood corpuscles or spermatozoa.

Schütze (6, XI. 1902, p. 804) found human lactoserum to precipitate solutions of human spermatozoa, and von Dungern (1899—1900 *Münchener med. Wochenschr.*) found lactosera to agglutinate red blood corpuscles. (See further Hamburger and others, under the Action of Precipitins upon different Albumins of the same animal, p. 104.)

Uhlenhuth (6, XII. 1902) found lactoserum for cows' milk to produce clouding in ox blood dilutions. He cites an observation of von Dungern to the effect that lactoserum for human milk exerts a toxic action on human milk-gland epithelium.

Although perhaps not identical with the haematosera, the lactosera for the same animal would appear to show but a very slight difference, if any. We see that lactosera and haematosera both act similarly. Although there may be quantitative differences there are none apparently of a qualitative character. Comparative quantitative experiments have still to be made.

As in haematosera, the precipitin contained in lactosera is very stable. Moro (31, x. 1901) found it still gave reactions with its homologous milk after being sealed for months in capillaries, thus confirming what I have found for some haematosera. Lactosera withstand a temperature of 56° C., in this resembling the haematosera, and the agglutinins. On warming lactoprecipitum in normal saline it almost completely redissolves. Moro obtained lactosera by treating animals with pure powdered human or cow's casein.

Milks of the same species show individual differences according to Moro (*loc. cit.*), in their reaction to lactosera. On the other hand, the reaction of each individual milk to lactoserum, as tested on the milk of wet-nurses, remained constant. Individual differences were less marked in cow's milk. The individual difference was shown by the reaction of each milk to its homologous antiserum taking place with higher dilutions than with a non-homologous antiserum. Personally I am inclined to be somewhat sceptical on this point in consequence of the results which I have obtained with Mr Strangeways with haematosera. To prove the presence of individual differences, and I do not question but that there may be such, it is necessary to examine a large series of antisera and homologous milks or sera, and this does not seem to have been done by Moro. His results have therefore to be accepted with caution. I will however state the reasons which led him to the conclusion that there are individual as well as specific groups in a given milk.

1. Antiserum for milk *A* added to milk *a* gives reaction in a high dilution, and *vice versâ*.
2. Antiserum for milk *A* added to milk *b* gives no reaction in a high dilution, and *vice versâ*.

The reaction is due in the first instance to the specific and individual groups entering into combination. In the second instance, using a stronger dilution, only the specific, not the individual groups combine. The deficit in the reaction *A* plus *b*, cannot be made good by adding more of milk *b*. He concludes, therefore, that the specific groups are neutralized, or destroyed, only the individual ones which are least numerous being left. These being unable to combine account for the precipitum-deficit. He represents this schematically by a figure analogous to the well-known diagrams of Ehrlich, the lactoprecipitin molecule being represented by a circle from which protrude an assumed number (6) of receptors, the majority of which (4) represent specific groups. Whereas the latter are alike, the individual groups (2) are different, and the various combinations of the two lactosera (*A* and *B*) and milks (*a* and *b*) are thus shown, the precipitable substance in milk being fixed, or not, as the case may be, to receptors. See further under "Individual Differences," p. 145.

II. Haematosera.

In the following pages the results of precipitin tests with haematosera are given in the zoological order of the antisera, the tests made by other observers being summarized in each case, the results of my tests following.

The number of tests, made by me with 30 antisera produced, is given in the following table, the total number of tests being 16,000.

Antiserum for	No. of tests therewith	Antiserum for	No. of tests therewith
Man	825	Ox	790
Chimpanzee	47	Sheep	701
Ourang	81	Horse	790
Cercopithecus	733	Donkey	94
Hedgehog	383	Zebra	94
Cat	785	Whale	94
Hyaena	378	Wallaby	691
Dog	777	Fowl	792
Seal	358	Ostrich	649
Pig	818	Fowl-egg	789
Llama	363	Emu-egg	630
Mexican Deer	749	Turtle	666
Reindeer	69	Alligator	468
Hog Deer	699	Frog	551
Antelope	686	Lobster	450
	7751		8249
		7751	
		8249	
		Total number of tests	16,000

In the short tables which follow, I have briefly summarized the general results of the tests given in the experimental tables at the end, the object of this being to render the conclusions clearer. The number of sera reacting in each case is given, the intensity of reaction being indicated by signs, as follows:—

.	No reaction	+	Marked clouding
*	Faint clouding	+	Full reaction
×	Medium clouding		

The number of *full reactions* obtained is indicated by the numbers printed in *black type*. Although in some cases the number of bloods examined has been small, it has been thought best to introduce the percentages of reactions under each class, as the numbers alone convey no clear impression to the mind as to the relative frequency of reactions. Percentages are frequently given alongside each class of reaction. The percentages of positive reactions " all told," *including the faintest*, are given on the right-hand margin of the tables[1].

I. Antisera for bloods of Primates.

(1) *Tests with Anti-Human Serum.*

Before proceeding to give the results of my own tests with this antiserum I shall briefly summarize what has been done by different workers who have used anti-human serum in making various tests. It may seem somewhat pedantic to give the exact day on which the authors, cited in the following pages, published their papers; my object in doing so is to dispose of certain claims to priority which have been made in various quarters. I wish to accentuate the fact that no new principle was discovered when an antiserum for human blood was found. The antisera for eel, fowl, horse, etc. had been found by Tchistovitch and Bordet, and the whole impulse given to work in this line undoubtedly emanates from them. Their papers were most suggestive, and it was quite natural, in reading them, that many should be seized by the same idea as to the *possible applications* of the methods they had pointed out. This was the more natural in view of the knowledge previously gained with regard to other classes of antibodies.

Leclainche and Vallée (25, I. 1901) injecting human albuminous urine into rabbits, found that they obtained an antiserum which precipitated albuminous urine, and pleuritic exudation, but not serum dilutions of man. This antiserum did not produce a reaction with albuminous urines of the horse and cow.

---

[1] I am indebted to Dr Graham-Smith for kindly calculating the percentages given in the succeeding tables. The percentages are given in round numbers, fractions of percentages being only included in the totals at the right-hand margins of the tables.

Uhlenhuth (7, II. 1901) was however the first to publish tests of a kind that awakened general interest, more especially regarding the medico-legal aspects of the test and its possibilities. He injected rabbits with human blood, and tested the antiserum on 19 bloods (1), finding *only human blood to react*. Normal rabbit serum had no such effect. He made the practically important observation that human, horse, or ox blood which had been *dried* 4 weeks on a board, could readily be distinguished by the antiserum when brought into solution.

A day later than Uhlenhuth, Wassermann (8, II. 1901) demonstrated experiments similar to Uhlenhuth's at the Meeting of the Physiological Society, Berlin. Outside of human blood only that of a "*monkey*" gave the reaction. A report of the experiments appeared under the names of Wassermann and Schütze (18, II. 1901) shortly after, it being stated therein that they had tested 23 bloods with this antiserum (2). I have ordered the bloods they tested here, and it will be seen that many of those examined were the same as in Uhlenhuth's list.

(1) Uhlenhuth's List		(2) Wassermann and Schütze's List	
Man	*Reacted*	Man, baboon	*Reacted*
Dog, cat	No reaction	Dog, cat	No reaction
Pig	,,	Pig	,,
Ox, sheep, deer, fallow-deer	,,	Ox, sheep, goat	,,
Horse, donkey	,,	Horse, donkey	,,
Hare, rabbit, guinea-pig, rat, mouse	,,	Rabbit, guinea-pig, rat, mouse	,,
Fowl, turkey, pigeon, goose	,,	Fowl, duck, goose, sparrow	,,
		Eel, pike, "Schlei"	,,

The latter authors also found that blood could be perfectly well tested when brought into solution after being dried for 3 months on various objects. The baboon's blood reacted much more slowly and less markedly than did human blood. This finding possessed peculiar interest to me, and, together with the similar observation of Stern, led me to plan an investigation of the bloods of Anthropoidea. Stern (28, II. 1901) also obtained anti-human serum from rabbits and found it to act on blood which had been dried, and also on albuminous urine. His antiserum did not act on horse, ox, sheep and pig bloods, whereas, it gave a feeble but distinct clouding with the blood of 3 species of monkeys "Meerkatze, Java-Affe, Kronen-Affe," the first being I suppose a *Cercopithecus*, the second *Macacus cynomolgus* L. He does not give their scientific names, so that, as in the case of Wassermann and Schütze, it is difficult to identify them.

Dieudonné (2, IV. 1901) obtained anti-human serum by treating rabbits with human albuminous urine, pleuritic exudation, and blood serum, injecting quantities of 10 c.c. every 3 to 4 days. He tested his antiserum on only four other bloods (rabbit, guinea-pig, pigeon, goose), finding that normal rabbit serum had not the precipitating effect of the antiserum on human blood, albuminous urine, pleuritic and peritoneal exudate.

At this stage I reported (11, v. 1901) having tested 24 bloods; only human, and to a lesser degree, 2 monkey bloods reacting. Soon after (1, vii. 1901; 36 bloods tested) I reported the blood of four species of monkey as having given a slight but distinct reaction, a very faint clouding appearing in solutions of the bloods of the horse, ox, and sheep," all the others remaining perfectly clear.

Ziemke (27, vi. 1901) obtained antiserum from Wassermann and Schütze, and tested it upon 12 bloods (1). A reaction in 15 minutes, and a precipitum after 24 hours only being found in human blood dilutions.

Schirokich (21, vii. 1901) injected rabbits with blood from the placenta and cord. He found that blood-stains 2 years old gave the reaction when dissolved and brought into contact with his anti-human serum. He tested 9 bloods (2).

(1) Ziemke's List		(?) Schirokich's List	
Man	*Reacted*	Man	*Reacted*
Dog, cat	No reaction	Cat	No reaction
Pig	,,	Pig	,,
Ox, calf, sheep, lamb	,,	Ox, goat	,,
Horse	,,	Horse	,,
Rabbit, guinea-pig, mouse	,,	Camel	,,
		Rabbit, guinea-pig	,,

In the meantime my collection of bloods had reached 140, including the bloods of 12 species of monkeys (some in duplicate) including 10 *Cercopithecidae*, and 2 *Cebidae*. I reported on these tests in my paper of 14, ix. 1901: "All the bloods of the Old World monkeys gave a very marked reaction, less powerful however than that of human blood. They also reacted to a weak antiserum. On the other hand the South American monkeys (*Cebidae*) gave but a slight reaction with human antiserum as compared to that of the other monkeys, and a weak antiserum produced no precipitation in the blood of *Mycetes seniculus*." I reported further upon this interesting observation in subsequent papers (21, xi. and 16, xii. 1901, p. 408), having in the meantime also tested the bloods of the Ourang-Outang and Chimpanzee with positive results. These bloods appeared to give about as much reaction as human blood to anti-human serum. Grünbaum (18, I. 1902, p. 143) also obtained this result with anti-human serum tested upon these bloods and that of the Gorilla. I shall refer to his results again further on. In my paper of 20, I. 1902, I remarked that the amount of reaction would appear to correspond with the degree of relationship amongst the Anthropoidea, the Lemurs, as already noted, having given negative results. Uhlenhuth seems to have tested a monkey blood of some kind lately, for he says (11—18, IX. 1902, pp. 661) that the weak reaction occurring in monkey blood upon the addition of anti-human serum cannot be confounded with the marked reaction occurring in human blood dilutions. In the remarks which follow the table below we see that even monkey blood may give a high degree of reaction under certain conditions. Schütze (6, XI. 1902, p. 805) was able to obtain precipitation with anti-human serum in solutions of human spermatozoa made from stains which had been dried 6 months.

Confirming what has gone before, Whittier (18, I. 1902) obtained negative results in testing anti-human serum on four bloods: of horse, cow, rabbit, guinea-pig. Butza (18, IV. 1902) tested anti-human serum upon 14 bloods (man, dog, cat, pig, ox, sheep, rabbit, guinea-pig, fowl, pigeon, turkey, duck, goose, and a fish) with negative result except the first. Biondi (1902) found anti-human serum to precipitate the bloods of *Cercopithecus flavus viridis* and *Macacus radiatus*. Anti-human serum precipitated human milk but not that of the cow, goat and donkey; human saliva, but not that of the dog, cat, horse and donkey. Human bloods in health and disease gave similar reactions.

Biondi found furthermore that human serum injected into a monkey did not lead to the formation of precipitins, as might be expected theoretically in consequence of the close relationship between man and monkey.

Lastly we find that Ewing (III. 1903) tested four specimens of monkey blood (baboon, rhesus, and two Java monkeys, species not stated) with anti-human sera obtained from rabbits and a fowl. He considers it possible to distinguish these bloods from human blood by means of higher dilutions of the test sera[1]. He omits to mention that the same thing can be accomplished by using higher dilutions of the bloods tested, as others have found. It must always be remembered that the precipitin should be present in excess.

[1] See page 142.

Layton (1903, pp. 219, 220) has treated rabbits according to Uhlen-huth's method and produced anti-human serum. He tested 28 bloods (species not stated) and was able to identify human and exclude monkey blood[1].

825 *Tests with Anti-Human Serum.*

The following tests were carried out by means of 5 different anti-human sera obtained by treating rabbits with the blood of Europeans. All the sera were powerful, being freshly prepared in succession, so that when the one began to weaken another was substituted[2].

			Reactions			
	.	*	×	+	**+**	%
97 Primates						
Anthropoidea						
34 Human (4 races)	.	.	3 (8 %)	7 (21 %)	**24** (71 %)	100
8 Simiidae (3 species)	.	.	.	.	**8**	100
36 Cercopithecidae (26 species)	3	.	26 (72 ,,)	3 (8 ,,)	**4** (10 ,,)	92
13 Cebidae (9 species)	3	2 (15 %)	5 (38 ,,)	3 (23 ,,)	.	78
4 Hapalidae (3 spec.)	2	1	1	.	.	50
Lemuroidea						
2 Lemuridae (2 spec.)	2	.	.	.	.	0
29 Chiroptera	26	3 (10 ,,)	.	.	.	10
15 Insectivora	13	2 (13 ,,)	.	.	.	13
97 Carnivora	70	13 (13 ,,)	14 (14 ,,)	.	.	27
65 Rodentia	53	7 (11 ,,)	5 (7 ,,)	.	.	18
70 Ungulata.................	40	19 (27 ,,)	11 (16 ,,)	.	.	43
3 Cetacea	.	3 (100 %)	.	.	.	100
13 Edentata	12	1 (7 ,,)	.	.	.	7
26 Marsupialia...............	25	1 (4 ,,)	.	.	.	4
1 Monotremata	1		.	.	.	0
320 Aves	319	1 (·3 ,,)	.	.	.	0·3
49 Reptilia	49				.	0
14 Amphibia	14					0
19 Pisces	19					0
7 Crustacea	7					0

(Ungulata group: 24 %)

(Aves and below: ·2 %)

The preceding table shows *maximum reactions only amongst the Hominidae, Simiidae, and Cercopithecidae.* In the case of the last only 4 bloods gave a maximum reaction, in two the precipitum was moderate in quantity, in two it was voluminous, in fact almost equal in quantity to that observed in the reactions with human

[1] Layton added ·5 % carbolic acid to his antiserum.
[2] See p. 144 regarding the strength of antisera.

and ape bloods. This great reaction with the blood of the two Cercopithecidae appears to find its explanation in the cause of death which was given for these two animals (Nos. 49 and 58) viz., intussusception and dysentery. In both of these affections the blood is liable to become concentrated, and therefore, taken volume for volume, the serum of such an animal will contain more of the reacting body than does normal serum. Within limits, moreover, the more concentrated a blood dilution, the more precipitum will it yield upon the addition of an antiserum. In point of fact then, the bloods of Cercopithecidae give under ordinary conditions a very moderate reaction as compared to those of Hominidae and Simiidae. Three negative results amongst the Cercopithecidae (Nos. 49, 59, 75) are due to some common cause, which I am not at present able to fully explain. All three bloods were sent to me by Dr Langmann from the New York Zoological Gardens. None of them went well into solution, a fact which was possibly due to their having been exposed too long in the sun, or affected in some other way. The bloods of the Cebidae and Hapalidae show a still further reduction in the amount of reaction.

If we include even the faint and medium cloudings (*   ×) amongst the positive reactions, we see that the *percentage of positive reactions falls as we read down the column of the reactions with the bloods of Primates, from man to the Hapalidae.* It must however not be lost sight of, that the great reactions in the two last columns are those which more clearly show the relationship. When we come to consider the other Mammalia, we see that there is a notable absence of larger reactions, the largest reactions noted being a few medium cloudings, and these occurred most frequently amongst the Carnivora and Ungulata. This will be better seen by reference to the percentages bracketed alongside the figures. The percentage of faint and medium cloudings in all the mammalian bloods examined, outside the Primates, amounts but to about a half of that noted for the Hapalidae, viz. 24%. Amongst the Primates we see that the result of the test with two lemurs was negative. I demonstrated a small series of primate bloods before the Royal Society (21 Nov. 1901), which showed this progressive decrease of reactions from the Hominidae downward, and no reaction with the lemur bloods. I stated at the time, that a weaker antiserum included the Hominidae, Simiidae and Cercopithecidae within its sphere of action. When the antiserum was stronger it included the Cebidae and Hapalidae, these reactions being comparatively very slight. On the other hand, if we use a very powerful anti-human serum, faint cloudings may also be observed in the dilutions of other mammalian bloods, and such an antiserum will also cloud the dilution of lemur blood. This is however a feeble reaction to which I have already given the name of mammalian reaction, for I have not found it to occur amongst non-mammalian bloods. This is very well seen by reference to the preceding table, where only one avian blood, out of 320 tested, is recorded as showing a faint clouding. On what such faint cloudings may depend, seems fairly clear. To begin with, they are such great exceptions that this alone suggests their being due to some error. At times they may be due to clouded solutions, or to clouded antiserum, at other times they may be due to the blood samples having been brought together by collectors. In an earlier paper I first showed that this reaction took place in blood mixtures. Consequently, if a blood sample, say from a bird, were brought in contact with one from a mammal, a positive reaction might take place with anti-avian as well as with anti-mammalian serum upon these being added to a dilution made from a sample of mixed bloods. Also, supposing that samples of dried bloods are stored in contact with one another, and are allowed to rub against each other, we may have the same result. This especially might occur where blood had been allowed to clot upon the paper, for blood-scales might break off and become attached to a paper saturated with another specimen of blood. Finally, a certain number of these results may be reasonably ascribed to the fact that in cutting out squares of filter-paper in succession from a number of different samples, a trace of a foreign blood may at times adhere to the scissors used and thus find its way into a solution to which it does not belong. It is remarkable however how few tests have given other than the result which experience taught one to expect. It is nevertheless well to remember these sources of error so as not to attach any special importance to exceptional reactions.

The anti-human sera used in the foregoing tests were usually powerful, and therefore we find more "mammalian reactions" noted. In contrast to this, we see on referring to the tests with a much weaker antiserum (anti-monkey, p. 171) that the action of the antiserum was practically limited to the Primates, for, even when we include the faint cloudings, reactions only occurred in 1% of all the other mammalian bloods examined.

(2—3) *Tests with Antisera for Simiidae.*

The first to obtain antisera for the Simiidae was Grünbaum (18, I. 1902, p. 143), who treated rabbits with the blood of the gorilla, ourang-outang, and chimpanzee, obtaining as many antisera, all of which acted upon their homologous bloods, as well as upon those of the related species, and man. Grünbaum was unable to "assert that there is any difference of reaction" amongst these bloods to the several antisera above-named. Dr Grünbaum, knowing of my investigations, very kindly sent me some of his anti-gorilla and anti-chimpanzee serum, so that I might test them upon the bloods of Anthropoidea which I had collected. Unfortunately his anti-gorilla serum had grown so weak that no results of value could be recorded for it. On the other hand, the following table gives the results of tests made with the anti-chimpanzee serum.

47 *Tests with Anti-Chimpanzee Serum.*

The anti-chimpanzee serum used in these tests was kindly sent to me by Dr A. S. F. Grünbaum of Liverpool. It bore the date 24, I. 1902, and the tests were made on its receipt four days later. The antiserum was weak and limited in quantity, and consequently but few bloods could be tested with it, the tests being confined to bloods of Primates. Although the amount of precipitum obtained was small, considerable differences could be noted.

	Reactions				
	•	*	×	**+**	%
3 HOMINIDAE (European)	•	•	•	**3**	100
3 SIMIIDAE (2 spec.)	•	•	•	**3**	100
23 CERCOPITHECIDAE (19 spec.)	8	9 (39 %)	6 (26 %)	•	65
12 CEBIDAE (9 spec.)	11	1	•	•	
4 HAPALIDAE (3 spec.)	1	3	•	•	
2 LEMURIDAE	2	•	•	•	0

It will be seen that the results are in accord with those obtained with anti-human serum. The proportion of negative results with the bloods of Cercopithecidae and Cebidae was naturally greater because a weaker antiserum was used. The results with Cebidae and Hapalidae do not agree with those obtained with the preceding and two succeeding series, where the Hapalidae gave lower values than the Cebidae.

81 *Tests with Anti-Ourang Serum.*

The antiserum used in these tests was produced by injecting a rabbit with the serum of *Simia satyrus*, which died at the Zoological Society's Gardens (No. 38). The antiserum was of moderate power. Here again the tests were limited to those upon the bloods of Primates, the bulk of the antiserum being reserved for studies by the quantitative method reported in Section VII. Standardized 6 weeks after these tests were made, this antiserum gave a precipitum of ·008 c.c.

	Reactions				
	•	*	×	+	%
23 HOMINIDAE (4 races)	3	7 (30 %)	13 (56 %)	•	86
8 SIMIIDAE (3 spec.)	1	•	2 (25 ,,)	5 (62 %)	87
32 CERCOPITHECIDAE (23 spec.)	5	2 (6 ,,)	23 (71 ,,)	2 (6 ,,)	84
12 CEBIDAE (8 spec.)	7	2 (17 ,,)	3 (25 ,,)	•	42
4 HAPALIDAE (3 spec.) ....,,,	4	•	•	•	0
2 LEMURIDAE (2 spec.)	2	•	•	•	0

We see here that the results are again mainly in accord with the preceding. It is possibly due to the fact of the human blood dilutions being somewhat weaker (old dried bloods mostly) that the human bloods gave lower values here. The quantitative tests made with fluid sera gave results similar to those in the two preceding tables. Owing to the fact that this antiserum was more powerful than the anti-chimpanzee serum the greater number of the bloods of Cercopithecidae gave medium reactions. Of the 5 that gave negative results 3 were samples from India, Borneo, and China respectively, 1 came from New York (gave a * with anti-monkey), and the fifth sample had been dried one year. Similarly the 2 giving feeble reactions (*) came respectively from India and New York. The 3 negative results with human blood were with 3 samples (Negro, Nos. 33—35) sent from W. Africa, these samples also showing less than full reactions even with anti-human serum. The negative result noted among Simiidae is due to the blood of a gorilla (No. 43) which Dr Grünbaum informed me a year ago gave but feeble reactions in his tests with anti-human and anti-chimpanzee serum.

(4) *Tests with Anti-Cercopithecus Sera.*

From my results with other anti-primate sera, it seemed of interest to repeat a series of tests with an antiserum for *Cercopithecidae*, and in due course for other families of Primates. The tests I have made with anti-monkey serum are given in the short table which follows. Before coming to these, I shall refer to a paper by Friedenthal (10, VII. 1902, p. 831). Pursuing my line of investigation, he sought to obtain an antiserum for *Cynocephalus hamadryas*, the blood of which animal he injected subcutaneously into rabbits in the huge doses of 26—51 c.c., which appears to have killed most of his rabbits outright, as might naturally be expected. It is almost impossible to attach any value to his results, because of the faulty manner in which he proceeded to immunify his animals, and especially because of the date at which he bled them after their last injection. In a protocol, which he gives, he states that he obtained an antiserum from a rabbit as follows: A rabbit received monkey serum subcutaneously in doses of 2, 2, 3, 5, 5, 5 c.c., the day intervals between inoculations being 3, 2, 5, 5, 4. Bled 3 days after the last injection, the rabbit's serum contained no precipitin for its homologous blood. Having waited seven days, he injected 25 c.c. (!) subcutaneously, and 48 hours (!) later, found precipitin (?) in the rabbit's serum, the serum being then used for tests. He made the tests by adding 0·2 c.c. of the serum to be tested to 5 c.c. of the "antiserum[1]."

Although Friedenthal's method was evidently bad, the results he claims to have obtained are not opposed to mine. His antiserum acted on its homologous blood, and also upon that of three other Cercopithecidae (*Cynocephalus* "*dschedala*," *Colobus guereza*, *Macacus cynomolgus*) to an equal degree. His antiserum, when first used, was certainly very weak (as might be expected from the very early date at which he bled his rabbit after its last injection), for he states that it failed to produce any effect upon human or anthropomorphic ape blood even after 24 hours. He adds, later (p. 833), that the rabbits which survived his (mal-) treatment, gave an antiserum which did also produce reactions, though weaker ones, with both human and chimpanzee blood. In this he confirms my results with the anti-primate sera described on the preceding pages, and also agrees with what follows. It is unfortunate that Friedenthal used such methods, for, as stated, they tend very much to vitiate his results.

---

[1] See further under "Sources of Error," p. 74.

733 *Tests with Anti-Monkey (Cercopithecus) Serum.*

The antiserum used in these tests was procured through the injection of a rabbit with the blood of *Cercopithecus (Papio) hamadryas.* The blood was kindly sent to me by Dr Lühe, the animal having died at the Zoological Gardens, Königsberg (No. 65, 14, II. 1902). The antiserum was feeble.

		Reactions			
	•	*	×	**+**	%
85 PRIMATES					
Anthropoidea					
23 Hominidae (4 races) ...	3	16 (70%)	4 (17%)	•	87
8 Simiidae (3 spec.).........	2	2 (25 ,,)	4 (50 ,,)	•	75
35 Cercopithecidae (24 spec.)*	•	11 (31 ,,)	21 (60 ,,)	**3** (8%)	100
13 Cebidae (9 spec.)	7	3 (23 ,,)	3 (23 ,,)	•	46
4 Hapalidae (3 spec.)	3	1 (25 ,,)	•	•	25
Lemuroidea					
2 Lemuridae (2 spec.)	2	•	•	•	0
29 CHIROPTERA	29	•	•	•	0
12 INSECTIVORA	12	•	•	•	0
95 CARNIVORA	93	2	•	•	2
62 RODENTIA ,,,....................	62	•	•	•	0
67 UNGULATA	66	1	•	•	1
3 CETACEA	3	•	•	•	0
13 EDENTATA.......................	13	•	•	•	0
26 MARSUPIALIA	24	2	•	•	8
1 MONOTREMATA	1	•	•	•	0
271 AVES (incl. 3 eggs)	271	•	•	•	0
45 REPTILIA	45	•	•	•	0
9 AMPHIBIA	9	•	•	•	0
14 PISCES	14	•	•	•	0
1 CRUSTACEA	1	•	•	•	0

The brace covering the rows from Chiroptera through Monotremata is annotated: 1 %

* No. 59 Cercopithecus is not included here, as it gave negative results throughout. See note to No. 49 in tables.

The results will be seen to correspond to those obtained with the preceding antisera, only that the reactions amongst the Cercopithecidae form the majority of the marked reactions. The bloods of Cebidae and Hapalidae again give less marked reactions than do those of Hominidae and Simiidae. The Lemuridae give a negative result. Owing to the weakness of this antiserum the mammalian reaction occurring in bloods not closely related is scarcely perceptible, for only 1% of faint reactions (slight clouds) were observed amongst all the other mammalian bloods examined outside of the Primates. The action of this antiserum very well demonstrates a fact to which I have already drawn attention, namely that a weak antiserum is more limited in its action.

II.　Antisera for bloods of Insectivora.

Uhlenhuth (25, VII. 1901) treated rabbits with the serum of the hedgehog (*Erinaceus europaeus*) and obtained an antiserum which only acted on the blood of the hedgehog, he tested 24 bloods as follows:—

Man, hedgehog, bat, dog, fox, cat, pig, deer, ox, sheep, goat, horse, donkey, rabbit, guinea-pig, rat, mouse.　Fowl, pigeon, owl, crow, sparrow, duck, goose.

383 *Tests with Anti-Hedgehog Serum.*

The following tests were made with a powerful antiserum (Precipitum $= ·022$ c.c.) obtained by injecting a rabbit with the mixed serum of four *Erinaceus europaeus* (No. 135). In view of the continued negative results with different anti-mammalian sera upon non-mammalian bloods, these tests were limited to mammalia.

	Reactions					
	•	*	×	+	%	
85 Primates	83	2	•	•	2	
27 Chiroptera	26	1	•	•	3	
15 Insectivora (4 families)						
6 Erinaceus europaeus	•	•	•	6	100	With other than its
1 Crocidura coerulea	1	•	•	•	0	homologous serum
6 Talpa europaea	6	•	•	•	0	3·5 %
2 Centetes ecaudatus	1	1	•	•	50	
92 Carnivora..............................	87	5	•	•	5	
58 Rodentia	58	•	•	•	0	
65 Ungulata	60	5	•	•	7	
3 Cetacea..................................	3	•	•	•	0	
12 Edentata	12	•	•	•	0	
25 Marsupialia............................	25	•	•	•	0	
1 Monotremata	1	•	•	•	0	

The foregoing table shows a very limited action on the part of this antiserum, only the blood of *Erinaceus europaeus* giving a full reaction. The tests with other Insectivora were negative, with the exception of one sample from *Centetes ecaudatus* which only gave a faint reaction such as was observed in a few instances amongst the other mammalia.

III. Antisera for bloods of Carnivora.

(1) *Tests with Anti-Cat Serum.*

In my paper of 1, VII. 1901 (p. 381) I reported having failed to obtain an anti-cat serum by treating rabbits in the usual manner by intraperitoneal injections with cat serum. The rabbits were treated until they had received as many as 8 and 9 injections respectively, but there was no appearance of precipitin. Two further attempts were made which failed for the reason that a fresh cat serum was used for continuing the treatment begun with old serum, the fresh serum proving toxic. Discouraged by these results, I made no further attempts in this direction. Uhlenhuth (25, VII. 1901) was more successful, however. He does not state that he had any particular difficulty in obtaining this antiserum. He tested it on the 24 bloods cited on page 172, with the result that only cat serum dilutions were found to react, the antiserum being obtained, as usual, from rabbits. Stimulated by Uhlenhuth's success, I again made an effort to obtain anti-cat serum, finally succeeding, although I only obtained a weak antiserum. My results are summarized in the following table:

	• .	*	×	+	%	
			Reactions			
74 Primates	67	6 (8 %)	1 (1 %)	•	9	
27 Chiroptera	27	•	•	•.	0	
13 Insectivora	13	•	•	•	0	
92 Carnivora						
19 Felidae (8 spec.)	2	10 (52 %)	5 (26 %)	**2** (10 %)	89	
14 Viverridae (11 spec.)	12	1 (7 ,,)	1 (7 ,,)	•	14	
1 Proteleidae	1	1	•	•	50	
2 Hyaenidae (1 spec.)	•	1	1	•	100	
20 Canidae (14 spec. & ? races)	17	3	•	•	15	31·5 %
4 Ursidae (3 spec.)	3	1	•	•	25	
7 Procyonidae (5 spec.)	5	1 (14 ,,)	1 (14 ,,)	•	28	
19 Mustelidae (10 spec.)	16	3	•	•	16	
4 Pinnipedia (3 spec.)	4				0	
63 Rodentia	61	1	1*	•	3	Other mammals than Carnivora 4 %
73 Ungulata	72	1	•	•	1	
2 Cetacea	2	•	•	•	0	
13 Edentata	12	1	•	•	7	
24 Marsupialia	24	•	•	•	0	
1 Monotremata	1	•	•	•	0	
324 Aves (incl. 4 eggs)	319	1	•	•	0·3	
41 Reptilia	41	•	•	•	0	
10 Amphibia	10	•	•	•	0	
22 Pisces	22	•	•	•	0	
6 Crustacea	6	•	•	•	0	

785 *Tests with Anti-Cat Serum.*

This antiserum was obtained by injecting a rabbit with the serum of *Felis domesticus*. The antiserum was weak, giving a precipitum of ·005 c.c.

It will be seen from the preceding table that this antiserum only produced full reactions amongst the Felidae. Many of the bloods had been dried for over a year when brought into solution and tested. This, but especially the weakness of the antiserum, accounts for most of the reactions being so slight. The two negative results obtained with Felidae were with one blood of *F. domesticus* (No. 155) dried 19 months, and one of a leopard (147) sent to me by Mr E. H. Hankin from Agra, India. The blood-relationship amongst the Carnivora is however fairly well shown, excepting in the case of the Pinnipedia. The other mammalia showed a much smaller number of faint reactions. One rodent blood (marked *, No. 288) gave a medium reaction, this was from an agouti and it behaved curiously with other antisera, as will be seen by reference to its number in the tables. The agouti blood, which had been preserved with chloroform, showed slight reactions with anti-wallaby and anti-Mexican-deer and also gave a medium reaction with anti-horse serum. I cannot as yet explain this result, as this blood did not react to 16 other antisera.

(2)　378 *Tests with Anti-Hyaena Serum.*

This antiserum was obtained by treating a rabbit with the serum of *Hyaena striata*, the animal having died at the Zoological Gardens, London. This antiserum was much more powerful than the preceding. It gave a precipitum measuring ·031 c.c. when ·1 c.c. of the antiserum was added to ·5 c.c. of a 1 : 100 dilution of hyaena serum.

A glance over the following table immediately shows that a considerable number of reactions took place with the bloods of mammalia other than those belonging to the Carnivora. *When we however confine our attention to the largest reactions, we find that they occur solely amongst the Hyaenidae and Felidae.* If we include all the reactions, faint and otherwise, as positive ones, we find on reading the percentages opposite each family of Carnivora, that they are higher in each of these families than in any other mammalia not belonging to the order Carnivora. The percentage of reactions amongst the Carnivora, taken as a whole, is also higher. The anti-hyaena serum is sufficiently powerful to include the Pinnipedia. On referring to the percentages of reactions all told, occurring between anti-cat serum and the bloods of Felidae and Hyaenidae, we find that even the weak anti-cat serum proved the close relationship of these two families, in a manner corresponding to what we have found with anti-hyaena serum. Moreover, in the case of both antisera, the lowest percentages of reactions all told are given by the bloods of Viverridae, Canidae, Ursidae and Mustelidae.

	•	*	×	+	%	
82 PRIMATES	41	35 (42 %)	12 (14 %)	•	56	
27 CHIROPTERA	22	5	•		18	
13 INSECTIVORA	6	7	•		54	
90 CARNIVORA						
18 Felidae (8 spec.)	•	1 (5 ,,)	11 (61 ,,)	6 (33 %)	100	
14 Viverridae (11 spec.)	2	3 (21 ,,)	9 (64 ,,)	•	85	
8 Proteleidae	•	•	1	•	100	
2 Hyaenidae (1 spec.)	•	•	•	2	100	
22 Canidae (13 spec. & ? var.)	4	4 (18 ,,)	14 (63 ,,)	•	81	87 %
4 Ursidae (3 spec.)	1	1 (25 ,,)	2 (50 ,,)	•	75	
6 Procyonidae (4 spec.)	•	2	4	•	100	
18 Mustelidae (10 spec.)	4	6 (33 ,,)	8 (44 ,,)	•	77	
5 Pinnipedia (3 spec.)	•	1 (20 ,,)	4 (80 ,,)	•	100	
59 RODENTIA	52	5 (8 ,,)	2* (3 ,,)	•	12	Other mammals than Carnivora (38 %)
65 UNGULATA	31	26 (40 ,,)	9 (14 ,,)	•	54	
3 CETACEA	2	1	•		33	
12 EDENTATA	8	3 (25 ,,)	1 (8 ,,)	•	33	
25 MARSUPIALIA	22	3	•		12	
1 MONOTREMATA	1	•	•	•	0	

* Rodent blood 257 (see text).

Of the 11 bloods of Carnivora giving negative results in the preceding table, there were two of Viverridae (168 from Borneo, 169 dried 18 months), four of Canidae (187, 189, 196, 204, the three first from India, the last from Paraguay), one of Ursidae (206 dried 17 months). These bloods have evidently been the cause of lowering the percentage of positive results with anti-hyaena serum, but they do not account entirely for the lower percentages obtained with these families. On the other hand, the four bloods of Mustelidae giving a negative reaction, had all been collected in England, and presumably had not been subjected to any conditions which would alter the bloods so that they would not give good reactions with a suitable antiserum.

One of the two rodent bloods(*) giving a medium clouding was No. 257

(see tables), which acted peculiarly with other antisera. The other rodent blood giving this slight reaction was No. 283, and no explanation can be offered for the result.

(3) *Tests with Anti-Dog Sera.*

Nolf (v. 1900) first produced anti-dog serum by treating rabbits with dog serum. He was unsuccessful when he treated them with dog blood corpuscles. The antiserum did not precipitate some other bloods tested. In my note of 11, v. 1901 I reported having obtained negative results with 23 bloods other than dog's. Similar results have been obtained by Uhlenhuth, who found the anti-serum to react with the blood of the fox. In my paper of 21, xi. 1901 I reported having only obtained reactions with the bloods of six species of Canidae, 196 non-canine bloods having given a negative result. By reference to the tables, it will be seen that a considerably larger number of canine bloods has since been examined. Gengou (25, x. 1902, p. 751) tested this antiserum upon four bloods (dog, horse, ox, guinea-pig) and found only that of the dog to react.

I will add here that Uhlenhuth (25, vii. 1901) has produced an *anti-fox* serum, which, tested on 24 bloods (see his list, p. 172) only reacted with the blood of the fox and dog, less with the latter.

Farnum (28, xii. 1901) prepared anti-dog serum and found it to precipitate solutions of serum of the dog, but not of man and bull.

777 *Tests with Anti-Dog Serum.*

These tests were carried out in the course of two years and necessitated the use of five different antisera obtained from as many rabbits which were treated with the serum of *Canis familiaris* (different breeds). Four of the sera were powerful, one weak. The weak antiserum was scarcely used. One of the powerful sera (the last used) was standardized, being found to give a precipitum of ·015 c.c.

We see from the following table that the dogs form a detached group amongst the Carnivora, for with the exception of one of the Mustelidae, no other bloods gave a full reaction. Three canine bloods gave negative or faint reactions, the two giving negative results were sent from India (196) and South America (204), the one giving a faint clouding only, from India (187). It may therefore be assumed that these bloods had become relatively insoluble. The proportion of medium, but especially of weak, reactions amongst the Carnivora is obviously much higher (note the percentages) than in the other mammalia when the number of bloods examined is taken into account.

	Reactions			+	%	
	•	*	×			
84 Primates	66	16 (19 %)	2 (2 %)	•	21	
25 Chiroptera	23	1 (4 ,,)	1 (4 ,,)	•	8	
14 Insectivora	12	1 (7 ,,)	1 (7 ,,)	•	14	
91 Carnivora						
18 Felidae (8 spec.)	12	3 (8 ,,)	3 (8 ,,)	•	16	
14 Viverridae (11 spec.)	11	3	•	•	21	
1 Proteleidae	1	•	•	•	0	
2 Hyaenidae (1 spec.)........	1	1	•	•	50	46 %
22 Canidae (12 spec. & ? races)	2	1 (7 ,,)	3 (13 %)	16 (72 %)	91	
3 Ursidae (3 spec.)	1	•	2	•	66	
7 Procyonidae (? 5 spec.) ...	4	2 (19 ,,)	1 (14 ,,)	•	43	
20 Mustelidae (9 spec.)	16	2 (10 ,,)	1 (5 ,,)	1 (5 ,,)	20	
4 Pinnipedia (3 spec.)........	2	2	•	•	50	
61 Rodentia	55	4 (6 ,,)	2 (3 ,,)	•	9	Other mammals than Carnivora 11 %
65 Ungulata	63	2	•	•	3	
2 Cetacea	2	•	•	•	0	
13 Edentata	12	1	•	•	8	
22 Marsupialia	21	1	•	•	4	
1 Monotremata	1	•	•	•	0	
322 Aves (incl. 4 eggs)...............	321	1	•	•	0·0	
40 Reptilia	40	•	•	•	0	
13 Amphibia	13	•	•	•	0	
18 Pisces	18	•	•	•	0	
6 Crustacea	6	•	•	•	0	

It is worthy of note, that the tests made with the three preceding antisera for Carnivora agree in causing a larger proportion of reactions amongst the Primates than amongst any of the other mammalia, excepting the results with anti-hyaena serum which show a large proportion of faint reactions amongst the Ungulata.

One of the two rodent bloods giving a × reaction again proved to be No. 257 referred to under anti-hyaena serum.

(4) 358 *Tests with Anti-Seal Serum.*

The antiserum used in the following tests was obtained by treating a rabbit with the serum of *Phoca vitulina* L., the Common Seal (No. 242). The antiserum was weak, only giving a precipitum measuring ·006 c.c. when ·1 c.c. of antiserum was added to ·5 c.c. of a 1 : 100 dilution of homologous serum.

	Reactions			
	•	*	×	%
75 PRIMATES	63	9 (12 %)	3 (4 %)	16
24 CHIROPTERA	21	3	•	12
13 INSECTIVORA	12	1	•	7
88 CARNIVORA				
18 Felidae (8 spec.)	14	1 (5 „)	3 (17 „)	22
13 Viverridae (10 spec.)	10	2 (15 „)	1 (7 „)	23
1 Proteleidae	•	1	•	100
2 Hyaenidae (1 spec.).........	2	•	•	0
22 Canidae (13 spec. & ? races)	11	7 (31 „)	4 (18 „)	50
4 Ursidae (3 spec.)	2	1	1	50
6 Procyonidae (? 4 spec.) ...	4	1	1	33
18 Mustelidae (10 spec.)	10	4	4	44
	53	17	14	
4 Pinnipedia (3 spec.).........	•	1 (25 „)	3 (75 „)	100
56 RODENTIA	54	1	1	3
63 UNGULATA	61	2	•	3
3 CETACEA.............................	3	•	•	0
12 EDENTATA	10	2	•	16
23 MARSUPIALIA........................	22	1	•	4
1 MONOTREMATA	1	•	•	0

(brace for Carnivora) 37 %

Other mammals than Carnivora 8 %

It is owing doubtless to a slight cloudiness and also to the feebleness of this antiserum that the results tabulated above appear at first sight less in accord with the preceding. It is obvious that it is more difficult to estimate different degrees of reaction when at most moderate cloudings occur and the precipitum observed after 24 hours is so small in quantity. It is evident nevertheless that the bloods of the Pinnipedia react more than do the others, numerically if not apparently quantitatively, and it would appear also that there is a preponderance of reactions amongst the other Carnivora. It is notable again that the Primates stand out amongst the other mammalia, in a manner corresponding (but less marked) to what has been stated with regard to the preceding three antisera for Carnivora.

IV. Antisera for Bloods of Rodents.

Antisera for rabbit blood have been obtained by only a few observers, the first of these being Nolf (v. 1900) who produced it by treating a fowl with rabbit serum. Bordet (III. 1899) failed to find any precipitin in guinea-pigs treated with rabbit serum, whereas Gengou (25, x. 1902, p. 743) claims that the serum of such guinea-pigs produced a "distinct opalescence" in rabbit serum dilutions, although it did not produce a precipitum. Two years ago I sought to discover if isoprecipitins were formed in rabbits treated with the serum of other rabbits, but the result was negative, and I have not mentioned it hitherto. Hamburger (6, XI. 1902, p. 1189, footnote) states that he obtained negative results in treating guinea-pigs with rabbit serum, consequently Gengou's statement requires further confirmation before it can be accepted. On the other hand Hamburger reports that he has succeeded in obtaining anti-rabbit serum by treating a goat with rabbit serum. De Lisle (XI. 1902, p. 399) states that he has obtained anti-rabbit serum by injecting rabbit blood into eels. Haemolysins were also formed in the eel, so that its serum after treatment was four times as toxic for rabbits as before. He does not state the strength of the blood dilutions he tested, nor is there mention of control tests on other sera, consequently his precipitin results are not as valuable as they might be.

I have only had time to make two attempts to obtain anti-rabbit sera. Following Nolf, I used a fowl and duck for treatment, but in neither case was I successful. Working with birds is not satisfactory in any case, as compared to mammals. The goat would appear to be the most likely animal for such purposes. I regret not to be able to give a series of tests with an anti-rodent serum at present, but shall perhaps have an opportunity of doing so later. It will be very interesting to see how the collection of rodent bloods will behave to such an anti-serum. On glancing through the tables, we see that the results are throughout practically negative with rodent bloods. Here and there a rodent blood reacts, but this can be put down in some cases to experimental error. Also where the most powerful anti-mammalian sera were used, we see a number of slight cloudings noted (anti-pig, anti-human, anti-hyaena) but they appear to have no significance excepting as slight indications of mammalian relationship. I would surmise that a powerful anti-rodent serum might possess a very generalized action in this group.

V. Antisera for Bloods of Ungulata.

a. Suidae.

(1) *Tests with Anti-Pig Serum.*

The first to produce anti-pig serum was Uhlenhuth (25, VII. 1901) who tested 24 bloods therewith (see list on p. 172); the only bloods he found to react were those of the pig and wild-boar. I subsequently (20, I. 1902) tested 250 bloods with this antiserum, with the result that

I found it to produce marked clouding in a number of mammalian bloods: that of man, several species of monkey, bear, dog, opossum, raccoon, cat, coati, genet, stoat, rat, mouse. It did not however produce a full reaction with any blood except that of the pig, no other suilline bloods being tested. Only once did the antiserum produce a slight clouding in an avian blood, this being attributed to experimental error, the sample in question having probably been in contact with some mammalian blood, through carelessness in collection. Still later, (5, IV. 1902), I observed faint cloudings to occur upon the addition of this antiserum to the bloods of the antelope and deer, and after 30 to 120 minutes in the bloods of bats (6 species), 3 species of Edentates, and 8 species of Australian Marsupials. I also drew attention to the interesting fact that the blood of the porpoise immediately gave a slight clouding with this antiserum, a well-marked though slight deposit being formed after 24 hours. I wrote at the time that "The more general action of this particular anti-pig serum on other mammalian bloods may at first appear to contradict what has hitherto been claimed with regard to the relatively specific character of these antisera; but this is actually not the case. I have already noted elsewhere the occasional occurrence of clouding in non-homologous bloods upon the addition of an antiserum." I have attributed these results to the great power of my antisera, giving what I have described elsewhere as a "mammalian reaction."

Besides Uhlenhuth and myself, only Kister and Wolff (18, XI. 1902, p. 422) are known to me as having experimented with anti-pig serum. They state that their antiserum did not act on human blood in any concentration, and this I attribute to their antiserum not being as powerful as mine. Under the same conditions the antiserum gave a reaction in 5 minutes with pig blood (followed by a large deposit in 1—2 hours), a cloud in the sera of ox and sheep after 20 minutes, and in the horse after 60 minutes.

Finally, Schütze (6, XI. 1902, p. 805) found pig haematoserum to precipitate solutions of pig spermatozoa, but not those of man.

818 *Tests with Anti-Pig Serum.*

Three different antisera were used in the following tests, the three having been obtained by injecting as many rabbits with the serum of the domestic pig. All three antisera were exceedingly powerful, two of them on being standardized giving a precipitum of ·045 and ·055 c.c. respectively.

		Reactions			+	%
	•	*	×	+		
89 Primates	36	16 (18 %)	19 (21 %)	18 (20 %)	•	60
29 Chiroptera	18	2 (6 ,,)	8 (27 ,,)	1 (3 ,,)	•	38
14 Insectivora	8	3 (21 ,,)	2 (14 ,,)	1 (7 ,,)	•	43
96 Carnivora	45	20 (21 ,,)	26 (27 ,,)	5 (5 ,,)	•	53
65 Rodentia	44	16 (24 ,,)	4 (6 ,,)	1* (1 %)	•	32
69 Ungulata						
a. *Artiodactyla*						
Suina						
4 Suidae	•	•	•	•	4	100
1 Dicotylidae	•	1	•	•	•	100
Tylopoda						
4 Camelidae (2 spec.)	1	1	1	1	•	75
Tragulina						
1 Tragulidae	1	•	•	•	•	0
Pecora						
17 Cervidae (12 spec.)	8	4 (23 ,,)	•	5 (18 ,,)	•	41
34 Bovidae (22 spec.)	23	6 (17 ,,)	5 (15 ,,)	12 (35 ,,)	•	67
b. *Perissodactyla*						
1 Tapiridae	•	1	•	•	•	100
6 Equidae (3 spec.)	3	1	2	•	•	50
c. *Hyracoidea*						
1 Hyracidae	•	1	•	•	•	100
3 Cetacea (2 spec.)	•	•	2	1	•	100
13 Edentata	3	5 (38 ,,)	5 (38 ,,)	•	•	77
26 Marsupialia	10	5 (19 ,,)	11 (42 ,,)	•	•	61
1 Monotremata	1	•	•	•	•	•
322 Aves (incl. 4 eggs)	321	1	•	•	•	0·3
47 Reptilia	47	•	•	•	•	
13 Amphibia	13	•	•	•	•	
24 Pisces	24	•	•	•	•	
7 Crustacea	7	•	•	•	•	

Artiodactyla group (Suidae through Bovidae) bracketed: 66 %

Perissodactyla group (Tapiridae, Equidae) bracketed: 57 %

Other mammals than Ungulata 51 %

* Rodent No. 257 which gave curious results with other antisera.

The antisera employed for these tests were amongst the most powerful used throughout the investigation. The general mammalian reaction is consequently well marked. Nevertheless we see that *full reactions only occurred amongst the Suidae*. The four bloods of Suidae tested included the blood of the domestic pig, of the wild boar (Europe), and of a wild boar, whose blood was collected at Singapore. Moderate reactions were observed with the bloods of Bovidae, Cervidae, Camelidae. Taking the Artiodactyla as a whole we find that 66 % of the bloods gave some sort of reaction. The Perissodactyla and Hyracoidea (1 sample only of

last) gave but slight reactions. Whether it is due or not to the
fact that the pig is a "generalized mammal" I do not know, but
certainly the results, especially with the Primates, seem somewhat
anomalous. We find in fact that nearly all the mammalia are more or
less connected, the only quite negative result being with the blood of a
Monotreme, which has shown no trace of reaction with any of my
antisera, although it went perfectly into solution. These results seem
to be confirmed by our quantitative tests as far as they have. gone.

This antiserum brings out the reaction with the bloods of Cetacea in
a very striking manner. It will be seen further that these bloods also
reacted to other anti-ungulate sera. The quantitative tests (see Section
VII), very clearly show a remarkable tendency of cetacean blood to
react with anti-ungulate sera, giving large perfectly measurable quantities of precipitum.

The following passage from Flower and Lydekker (p. 233)[1] would
seem to offer the clearest explanation of this action of anti-ungulate sera
upon cetacean bloods. They write, "But the structure of the Cetacea is,
in so many essential characters, so unlike that of the Carnivora that the
probabilities are against these orders being nearly related. Even in the
skull of the Zeuglodon, which has been cited as presenting a great
resemblance to that of a seal, quite as many likenesses may be traced to
one of the primitive Pig-like Ungulates (except in the purely adaptive
character of the form of the teeth), while the elongated larynx, complex
stomach, simple liver, reproductive organs both male and female, and
foetal membranes of the existing Cetacea are far more like those of that
group than of the Carnivora. Indeed it appears probable that the old
popular idea which affixed the name of "Sea-Hog" to the Porpoise
contains a larger element of truth than the speculations of many
accomplished zoologists of modern times......"

b. Camelidae.

(2)   363 *Tests with Anti-Llama Serum.*

This antiserum was produced by injecting a rabbit with the serum of
Auchenia huanacos Molina. The antiserum was weak, giving a
precipitum of ·005 c.c. when standardized, six weeks later.

---

[1] Flower, W. H., and Lydekker, R. (1891), *An Introduction to the Study of Mammals
Living and Extinct*, London, Adam and Charles Black.

	Reactions					
	•	*	×	+	**+**	%
75 PRIMATES	72	2	1	.	.	4
25 CHIROPTERA	25	.	.	.	.	0
13 INSECTIVORA	13	.	.	.	.	0
89 CARNIVORA	85	3	1	.	.	5
57 RODENTIA	57	.	.	.	.	0
62 UNGULATA						
a. *Artiodactyla*						
Suina						
4 Suidae	3	1	.	.	.	25
1 Dicotylidae	.	.	1	.	.	100
Tylopoda						
4 Camelidae (3 spec.) ...	.	.	2	.	**2**	100
Tragulina						
1 Tragulidae	1	.	.	.	.	0
Pecora						
17 Cervidae (12 spec.) ...	13	4	.	.	.	23
31 Bovidae (22 spec.) ...	22	9	.	.	.	20
b. *Perissodactyla*						
1 Tapiridae	1	.	.	.	.	0
2 Equidae (1 spec.)	1	1	.	.	.	50
c. *Hyracoidea*						
1 Hyracidae	1	.	.	.	.	0
2 CETACEA	2	.	.	.	.	0
12 EDENTATA	12	.	.	.	.	0
27 MARSUPIALIA	26	1	.	.	.	3
1 MONOTREMATA	1	.	.	.	.	0

60 % (Artiodactyla Suina + Tylopoda + Tragulina bracket)

25 % (Pecora bracket)

Other mammals than Ungulata 2·6 %

The preceding table shows that large reactions only occurred with two samples of the blood of two animals of the species *Auchenia huanacos,* the blood of one of which was used for the treatment of the rabbit yielding the antiserum. The blood of *Auchenia glama* and *Camelus dromedarius* gave × reactions. The percentage of reactions all told, is larger amongst the Ungulata than among the other mammalia. Again the Primates and Carnivora stand out among these.

c. Cervidae.

(3) 749 *Tests with Anti-Mexican-Deer Serum.*

The following tests were made with one antiserum obtained by injecting a rabbit with the serum of *Cariacus mexicanus,* H. Smith (No. 338, belonging to the Family Cervidae). This antiserum was powerful, giving a precipitum of ·015 when standardized.

	·	*	×	+	✚	%	
73 Primates	40	27 (37 %)	4 (5 %)	2 (2 %)	·	47	
25 Chiroptera	15	10	·		·	40	
11 Insectivora	7	3 (27 ,,)	1* (9 ,,)		·	36	
84 Carnivora	62	15 (18 ,,)	7 (8 ,,)		·	27	
61 Rodentia	57	3	1		·	6	
61 Ungulata							
a. *Artiodactyla*							
Suina							
3 Suidae	·	·	2	1	·	100	⎫
1 Dicotylidae	·	·	1	·	·	100	⎪
Tylopoda							⎪
2 Camelidae (2 sp.)	·	2	·	·	·	100	⎬ 100 %
Tragulina							⎪
1 Tragulidae	·	1	·	·	·	100	⎭
Pecora							
15 Cervidae (11 sp.)	1†	1 (6 ,,)	8 (53 ,,)	2 (13 ,,)	3 (20 %)	93	⎫
34 Bovidae (22 sp.)	2‡	7 (20 ,,)	17 (50 ,,)	5 (14 ,,)	3 (8 ,,)	94	⎬ 94 %
b. *Perissodactyla*							
1 Tapiridae	·	1	·	·	·	100	
3 Equidae (1 spec.)	2	·	·	1§	·	33	
c. *Hyracoidea*							
1 Hyracidae	1	·	·	·	·	0	
2 Cetacea	1	·	·	1‖	·	50	Other
13 Edentata	10	3	·	·	·	23	mammals
23 Marsupialia	21	2	·	·	·	8	than
1 Monotremata	1	·	·	·	·	0	Ungulata
314 Aves (incl. 4 eggs)	314	·	·	·	·	0	27 %
40 Reptilia	40	·	·	·	·	0	
11 Amphibia	11	·	·	·	·	0	
23 Pisces	23	·	·	·	·	0	
7 Crustacea	7	·	·	·	·	0	

* Fluid serum No. 143.

† No. 332 test not repeated, result doubtful as gave good reactions with other antisera.

‡ No. 347 from Central Africa, 369 dried 15 months when tested.

§ No. 380 a fluid serum 17 days on ice tested with antiserum 19 days old also on ice.

‖ No. 385 Balaenoptera rostrata. This and another sample of fluid blood showed reactions with other anti-ungulate sera.

It will be noted in the preceding table that large reactions occur only amongst the Pecora, forming 20 % of such reactions amongst the Cervidae, 8 % among the Bovidae. The bloods of both of these families all reacted if we except the three giving a negative result (about which see footnotes). Second-class reactions occurred chiefly amongst the Pecora, as also in 1 blood each of Suidae, Equidae, Cetacea, and in only 2 (2 %) of the Primates. All bloods of Artiodactyla, outside the Pecora, reacted, though weakly. Of 3 equine bloods two gave no reaction, one gave a moderate reaction; the blood of Hyrax gave a negative result. Of the non-ungulate bloods those of the Primates gave the highest percentage of reactions all told.

(4) 69 *Tests with Anti-Reindeer Serum.*

This antiserum was obtained by injecting a guinea-pig with the blood and clouded serum of *Rangifer tarandus*, kindly sent me by Privatdocent Dr Max Lühe from the Zoological Gardens at Königsberg. The antiserum was unfortunately very weak, yielding only a precipitum of ·004 when standardized with its homologous blood. Owing to its limited quantity it was impossible to carry out more than a very small number of tests, especially as some antiserum was used for the quantitative tests reported in Section VII.

	Reactions			
	•	*	×	%
4 CHIROPTERA	4	•	•	0
6 CARNIVORA	6	•	•	0
1 RODENTIA	1	•	•	0
9 UNGULATA				
a. *Artiodactyla*				
Suina				
1 Suidae	•	1	•	100
Pecora				
4 Cervidae (4 spec.)	1	•	3	75
4 Bovidae (4 spec.)	•	2	2	100
b. *Perissodactyla*				
1 Equidae	•	1	•	100
1 CETACEA	•	1	•	100
1 MARSUPIALIA	1	•	•	0
41 AVES	41	•	•	0
5 REPTILIA	5	•	•	0

(Cervidae, Bovidae, and Equidae bracketed: 90 %)

Other mammalian bloods than those noted were not tested for the reason that the antiserum had become exhausted. Here, as in the preceding table, the Pecora only show greater reactions, the Cervidae giving more of the marked reactions than do the Bovidae. Of the other mammalian sera tested those of a pig, horse, Rorqual (385 *Balaenoptera rostrata*) gave a faint clouding. All other bloods showed no trace of reaction.

(5)   699 *Tests with Anti-Hog-Deer Serum.*

This antiserum was obtained by treating a rabbit with the serum of *Cervus porcinus* Zimm. (No. 326). The antiserum gave a precipitum of ·0063 c.c. when standardized.

		Reactions				
	•	*	×	+	✚	%
74 Primates	70	3	1	•	•	5
22 Chiroptera	22	•	•	•	•	0
11 Insectivora	11	•	•	•	•	0
86 Carnivora	80	3	3	•	•	7
57 Rodentia	56	1	•	•	•	1
62 Ungulata						
a. *Artiodactyla*						
Suina						
3 Suidae (3 spec. ?)	•	3	•	•	•	100
1 Dicotylidae.........	1	•	•	•	•	0
Tylopoda						
2 Camelidae (2 sp.)	1	1	•	•	•	50
Tragulina						
1 Tragulidae	1	•	•	•	•	0
Pecora						
17 Cervidae (12 sp.)	1	1 (6 %)	7 (41 %)	2 (12 %)	6 (35 %)	94
34 Bovidae (22 sp.,						
4 races)	2	9 (26 ,,)	15 (44 ,,)	2 (6 ,,)	6 (17 ,,)	94
b. *Perissodactyla*						
1 Tapiridae	1	•	•	•	•	0
2 Equidae (1 spec.)	2	•	•	•	•	0
c. *Hyracoidea*						
1 Hyracidae	1	•	•	•	•	0
2 Cetacea	1	1	•	•	•	50
12 Edentata	11	1	•	•	•	8
17 Marsupialia	17	•	•	•	•	0
1 Monotremata	1	•	•	•	•	0
290 Aves (incl. 4 eggs) ...	290	•	•	•	•	
33 Reptilia	33	•	•	•	•	
11 Amphibia	11	•	•	•	•	
15 Pisces.....................	15	•	•	•	•	
6 Crustacea	6	•	•	•	•	

Bracket groupings at right: Suidae through Tragulidae group 57%; Cervidae and Bovidae group 94%. "Other mammals than Ungulata 4%".

It will be seen from the above table that the bloods of the Pecora only give large reactions, and that the percentage of large and moderate reactions amongst the Cervidae is higher than amongst the Bovidae. In some cases the bloods from both of these families appear to react equally. One sample of cervine blood failed to react: 322, from India. Two bovine bloods failed to react: 347 from Central Africa, and

369, dried over a year when tested; if we exclude these then 100 % of the bloods of Pecora reacted.   Next in order come the Suidae and Camelidae and the Cetacea.

d.   Bovidae.

(6)   686 *Tests with Anti-Antelope Serum.*

This antiserum was obtained by injecting a rabbit with the serum of *Cobus unctuosus* Lorrill., or Sing-Sing antelope (340) which died at the

		Reactions				
	•	*	×	+	✚	%
68 Primates	62	4	2	•	•	9
23 Chiroptera	23	•	•	•	•	0
10 Insectivora	10	•	•	•	•	0
82 Carnivora	77	4	1	•	•	6
58 Rodentia	56	1	1*	•	•	3
60 Ungulata						
a. *Artiodactyla*						
Suina						
3 Suidae,..............	1	2	•	•	•	66
1 Dicotylidae.........	•	1	•	•	•	100
Tylopoda						
2 Camelidae (2 sp.)	1	1	•	•	•	50
Tragulina						
1 Tragulidae	•	1	•	•	•	100
Pecora						
16 Cervidae (11 sp.)	1†	3 (18 %)	7 (43 %)	5 (31 %)	•	94
33 Bovidae (22 sp.,						
?var.)	1‡	1 (3 ,,)	19 (57 ,,)	7 (21 ,,)	5 (15 %)	97
b. *Perissodactyla*						
3 Equidae (1 sp.) ...	3	•	•	•	•	0
c. *Hyracoidea*						
1 Hyracidae	•	1	•	•	•	100
2 Cetacea	2§	•	•	•	•	0
11 Edentata	11	•	•	•	•	0
16 Marsupialia	16	•	•	•	•	0
1 Monotremata	1	•	•	•	•	0
292 Aves (incl. 4 eggs) ...	292	•	•	•	•	
33 Reptilia	33	•	•	•	•	
10 Amphibia	10	•	•	•	•	
15 Pisces	15	•	•	•	•	
5 Crustacea	5	•	•	•	•	

Bracketed annotations at right: Suina–Tragulina grouped as 71 %; Cervidae–Bovidae grouped as 96 %; "Other mammals than Ungulata 5 %".

* 257, see tables, this sample reacted most peculiarly.   Collected properly?

† 336, roebuck, gave feebler reactions than another sample from same species when tested with other anti-ungulate sera.

‡ 347 from Central Africa.

§ 385–6 showed trace of deposit after 24 hours.

Zoological Gardens, London.   The antiserum was very powerful, giving
a precipitum of ·055 when standardized.

We see from the above table that the large reactions are limited to
the Bovidae, the second-class reactions to Bovidae (21 %) and Cervidae
(31 % of their totals).   Two negative results amongst the Pecora are
referred to in footnotes.   Other Artiodactyla than the Pecora gave 71 %
total reactions.   Hyrax gave a faint clouding, the Equidae negative
results.   The Cetacea also gave negative results, although it is stated in
the tables that a trace of deposit was noted in the tubes after 24 hours.
Of the non-ungulate bloods those of the Primates and Carnivora give
most reaction.

(7)   *Tests with Anti-Ox Serum.*

The first to produce anti-ox serum was the late Dr Walter Myers
(14, VII. 1900) who treated rabbits with ox serum-globulin, obtaining
an antiserum which precipitated solutions of ox globulin, besides having
a slight action on a similar solution from sheep.   Uhlenhuth (1, II. 1901)
produced it next by subjecting rabbits to intraperitoneal injections of
defibrinated ox blood, doses of about 10 c.c. being administered every
6—8 days until 5 injections had been made.   He tested 19 different
bloods therewith in 1 : 200 dilutions.   The bloods tested were the
following, which I have grouped in order:

Ox	Man	Fowl
Sheep		Turkey
Deer	Dog	Pigeon
Fallow Deer	Cat	Goose
Pig	Hare	
	Rabbit	
Horse	Guinea-pig	
Donkey	Rat	
	Mouse	

Presumably his anti-ox serum was weak, for he states that he
obtained a reaction *only* with ox blood.   In my preliminary note of
11, V. 1901, in which tests on 24 bloods were reported, I noted that
anti-ox serum also produced a weak reaction with the blood of the sheep.
As the number of bloods tested increased, I found that those of the
gazelle and axis-deer gave distinct reactions (1, VII. 1901), as did also
those of the goat, roebuck, and Burrhel sheep (14, IX. 1901), and in

a later paper (21, XI. 1901) I stated that the tests conducted both with anti-ox and anti-sheep sera had "given reactions, which indicate the existence of a 'blood relationship' between certain of the true ruminants," positive reactions having further been obtained with the bloods of the deer, antelope and gnu. The bloods of the Tragulidae and Camelidae (20, I. 1902) gave no indication of relationship with the true ruminants. Subsequently I found what appeared to be differences in the degree of the reaction, the antisera for ox and sheep acting "to a greater degree upon the bloods of more closely allied species" (5, IV. 1902).

In the meantime Uhlenhuth (25, VII. 1901) tested 24 bloods (see his list on page 172) with this antiserum, and found it to give much precipitum with ox blood, and a small precipitum with the bloods of the goat and sheep. Farnum (28, XII. 1901), in America, treated rabbits with the semen of the bull and found the antiserum thus obtained to act upon homologous solutions, not upon those of goat semen.

Michaëlis and Oppenheimer (1902, p. 342) found anti-ox serum to act on ox and on sheep blood, not on that of the horse. Michaëlis (9, X. 1902, p. 734), states that this antiserum acts very much less upon sheep blood than upon that of the ox. Kister and Wolff (18, XI. 1902, p. 422) publish results which certainly suggest some experimental error, probably, as Uhlenhuth suggests, the use of "milky antisera" (see p. 72). Correctly enough, according to the results above cited, deposits were found to occur upon the addition of this antiserum to ox and sheep blood, but they found a greater reaction with human blood than with that of the pig and horse.

790 *Tests with Anti-Ox Serum.*

Four different antisera were used in the following tests, all being obtained from rabbits injected with the serum of the ox. The antisera were fairly powerful, one of them, when standardized, giving a precipitum of ·011 c.c.

The larger reactions are limited here to the bloods of the Bovidae, the second-class and weaker reactions being about equally divided amongst the Bovidae and Cervidae, medium cloudings occurred in a horse, a pig, and a whale blood and in a few Primates (11 %) and Carnivora (4 %). All three of the cetacean bloods gave some reaction. Two of the four cervine bloods sent from India, which gave no reaction, should be excluded.

	•	*	×	+	‡	%	
94 Primates	84	6 (6%)	4 (4%)	•	•	1	
29 Chiroptera	29	•	•	•	•	0	
12 Insectivora	12	•	•	•	•	0	
91 Carnivora	87	3	1	•	•	4	
63 Rodentia	62	1	•	•	•	1	
67 Ungulata							
a. *Artiodactyla*							
Suina							
4 Suidae	3	•	1	•	•	25	
1 Dicotylidae	1	•	•	•	•	0	
Tylopoda							
4 Camelidae	4	•	•	•	•	0	} 10%
Tragulina							
1 Tragulidae	1	•	•	•	•	0	
Pecora							
17 Cervidae (12 sp.)	4*	2 (11%)	3 (17%)	8 (47%)	•	77	} 88%
35 Bovidae (22 sp., etc.)	2†	4 (11 ,,)	6 (17 ,,)	17 (48 ,,)	6 (17%)	94	
b. *Perissodactyla*							
1 Tapiridae	1	•	•	•	•	0	
3 Equidae	2	•	1	•	•	66	
c. *Hyracoidea*							
1 Hyracidae	1	•	•	•	•	0	
3 Cetacea (2 spec.)	•	2	1	•	•	100	Other
13 Edentata	13	•	•	•	•	0	mammals
26 Marsupialia	26	•	•	•	•	0	than
1 Monotremata	1	•	•	•	•	0	Ungulata
301 Aves (incl. 4 eggs)	299	2	•	•	•	0·6	5 %
51 Reptilia	51	•	•	•	•	0	
13 Amphibia	13	•	•	•	•	0	
19 Pisces	19	•	•	•	•	0	
7 Crustacea	7	•	•	•	•	0	

* Two of these, 322 and 323, from India.   † ?

(8) *Tests with Anti-Sheep Serum.*

The first to prepare this antiserum was Myers (14, VII. 1900). He treated rabbits with sheep serum-globulin, and found that their serum contained a "specific" precipitin for sheep globulin solutions, and also exerted a slight precipitating action upon ox globulin. Nuttall (17, V. 1901) treated rabbits with sheep serum and obtained a similar result, when testing 24 bloods. Subsequently he found (1, VII. 1901) the blood of the gazelle and axis-deer to give "distinct but less marked" reactions than did the blood of the sheep. Uhlenhuth (25, VII. 1901) tested 24 bloods (see list p. 172) with this antiserum, finding that it gave almost as much reaction with the blood of the goat as with that of the sheep, less with that of the ox. He observed no reaction in the blood of a deer, nor in any of the other bloods tested. With an increasing number of bloods tested, I noted more and more the tendency of this antiserum to include other Bovidae in its positive reactions, and not infrequently Cervidae. In several short papers these findings were briefly dwelt upon as in the preceding tests with anti-ox serum.

Recently Schütze (6, XI. 1902, p. 805) has found anti-sheep serum to act less upon ox spermatozoa solutions than upon corresponding ones of the sheep. The results of Kister and Wolff (18, XI. 1902, p. 421) with this antiserum are in accord with mine, and of themselves indicate that an experimental error lay at the bottom of those they obtained with anti-ox serum referred to above. Anti-sheep serum clouded both sheep and ox serum dilutions in 5 minutes, leading to a large deposit in the former after 2 hours. After 20 to 30 minutes slight clouds appeared in pig and horse serum. No effect was exerted on human blood after 2 hours. As the proportion of antiserum added to the dilution decreased, the blood of the pig and horse ceased to react, reactions still being obtained with the bloods of sheep and ox.

701 *Tests with Anti-Sheep Serum.*

Six different antisera, obtained from as many rabbits, were used for these tests. The rabbits received injections of the serum of the domesticated sheep. One of the antisera had moderate power, whereas the others were powerful. One of the latter gave a precipitum of ·02 when standardized.

Here again the large reactions are limited to the bloods of Bovidae (23 %), the second-class reactions occur amongst Bovidae and Cervidae, forming a higher percentage of the total reactions in the latter group. Of the bloods of other Ungulata, those of Suina and Equidae gave faint and medium cloudings. One of two cetacean bloods tested gave a medium clouding. Outside the Ungulata, those of Primates and Carnivora gave most of the "mammalian reactions" recorded, viz. 5 and 10 % respectively. Two of the four bloods of Pecora giving negative results were from India.

	Reactions					%	
	·	*	×	+	✚	%	
75 Primates	71	1	3	·	·	5	
24 Chiroptera	24	·	·	·	·	0	
9 Insectivora	9	·	·	·	·	0	
78 Carnivora	70	3	5	·	·	10	
58 Rodentia	57	·	1	·	·	2	
61 Ungulata							
a. *Artiodactyla*							
Suina							
3 Suidae	2	·	1	·	·	66	
1 Dicotylidae	·	1	·	·	·	100	
Tylopoda							28 %
2 Camelidae (2 sp.)	2	·	·	·	·	0	
Tragulina							
1 Tragulidae	1	·	·	·	·	0	
Pecora							
16 Cervidae (11 sp.)	2*	1 (6 %)	9 (56 %)	4 (25 %)	·	88	
34 Bovidae (21 sp., 4 races)	2†	2 (6 ,,)	12 (35 ,,)	10 (29 ,,)	8 (23 %)	94	92 %
b. *Perissodactyla*							
3 Equidae	1	1	1	·	·	66	
c. *Hyracoidea*							
1 Hyracidae	1	·	·	·	·	0	
2 Cetacea	1	·	1	·	·	50	Other
12 Edentata	12	·	·	·	·	0	mammals
22 Marsupialia	22	·	·	·	·	0	than
1 Monotremata	1	·	·	·	·	0	Ungulata
289 Aves (incl. 4 eggs)	287	2	·	·	·		5 %
33 Reptilia	33	·	·	·	·		
13 Amphibia	13	·	·	·	·		
17 Pisces	17	·	·	·	·		
7 Crustacea	7	·	·	·	·		

    * One from India (323), the other (329) from Germany.
    † One from China (393), the other (342) from London Zoo.

e. Equidae.

(9)　*Tests with Anti-Horse Serum.*

Anti-horse serum was first obtained by Tchistovitch (v. 1899) by treating rabbits with horse serum, injections of 3 c.c. being made 5—6 times, at intervals. We may conclude that his antiserum was very weak indeed, for the reason that he states it produced a reaction with the serum of the horse, but *not* with that of the donkey. Tests upon normal rabbit serum also proved negative. Nolf (v. 1900), and several observers since, have prepared this antiserum. In a series of 24 bloods tested by me and reported in my preliminary note of 11, v. 1901, only horse serum was found to react to its antiserum. Numbers of tests were reported on in my subsequent papers, wherein it was stated that the only other blood to react was that of the donkey. Uhlenhuth (25, VII. 1901) reported having tested 24 bloods with anti-horse serum (see list p. 172) and he found no other bloods outside those of the horse and donkey to react. Schütze (6, XI. 1902, p. 805) found this antiserum to precipitate spermatozoa solutions of the horse, but not those of man, and ox. Uhlenhuth (7, XI. 1901) suggests the use of this antiserum for the detection of horse-meat in sausages (see under practical applications), and, using the test on meats, Nötel (13, III. 1902) found horse and donkey meat indistinguishable. Kister and Wolff (18, XI. 1902, pp. 414—416) published seven tables of tests made by them with anti-horse serum upon 5 bloods. They do not state whether they were all conducted with one, or with several antisera, so it is to be presumed that they used one, which they say "surprised" them by the general reactions which it gave. I have no doubt, in view of my own results, that Uhlenhuth is right in stating that these authors probably had to deal with what is well described as a "milky antiserum" (see p. 72), for their results are somewhat anomalous. Possibly their antisera were preserved with chloroform (see p. 76). They tested the blood of the horse, pig, ox, sheep, and man. On adding antiserum in the proportion 1 : 5 blood-dilution (1 : 10 to 1 : 320) all the bloods clouded, the greatest clouding occurred with horse blood, where a large deposit soon formed. Deposits did not occur in the other bloods (except man, tests of 1 : 10 and 1 : 40). When the amount of antiserum added was less than 1 : 10, 1 : 30, 1 : 50, the reactions amongst non-homologous bloods took place more and more slowly, all the bloods outside that of the horse practically ceasing to react with tests of 1 : 100. Beginning with their tests of 1 : 5, and following the series of increasing amount of dilution to which antiserum was added to 1 : 100, we find that the blood of the pig first ceased to react, then that of man and ox. Strong sheep blood dilutions still gave clouds with antiserum added in the proportion of 1 : 100, the clouding only appearing after two hours; though the homologous horse serum reacted in five minutes. Next to horse, most positive reactions were obtained with man. These results are entirely in disaccord with mine and others.

		Reactions				
	•	*	×	+	**+**	%
97 PRIMATES	93	3*	1††	•	•	
28 CHIROPTERA	29	•	•	•	•	
13 INSECTIVORA	13	•	•	•	•	
88 CARNIVORA	86	2†	•	•	•	
63 RODENTIA	61	2‡	•	•	•	
71 UNGULATA						
a. *Artiodactyla*						
Suida						
4 Suidae	3	•	1§	•	•	25
1 Dicotylidae	1	•	•	•	•	
Tylopoda						
4 Camelidae (3 sp.)	4	•	•	•	•	
Tragulina						
1 Tragulidae	1	•	•	•	•	
Pecora						
17 Cervidae (12 sp.)	15	•	2	•	•	
36 Bovidae (22 sp.)	36	•	•	•	•	
b. *Perissodactyla*						
1 Tapiridae	1	•	•	•	•	
6 Equidae (3 sp.)	•	•	1‖	•	**5**	100
c. *Hyracoidea*						
1 Hyracidae	1	•	•	•	•	
3 CETACEA	2	1	•	•	•	33
13 EDENTATA	13	•	•	•	•	
26 MARSUPIALIA	26	•	•	•	•	
1 MONOTREMATA	1	•	•	•	•	
296 AVES (incl. 4 eggs)	296	•	•	•	•	
51 REPTILIA	51	•	•	•	•	
13 AMPHIBIA	13	•	•	•	•	
20 PISCES	20	•	•	•	•	
6 CRUSTACEA	6	•	•	•	•	

* The 3 bloods giving * reactions were Nos. 38, 45, 58, all fluid. No. 45 reacted peculiarly with other antisera. The other two were not tested with many antisera.

† One, a fluid serum No. 162, gave faint clouding with a number of other antisera. See tables. The other, No. 168, a sample from Borneo, also gave * with anti-human, and × with anti-hog-deer.

‡ One, a clouded, chloroform-preserved serum (No. 288) rejected in some of the tests because it gave clouded dilutions. Gave × with anti-wallaby and anti-cat, both weak antisera. Also * with anti-Mexican-deer. The other, No. 292, was a dried sample from S. America, and this also reacted, with faint clouding, with 4 different antisera. Owing to the peculiar behaviour of these two bloods they should be excluded.

§ No. 313, fluid serum. The negative results were with samples dried on filter-paper, but perfectly soluble.

‖ Anti-diphtherial serum 5 years old, fluid, bottled (No. 378). Two sets of tests on Nos. 376 and 377 (anti-diphtherial horse serum, dried in scales, bottled since 1895 and 1897 respectively) gave negative results with this antiserum, doubtless because of their insolubility; these are excluded from the above table as also from the other tables.

†† No. 46, fluid serum. The value of this reaction is open to doubt because of the behaviour of this blood to other non-homologous antisera, notably, weak anti-cat and anti-wallaby.

790 *Tests with Anti-Horse Serum.*

Four antisera were used in the tests noted on p. 194, the antisera having been obtained from as many rabbits treated with the serum of horses slaughtered at Cambridge. Three of the antisera were powerful, one of moderate strength, giving, when standardized, a precipitum of ·008 c.c.

The reactions with anti-horse serum, and with the antisera for the donkey and zebra, are practically in accord. The large reactions are in all three cases confined to equine bloods. In the case of the last two antisera, a time limit was put upon the reactions recorded, those only being included which took place within 40 minutes. For this reason the effects of these antisera seem more limited, although the difference is trivial.

Although the reactions with the bloods of Equidae are put down as equal, there is a very distinct difference both with regard to the time in which they take place, and the amount of precipitum formed. Anti-horse serum acts more rapidly on horse blood than upon the others, anti-donkey more rapidly upon that of the donkey, anti-zebra more rapidly upon that of the zebra. In the quantitative tests to be reported in Section VII. these differences are stated in figures, and are dwelt upon more in detail.

(10) *Tests with Anti-Donkey Serum.*

This antiserum was first obtained by Uhlenhuth (25, VII. 1901) who tested 24 bloods therewith (see list p. 172) and found it only to react with 2 bloods, those of the donkey and horse, less with the latter.

94 *Tests with Anti-Donkey Serum.*

The following tests were made on one day, and with one antiserum obtained from a rabbit which had been treated with the serum of *Equus asinus*. Only the reactions occurring within 40 minutes are recorded. The antiserum was fairly powerful, giving a precipitum of about 0·01 c.c. when standardized.

The tests on the bloods of Ungulata and Cetacea are included in the extended tables at the end, but not the others, as they were made in such limited numbers. Only one test on Ungulata is not included in the tables at the end, the test having been made on a fluid ox-serum (1, I. 1902) not included amongst the numbered samples   The following numbers, referring to the bloods tested, are to be found in the first column of the tables in question.

		Reactions			
	.	*	×	+	**+**
5 Primates	5	.	.	.	.
5 Chiroptera	5	.	.	.	.
2 Insectivora	2	.	.	.	.
6 Carnivora	6	.	.	.	.
6 Rodentia	6	.	.	.	.
46 Ungulata					
a. *Artiodactyla*					
Suina					
4 Suidae	2	2	.	.	.
1 Dicotylidae	.	1	.	.	.
Tylopoda					
3 Camelidae	3	.	.	.	.
Tragulina					18 %
1 Tragulidae	1	.	.	.	.
Pecora					
13 Cervidae	12	1	.	.	.
18 Bovidae	15	3	.	.	.
b. *Perissodactyla*					
1 Tapiridae	1	.	.	.	.
4 Equidae (3 spec.)	.	.	.	.	**4** 100 %
c. *Hyracoidea*					
1 Hyracidae	1	.	.	.	.
3 Cetacea	3	.	.	.	.
5 Edentata	5	.	.	.	.
5 Marsupialia	5	.	.	.	.
1 Monotremata	1	.	.	.	.
8 Aves	8	.	.	.	.
1 Reptilia	1	.	.	.	.
1 Amphibia	1	.	.	.	.

The bloods tested in the above table were Primates, Nos. 38, 44, 45, 48, 57, all fluid; Chiroptera, Nos. 99, 100, 105, 114, 119, all samples dried on filter-paper; Insectivora, Nos. 133, 141, both fluid; Carnivora, Nos. 145, 162, 205 a, 220, 242, and unnumbered dog's serum (4, III 1902), all fluid; Rodentia, Nos. 247, 271, 280, 284, 286, 289, all dried on filter-paper; Edentata, Nos. 388, 389, 390, 391, 395, all fluid; Marsupialia, Nos. 406, 412, 415, 417, 425, all fluid; Monotremata, No. 427, dry; Aves, Nos. 428, 448, 461, 463, 477, 502, and two unnumbered samples, fowl (12, III. 1902) and pigeon (6, VI. 1902), all fluid; Reptilia (813), Amphibia (862), both fluid.

(11) 94 *Tests with Anti-Zebra Serum.*

These tests were all made on one day with one antiserum obtained from a rabbit treated with the serum of *Equus greyvi* Günther, Grevy's zebra, which died at the Zoological Society's Gardens, London. This antiserum was moderately powerful, giving a precipitum of ·012 c.c. with its homologous blood.

		Reactions			
	·	*	×	+	**+**
5 PRIMATES	5	·	·	·	·
5 CHIROPTERA	5	·	·	·	·
2 INSECTIVORA	2	·	·	·	·
6 CARNIVORA	6	·	·	·	·
6 RODENTIA	6	·	·	·	·
46 UNGULATA					
a. *Artiodactyla*					
Suina					
4 Suidae	4	·	·	·	·
1 Dicotylidae	1	·	·	·	·
Tylopoda					
3 Camelidae	3	·	·	·	·
Tragulina					
1 Tragulidae	1	·	·	·	·
Pecora					
13 Cervidae	12	1	·	·	·	8 %
18 Bovidae	11	4	3	·	·	39 %
b. *Perissodactyla*					
1 Tapiridae	·	1	·	·	·
4 Equidae (3 spec.)	·	·	·	·	**4**	100 %
c. *Hyracoidea*					
1 Hyracidae	1	·	·	·	·
3 CETACEA	3	·	·	·	·
5 EDENTATA	5	·	·	·	·
5 MARSUPIALIA	5	·	·	·	·
1 MONOTREMATA	1	·	·	·	·
8 AVES	8	·	·	·	·
1 REPTILIA	1	·	·	·	·
1 AMPHIBIA	1	·	·	·	·

Note. A number of the tests here recorded are not included in the tables at the end. The footnote to the preceding tests with anti-donkey serum applies also to these bloods. In addition, the three tests with anti-zebra serum upon the 3 cetacean blood-samples are not included in the tables.

The close relationship existing between *Equus caballus* and a species of zebra (*Equus burchelli*) has been demonstrated by crossing.

See J. C. Ewart (1889), *The Penycuik Experiments*. London (Adam and Charles Black), 177 pp.

VI. Antisera for bloods of Cetacea.

94 *Tests with Anti-Whale Serum.*

These tests are not included in the tables at the end, for the reason that the antiserum had not been obtained when they were written. The tests were all made with one antiserum obtained from a rabbit treated with the serum and blood of *Balaenoptera rostrata*, the Rorqual (No. 385 a), the blood having been kindly sent me by Professor Torup of Christiania.

When standardized, the antiserum gave a precipitum of ·009 c.c.

	Reactions					
	·	*****	**×**	**+**	**✚**	
5 Primates	5	·	·	·	·	
5 Chiroptera	1	4	·	·	·	80 %
2 Insectivora	2	·	·	·	·	
6 Carnivora	6	·	·	·	·	
6 Rodentia	6	·	·	·	·	
46 Ungulata						
a. *Artiodactyla*						
5 Suina	1	3	1	·	·	80 %
3 Tylopoda	3	·	·	·	·	
1 Tragulina	1	·	·	·	·	
31 Pecora	10	14	7	·	·	68 %
b. *Perissodactyla*						
1 Tapiridae	1	·	·	·	·	
4 Equidae	4	·	·	·	·	
c. *Hyracoidea*						
1 Hyracidae	1	·	·	·	·	
3 Cetacea (2 spec.)	·	·	1*	1	1	100 %
5 Edentata	3	2	·	·	·	40 %
5 Marsupialia	5	·	·	·	·	
1 Monotremata	1	·	·	·	·	
8 Aves	8	·	·	·	·	
1 Reptilia	1	·	·	·	·	
1 Amphibia	1	·	·	·	·	

* The weak reaction due to a test carried out upon a solution from a dried sample of blood of *Balaenoptera rostrata* (No. 285) of older date.

The bloods tested were the same as in the list appended to the tests with anti-donkey serum (see p. 196), and the same as those tested with anti-donkey serum (Ungulata, Cetacea) given in the tables at the end.

The reactions noted occurred within 40 minutes.

The reactions of the first and second class were confined to two cetacean bloods, those of *Balaenoptera rostrata*, and *Phocaena communis*. Some medium cloudings are noted amongst the Artiodactyla, some faint cloudings amongst these and other mammalia.

VII. Antiserum for Marsupialia.

691 *Tests with Anti-Wallaby Serum.*

The antiserum used in these tests was obtained from a rabbit treated with the serum of *Onychogale unguifera* Gould, the Nail-tailed Wallaby, the animal having died at the Zoological Society's Gardens, London. The antiserum possessed but moderate power, yielding a precipitum of ·007 c.c., on being standardized.

	Reactions				
	•	*	×	+	%
66 PRIMATES	59	6	1	•	11
24 CHIROPTERA	24	•	•	•	0
10 INSECTIVORA	10	•	•	•	0
79 CARNIVORA	75	2	2	•	5
56 RODENTIA	55	*	1*	•	2
61 UNGULATA	58	3	•	•	5
2 CETACEA	2	•	•	•	0
13 EDENTATA	13	•	•	•	0
25 MARSUPIALIA					
Suborder *Polyprotodontia*					
6 Didelphyidae (3 spec.)	5	1	•	•	17
1 Dasyuridae	•	•	1	•	100
1 Peramelidae	1	•	•	•	0
Suborder *Diprotodontia*					
2 Phalangeridae	2	•	•	•	0
15 Macropodidae (8 spec.)	•	1 (6%)	8 (53%)	6 (40%)	100
1 MONOTREMATA	1	•	•	•	0
278 AVES (incl. 4 eggs)	278	•	•	•	
38 REPTILIA	38	•	•	•	
11 AMPHIBIA	11	•	•	•	
21 PISCES	21	•	•	•	
6 CRUSTACEA	6	•	•	•	

5% (Primates–Edentata group); 68% (Marsupialia group)

* No. 289 gave curious results with other antisera. See tables.

It will be seen from the above that the action of this antiserum was limited, as far as the large reactions go, to the Macropodidae. The bloods of the latter gave 40% of the large reactions, 53% of the second-class reactions, one gave a faint clouding; in other words, taking the reactions all told, 100 % reacted. Two Phalangeridae bloods gave no reactions. Of the Polyprotodontia bloods one of Dasyuridae gave × and one out of 6 of Didelphyidae gave a faint clouding. The majority of the last gave a negative result, perhaps for the reason that they were not tested until my antiserum had grown still weaker than it was at the start. It is true that some × reactions are noted amongst other mammalia, but their number is very small if their percentage is reckoned. The Didelphis blood which showed a faint clouding, was the only one, outside that of other Marsupialia, which reacted at all in a long series of bloods tested on one day, the faint clouding in this case was therefore certainly suggestive. These tests will however have to be repeated with stronger antisera.

VIII. Antisera for bloods and egg-whites of Aves.

A. *Tests with Antisera for Avian Bloods.*

(1) *Anti-fowl serum.*

Anti-fowl serum was first obtained by Bordet (III. 1899) by treating a rabbit intraperitoneally with defibrinated fowl's blood. He observed that the rabbit's serum acquired agglutinating and haemolyzing, as well as precipitating, power for fowl and pigeon blood. As in Tchistovitch's experiments with eel serum, the precipitum Bordet obtained was soluble in dilute alkaline solutions. Nolf (v. 1900) obtained this antiserum by injecting fowl serum, but not when he injected the blood corpuscles. Uhlenhuth (15, x. 1900, p. 735) obtained anti-fowl serum in the same manner as Bordet. He found it to precipitate solutions of fowl blood, but not those of the pigeon, horse, donkey, ox and sheep. Evidently his antiserum was weak, otherwise it would have acted on pigeon blood. In my paper of 20, I. 1902, I reported having tested 250 bloods with anti-fowl serum. Amongst birds the results were different to those I had obtained with anti-mammalian sera amongst mammals, the reactions in the latter case being mostly restricted to groups of closely related animals. "Anti-fowl serum was found to produce a reaction not only with solutions of fowl blood, and that of the closely related pheasant, turkey, etc., but also with the bloods of widely divergent species, such as the parrot, various species of duck, the woodcock, sheathbill, heron, eagle, owl, condor, pigeon, a number of small Passerines, and American rhea. A marked clouding was moreover produced in the blood of the swallow, rook, landrail, stork, swan and

African ostrich. What I have termed a 'marked clouding' is probably to be regarded as an indication of a more remote relationship." Anti-fowl serum produced a slight cloud in only one mammalian blood of those tested, this being put down to some experimental error, the sample having been sent me from abroad. It may have been collected carelessly, viz. brought in contact with avian blood collected on the same day when out shooting. These results are incorporated in the following table. Subsequently (5, IV. 1902) I succeeded in obtaining an anti-ostrich serum, the tests with which follow those with anti-fowl serum.

It is interesting to note that but a slight action is exerted by anti-fowl serum upon fowl egg-white dilutions, compared to a very powerful action upon fowl blood. Gengou (25, X. 1902, p. 753) notes that this antiserum acts on fowl egg-white, but does not record the remarkable difference in the amount of the reaction, which I have repeatedly observed. A fairly powerful antiserum may in fact give no reaction with its corresponding egg-white.

792 *Tests with Anti-Fowl Serum.*

The following tests were made with two antisera obtained from two rabbits treated with the serum of *Gallus domesticus*. Both antisera were very powerful, giving, when standardized, the one a precipitum of ·05 c.c. with ·5 c.c. of a 1 : 40 fowl serum dilution, the other a precipitum of ·035 c.c., with ·5 c.c. of a 1 : 100 dilution.

The following table shows a striking difference from all of the preceding, and a great resemblance to the succeeding one with anti-ostrich serum. We see here an absence of reactions outside of the bloods of Aves, all of which react, whether they be of Ratitae or of Carinatae. Footnotes refer to some of the bloods which gave negative results. It is evident that not much importance can be attached to these when other bloods belonging to the same family have given even the largest reactions. It will be noted that the egg-white of the fowl, whose serum was used for treating a rabbit, gave a much smaller reaction than did the fowl's blood. Note that Galliformes gave 27 %, Anseriformes 26 %, of the largest reactions in their respective families.

	Reactions					
	•	*	×	+	**+**	%
85 PRIMATES	85	•	•	•	•	
28 CHIROPTERA	28	•	•	•	•	
12 INSECTIVORA	12	•	•	•	•	
88 CARNIVORA	87	1	•	•	•	385 mammals 0
61 RODENTIA	61	•	•	•	•	
70 UNGULATA	70	•	•	•	•	
3 CETACEA	3	•	•	•	•	
12 EDENTATA	12	•	•	•	•	
25 MARSUPIALIA	25	•	•	•	•	
1 MONOTREMATA	1	•	•	•	•	
320 AVES						
9 Ratitae						
Struthionidae, Rheidae						
(5 sp.), Casuaridae ...	•	1	6 (66 %)	1	**1**	100
311 Carinatae						
8 Colymbiformes	•	3	4	1	•	100
1 Procellariformes	1	•	•	•	•	0
14 Ciconiiformes (14 sp.)	2*	1	7 (50 %)	2	**2**	85
34 Anseriformes (18 sp.)	1†	3 (9 %)	13 (38 ,,)	8 (23 %)	**9** (26 %)	97
44 Falconiformes (27 sp.)	12‡	9 (20 ,,)	12 (27 ,,)	7 (16 ,,)	**4** (9 ,,)	72
30 Galliformes (17 sp.) ...	1§	5 (17 ,,)	8 (27 ,.)	8 (27 ,,)	**8** (27 ,,)	97
6 Gruiformes (4 sp.) ...	•	2	1	3 (50 ,,)	•	100
53 Charadriiformes (36 s.)	8‖	10 (19 ,,)	24 (45 ,,)	8 (15 ,,)	**3** (5 ,,)	84
15 Cuculiformes (12 sp.)	•	3 (20 ,,)	5 (33 ,,)	7 (46 ,,)	•	100
21 Coraciiformes (8 sp.)	1	5 (23 ,,)	10 (47 ,,)	3 (14 ,,)	**2** (9 ,,)	95
85 Passeres (60 sp.)	12	24 (28 ,,)	46 (54 ,,)	3 (4 ,,)	•	86
4 Eggs						
1 Ratitae	1	•	•	•	•	0
3 Carinatae	2	•	1††	•	•	33
42 REPTILIA	41	1	•	•	•	2
13 AMPHIBIA	13	•	•	•	•	0
22 PISCES	22	•	•	•	•	0
7 CRUSTACEA	7	•	•	•	•	0

87 % (bracket for the Aves rows)

* 449 from India, 453 smeary.
† 487 from India.
‡ 6 from India, 1 from Japan, rest England.
§ From India.
‖ 2 from China and India, rest England.
†† Fowl's egg-white.

(2)   649 *Tests with Anti-Ostrich Serum.*

The antiserum used in the following tests was obtained from a rabbit by the injection of the serum of *Struthio molybdophanes* Reichenow

(No. 428), the bird having died at the Zoological Society's Gardens, London. The antiserum was very powerful, giving, when standardized, a precipitum of ·042 c.c.

	Reactions					%
	·	*	×	+	**+**	
64 PRIMATES	63	1	·	·	·	1
24 CHIROPTERA	24	·	·	·	·	0
10 INSECTIVORA	10	·	·	·	·	0
69 CARNIVORA	68	1	·	·	·	1
51 RODENTIA	51	·	·	·	·	0
51 UNGULATA	51	·	·	·	·	0
2 CETACEA	2	·	·	·	·	0
12 EDENTATA	12	·	·	·	·	0
22 MARSUPIALIA	22	·	·	·	·	0
1 MONOTREMATA	1	·	·	·	·	0

·6 %

	·	*	×	+	**+**	%
276 AVES						
8 Ratitae						
3 Struthionidae (2 sp.)						
2 Casuaridae (2 sp.)	·	·	5 (62 %)	·	3*(37 %)	100
3 Rheidae (1 sp.)						
268 Carinatae						
6 Colymbiformes (5 sp.)	2	3	1	·	·	66
11 Ciconiiformes (11 sp.)	1	3	4	3	·	91
29 Anseriformes (16 sp.)	6	12 (41 %)	7 (24 %)	3 (10 %)	1 (4 %)	79
36 Falconiformes (21 sp.)	8	11 (30 ,,)	13 (36 ,,)	4 (11 ,,)	·	77
24 Galliformes (15 sp.) ...	1	13 (54 ,,)	9 (37 ,,)	1 (4 ,,)	·	96
5 Gruiformes (3 sp.)	·	3	1	·	1	100
46 Charadriiformes (34 s.)	9	18 (39 ,,)	14 (30 ,,)	4 (9 ,,)	1 (2 %)	80
13 Cuculiformes (12 sp.)	1	3 (23 ,,)	8 (61 ,,)	1 (8 ,,)	·	92
22 Coraciiformes (9 sp.)...	4	9 (40 ,,)	6 (27 ,,)	3 (13 ,,)	·	81
76 Passeres (56 sp.)	25	32 (42 ,,)	18 (24 ,,)	1 (1 ,,)	·	67
4 Eggs						
1 Ratitae	·	1	·	·	·	100
3 Carinatae	2	1	·	·	·	33

79 %

	·	*	×	+	**+**	%
33 REPTILIA						
12 Chelonia	10	2	·	·	·	17
2 Crocodilia	1	1	·	·	·	50
19 Sauria	18	·	1	·	·	5
10 AMPHIBIA	10	·	·	·	·	0
14 PISCES	14	·	·	·	·	0
6 CRUSTACEA	6	·	·	·	·	0

12 %

* All 3 Struthionidae.

The general results correspond to those given in the preceding table, nevertheless, we see that the bloods of Ratitae (three Struthionidae) give a higher percentage of large reactions than do those of

Carinatae. Here we also observe faint cloudings occurring in dilutions of egg-white both of Ratitae (Emu) and Carinatae. A slight action would seem to be exerted upon reptilian bloods, all other non-avian bloods giving practically negative results.

B. *Tests with Antisera for Avian Egg-White.*

(1) *Anti-fowl's Egg.*

Myers (14, VII. 1900) treated rabbits with crystallized egg-albumin of the fowl and obtained an antiserum which precipitated the egg-albumin of the fowl, and to a slight extent that of the duck. No reaction was obtained when the antiserum was added to normal rabbit, bullock, or sheep blood dilutions. Subsequently Uhlenhuth (15, XI. and 1, XII. 1900) injected fowl egg-white in salt solution intraperitoneally into rabbits, and obtained a powerful antiserum, which produced reactions with egg-white diluted to 1 : 100,000, and weaker reactions with pigeon egg than with that of the fowl. He concluded from this that the albuminous constituents of both species of eggs are closely allied. Mertens (14, III. 1901) confirmed the foregoing results. Uhlenhuth (25, IV. 1901) reported later that he had obtained a powerful anti-egg serum which acted on fowl blood dilutions, and produced a slight cloud in goose blood. He obtained an immediate precipitation upon adding the antiserum to 2·5 % fowl blood dilution. It produced almost as powerful reactions with egg-white dilutions of goose, duck, guinea-fowl, though less with the pigeon's than with its homologous (fowl) egg-white. A rabbit treated with goose egg yielded a serum which produced a precipitum in goose blood, less clouding in fowl blood, dilutions. It gave a great and immediate precipitation with goose and duck, a considerable clouding with fowl, guinea-fowl, and pigeon eggs. He concludes it will not be possible to differentiate eggs as has been possible for bloods.

Nuttall (16, XII. 1901, p. 408) found anti-fowl egg serum, when powerful, to produce a reaction with fowl blood, but *never so intense an action as upon its homologous substance.* Levene (21, XII. 1901) treated animals for two months with egg-white and found their serum to precipitate egg-albumin, egg-globulin, yolk of egg and fowl and turkey serum. Nearly all the preceding observers state that normal rabbit or other sera produce no such effect on the dilutions, these being used for control purposes.

Nuttall (20, I. 1902) reported that this antiserum produced clouds with a variety of avian bloods (parrot, swan, heron, stork, conure, crow, emu) also with egg-white of emu, and in addition with blood dilutions of *Alligator sinensis, A. mississippiensis, Chelone midas,* and *Testudo ibera,* and it might therefrom appear " that the egg possesses a vestige of reptilian character." In view of the continued negative results with other bloods I referred to this observation (5, IV. 1902) as " very suggestive in view of the reptilian origin of birds," and the reaction as possessing possibly a " reptilian-avian " character. Dr Graham-Smith has since been pursuing investigations in this direction, and reports upon them in Section VIII.

Pursuing a different line of research (see p. 94) Gengou (25, X. 1902, p. 750) found he was unable, according to his method, to find any difference in the action of anti-fowl egg serum as tested upon eggs of the fowl, turkey, pigeon, and duck. He was unable (p. 753) to determine that anti-egg serum possessed an action on fowl blood. As we have seen, anti-egg only acts upon the corresponding blood when powerful. Finally Obermayer and Pick (1902) remark that fowl serum pseudoglobulin contains a body which reacts to anti-egg (from rabbit) in a similar manner to egg-white.

789 *Tests with Anti-Fowl's Egg.*

The following tests were conducted with two antisera obtained from two rabbits treated with the white of egg of the domestic fowl. The one antiserum was moderately powerful, the other exceedingly powerful. The antisera were standardized, but I should estimate the precipitum of the stronger antiserum at ·06 c.c.

In this table it will be seen that the largest reaction occurs with the egg-white of the fowl, several eggs having been tested with the same result. Second-class reactions were obtained with another carinate egg, with the egg of the emu, with the blood of a heron, and the blood of two Reptilia (Chelonia and Crocodilia). Otherwise all the medium and slight cloudings occurred amongst the bloods of Aves and Reptilia, if we except a single slight clouding in one out of 388 mammalian bloods tested.

	Reactions					%
	·	*	×	+	**+**	
389 MAMMALIA	388	1	·	·	·	0·2
312 AVES						
8 Ratitae						
3 Struthionidae (2 sp.)						
4 Rheidae (1 sp.)	4*	2	2	·	·	50
1 Casuaridae						
304 Carinatae						
7 Colymbiformes	5	1	1	·	·	28
1 Procellariformes	1	·	·	·	·	0
13 Ciconiiformes	10	1	1	1	·	23
30 Anseriformes............	17	10 (33 %)	3 (10 %)	·	·	43
42 Falconiformes	41	1	·	·	·	2
30 Galliformes	18	3 (10 ,,)	9 (30 ,,)	·	·	40
6 Gruiformes..............	5	1	·	·	·	16
50 Charadriiformes	49	1	·	·	·	2
16 Cuculiformes............	11	2	3	·	·	31
21 Coraciiformes	18	3	·	·	·	14
88 Passeres	80	7 (8 %)	1	·	·	9
4 Eggs						
1 Ratitae	·	·	·	1	·	100
3 Carinatae	·	·	1	1	1†	100
46 REPTILIA						
13 Chelonia	10‡	·	2	1	·	23
3 Crocodilia	1§	·	1	1	·	66
30 Sauria....................	29	1	·	·	·	3
10 AMPHIBIA	10	·	·	·	·	0
23 PISCES	23	·	·	·	·	0
5 CRUSTACEA	5	·	·	·	·	0

(Marginal braces indicate 17 % for the Ratitae group and for the Carinatae orders from Colymbiformes to Passeres.)

* 4 Rheidae.

† Egg-white of fowl, several specimens of which gave same result.

‡ 8 of these dried on filter-paper. (See text, p. 207, top.)

§ No. 815 dried on paper, sent from Central Africa.

(2)   630 *Tests with Anti-Emu's Egg-White.*

The antiserum used in the following tests was obtained by treating a rabbit with the egg-white of *Dromaeus novae-hollandiae* Vieill., the emu (No. 799), received from the Zoological Society's Gardens, London. This antiserum was fairly powerful. Standardized by Dr Graham-Smith, it gave a precipitum of ·028 c.c. when tested after being stored 5 months on ice, ·1 c.c. of antiserum being added to ·5 c.c. of 1 : 21 solution of emu egg-white.

The preceding table shows that the antiserum for the egg-white of the emu only produced a large reaction with the dilution of the same egg-white with which the rabbit had been treated, viz. with that of the emu. A less marked reaction occurred with the egg-white of one of the Carinatae, whereas medium and faint cloudings occurred when the antiserum was added to the bloods of a number of Aves, as well as in two out of six bloods of Reptilia (Chelonia). The negative results with other bloods of this class may be due to their partial insolubility when dried.

	Reactions					
	•	*	×	+	**+**	%
309 MAMMALIA	309	.	.	.	.	0
258 AVES						
8 Ratitae	2	4	2	.	.	75
250 Carinatae						
5 Colymbiformes	5	.	.	.	.	0
11 Ciconiiformes	7	1	3	.	.	36
27 Anseriformes	25	2	.	.	.	7
32 Falconiformes	29	.	3	.	.	9
28 Galliformes	25	3	.	.	.	11
5 Gruiformes	4	1	.	.	.	20
43 Charadriiformes	35	4	4	.	.	18
20 Coraciiformes	18	1	1	.	.	10
12 Cuculiformes	12	.	.	.	.	0
67 Passeres	61	6	.	.	.	9
4 Eggs						
1 Ratitae	.	.	.	.	1*	100
3 Carinatae	.	.	2	1	.	100
31 REPTILIA						
6 Chelonia	4†	.	2	.	.	66
2 Crocodilia	2	.	.	.	.	0
23 Sauria	23	.	.	.	.	0
7 AMPHIBIA	7	.	.	.	.	0
15 PISCES	15	.	.	.	.	0
6 CRUSTACEA	6	.	.	.	.	0

The Carinatae rows are bracketed together as 14 %.

* Emu's egg-white with which rabbit was treated.

† Three out of the four were dried samples, on paper (see text above).

IX. Antisera for bloods of Reptilia.

(1) 666 *Tests with Anti-Turtle Serum.*

The following tests were conducted with two antisera obtained from two rabbits treated with the serum of (No. 809) *Chelone midas*, the Green Turtle, a specimen of which was killed in Cambridge. Both antisera were very powerful, one giving a precipitum of ·03 c.c.

		Reactions				%
	•	*	×	+	**+**	
323 Mammalia	323	•	•	•	•	0
262 Aves	262	•	•	•	•	0
4 Eggs of Birds	3	•	1	•	•	25
34 Reptilia						
7 Subclass Chelonia	2*	1	1	1	**2**	71
3 Subclass Crocodilia	2†	1	•	•	•	33
24 Subclass Sauria	24	•	•	•	•	0
14 Amphibia	14	•	•	•	•	0
22 Pisces	22	•	•	•	•	0
7 Crustacea.................................	7	•	•	•	•	0

* Both dried, one from Central Africa. † One dried.

The reactions of the first and second class will be seen to be confined to bloods belonging to Chelonia, medium and faint reactions were observed only amongst Crocodilia and avian egg-white dilutions. In my paper of 20, I. 1902 I reported the first series of 250 bloods tested with this antiserum, having found that the blood of *Testudo ibera* gave a marked clouding, and of *A. mississippiensis* a faint clouding. These tests are included in the above table. For further bloods tested see the large tables at the end.

(2) *Tests with Anti-Ophidian Sera.*

In view of the extensive work involved in the production of other antisera, I have been unable to produce anti-snake sera. Dr Graham-Smith has been carrying on a series of tests with these, which he reports upon in Section VIII. The observations of Lamb (16, VIII. 1902) on *precipitins for cobra venom* fall naturally under this head.

Lamb treated rabbits subcutaneously for 4—5 months with injections of pure unheated cobra venom, the injections being made at intervals of about 10 days. At the time when precipitins appeared, the rabbit could withstand four times the lethal dose of venom. Using Wright's sedimentation tubes, he added 4 volumes of antiserum to 1 volume of 1% cobra venom solution, and obtained a reaction leading to a marked precipitum after 24 hours. An excess of antiserum added to weak venom solutions gave the most precipitum. The precipitin was active for those proteids in venom which are incoagulable by heat. Moderate heating of the antiserum does not affect its action. A 0.2% solution of cobra venom, heated to 75° C. for 30 minutes, after which the coagulated proteids were removed by filtration, gave as much precipitum as unheated venom solutions which were used as controls. *Daboia* venom gave the same result. Cobra antiserum heated for 30 minutes at 55° remained unaffected, as did also daboia venom. Six parts of the antiserum added to one part of ·05 % of venom solution used up all the precipitin. Lamb considers that the method might be used for the standardization of anti-sera for therapeutic purposes, the test being more exact than in animals.

Cobra venom antiserum was tested upon the venoms of

Daboia rusellii (Indian Chain viper)	Reacted
Echis carinata (Fhoorsa, small Indian Viper)	No reaction
Bungarus fasciatus (Banded Krait, India, colubrine family)	,,
Hoplocephalus curtus (Australian Tiger Snake, colubrine family)	,,

The similar reactions obtained with cobra venom and that of daboia would, according to Lamb, indicate that the venoms are alike, but he and Hanna (*Journ. Pathol. and Bacteriol.* 1902, VIII. p. 1) have shown their physiological action to be quite different, and they found Calmette's antivenine to protect against cobra but not daboia venom. If the proteids are identical then the venom action will not depend on these, or else they possess the same haptophoric group. Nevertheless, Lamb adds that, on inspection of the protocols, it is seen that cobra venom gave a precipitum in higher dilutions than did that of daboia. These investigations are of considerable interest in relation to my work. It will be doubtless found that the injection of the serpents' bloods will yield precipitins having a similar action. Such a line of investigation appears to me indicated in view of the small amount of precipitin-producing substance present in venom compared with its toxic action.

(3) 468 *Tests with Anti-Alligator Serum.*

The following tests were made with one antiserum obtained from a rabbit treated with the serum of *Alligator mississippiensis* (No. 813), the reptile having died at the Zoological Society's Gardens, London. The antiserum was moderately powerful.

We see from the table (p. 210) that the action of anti-alligator serum was confined, so far as large reactions go, to Crocodilia; medium reactions took place with the bloods of a large proportion of Chelonia, a bird's blood, and egg-white. I may add here that the interaction of anti-avian egg sera and reptilian bloods, and *vice versâ*, has been amply confirmed by Dr Graham-Smith, who at my suggestion has investigated this interesting avian-reptilian reaction, as I termed it elsewhere (p. 205). Dr Graham-Smith reports upon his results, which are mainly quan-

titative, in Section VIII. In my preliminary communication of 20, 1.
1902, I reported that I had tested 250 bloods with the result that those
of *Alligator mississippiensis* and *A. sinensis* gave reactions, a clouding
occurring in the blood of *Chelone midas*. The tests with the alligator
and turtle sera were repeated several times with uniform results, these
being included in the tables.

		Reactions			
	·	*	×	+	✚	%
237 PRIMATES	237	·	·	·	·	0
174 AVES	173	·	1	·	·	·5
3 ,, eggs	3	·	1	·	·	25
20 REPTILIA						
6 Chelonia	2†	·	4	·	·	66
3 Crocodilia	1‡	·	·	·	**2**	66
11 Sauria	11	·	·	·	·	0
12 AMPHIBIA	12	·	·	·	·	0
15 PISCES	15	·	·	·	·	0
7 CRUSTACEA	7	·	·	·	·	0

† One from Central Africa, dried. The other dried.
‡ From Central Africa.

X. Antiserum for bloods of Amphibia.

551 *Tests with Anti-Frog Serum.*

The following tests were made with two antisera obtained from two
rabbits treated with the blood of freshly killed frogs (*Rana temporaria*).
Both antisera were weak. They were not standardized.

		Reactions			
	·	*	×	+	%
247 MAMMALIA	247	·	·	·	0
239 AVES (incl. 4 eggs)	239	·	·	·	0
27 REPTILIA	27	·	·	·	0
13 AMPHIBIA					
3 Order Urodela	3	·	·	·	0
1 Order Anura, Suborder Aglossa	1	·	·	·	0
Suborder Phaneroglossa					
2 Bufonidae (2 spec.)	2	·	·	·	0
1 Engystomatidae	1	·	·	·	0
6 Ranidae (2 spec.)	·	1	2	**3**§	100
18 PISCES	18	·	·	·	0
7 CRUSTACEA	7	·	·	·	0

§ *Rana temporaria.*

The foregoing table shows that the antiserum for *Rana temporaria* only produced a large reaction in the blood dilutions of this species, and that it also produced a reaction with the blood of *Rana tigrina*, the results with all other bloods being absolutely negative.

In my paper of 5, IV. 1902 I reported upon tests made on 508 bloods with this antiserum, these bloods being now included in the above table. The only investigator who has worked with anti-amphibian sera is Philippson (about Sept. 1902), who confined himself to a few tests. He cites my paper, and states that he has been able to confirm my results. He injected 1 c.c. of defibrinated frog's blood (*Rana viridis* var. *esculenta*) into a rabbit, making two such injections with a week's interval. He found his antiserum to precipitate the serum of *Rana viridis* var. *typica*, and that of *R. fusca*, but it had no effect on that of *Hyla arborea*, *Bufo vulgaris* and *Salamandra maculosa*. The antiserum had also no effect upon the serum of the pig or calf. I would remark that his antiserum must doubtless have been weak after only making two injections and, what is more, introducing such small amounts of frog blood. In my experience, I have not noted the presence of pre-cipitins in the serum of animals prior to the third injection, and then only in small amount. I did not however make a note as to the earliest period at which my rabbit showed precipitin when treated with frog's blood.

XI. Antisera for bloods of Pisces.

Tchistovitch (v. 1899) made his fundamental discovery upon the blood precipitins by treating the rabbit, guinea-pig, goat, or dog with eel serum. Tests upon different fish bloods are being made the subject of study by Dr Graham-Smith at my suggestion. The results should be of considerable interest. With the exception of de Lisle (XI. 1902, p. 399) who, following the method of Tchistovitch, also obtained anti-eel serum from rabbits, no further work has as yet been done. De Lisle (p. 403) obtained precipitins by injecting either eel serum or eel corpuscles.

XII. Antiserum for bloods of Crustacea.

450 *Tests with Anti-Lobster Serum.*

These tests were made with one antiserum derived from a rabbit treated with the blood of the lobster (*Homarus vulgaris*). The anti-serum was fairly powerful.

	Reactions					
	•	*	×	+	**+**	%
210 MAMMALIA	210	•	•	•	•	0
191 AVES (incl. 4 eggs)	191	•	•	•	•	0
17 REPTILIA..	15	2 (?)	•	•	•	0
12 AMPHIBIA	10	2 (?)	•	•	•	0
14 PISCES..	14	•	•	•	•	0
6 CRUSTACEA, all of the order Decapoda	•	1	•	3	**2**	100

Note: One sample of crustacean blood noted in the tables is excluded (No. 897), as the foam-test of the dilution was doubtful.

From the above table it will be seen that the anti-lobster serum only produced first and second-class reactions with crustacean bloods. The faint clouding in one of the six examined was doubtless due to the small quantity which went into solution from filter-paper. A faint clouding was observed to take place in two reptilian and two amphibian blood dilutions, the test in this case being unfortunately not repeated owing to lack of antiserum. Dr Graham-Smith has extended the studies in this relation also (see Section VIII). Otherwise these are the only investigations made with anti-crustacean sera. I might add that I first reported having produced such an antiserum in my paper of 20, I. 1902, and briefly referred to the results above given in my paper of 5, IX. 1902.

The Number of Species of Blood tested.

Of the 900 specimens of blood collected by me, about one-third were duplicates, the number of species represented being 586 or more, including four races of man under the species, but not several breeds of dog and sheep, as follows:—

	No. of Samples	No. of Species	
MAMMALIA			
Primates			
Anthropoidea			
Hominidae	35	4 (races)	
Simiidae	8	3	
Cercopithecidae	36	26	3 doubtful species
Cebidae	13	9	
Hapalidae	4	3	
Lemuroidea			
Lemuridae	2	2	
Chiroptera	30	25	1 doubtful
Insectivora	15	4	
Carnivora, incl. Pinnipedia	99	55	5 doubtful
Rodentia	69	39	
Ungulata	73	50	3 doubtful
Cetacea	9	9	
Edentata	13	10	
Marsupialia	26	16	
Monotremata	1	1	
AVES			
Ratitae, incl. 3 families	9	6	1 doubtful
Carinatae, incl. 11 Orders	319	213	
Avian Egg-white	4	4	
REPTILIA			
Chelonia	13	9	
Crocodilia	3	3	
Sauria	37	29	
AMPHIBIA			
Urodela	3	3	
Anura	11	6	
PISCES			
Elasmobranchia and *Teleostei* (over 14 families)	25	24	
CRUSTACEA			
Decapoda (Macrura and Brachyura)	7	7	

Bloods which proved insoluble have not been included in the foregoing list.

Note regarding the Classification of Animals adopted in the Tables.

As will be seen, the bloods tested have been ordered according to the zoological classification of the animals which yielded them. For the Mammalia, I have followed Flower and Lydekker; for Birds, see Evans; for Reptilia and Amphibia, see Gadow; for Fishes, see Günther; for Crustacea, see Claus: the works referred to being as follows:—

FLOWER, W. H., and LYDEKKER, R. (1891). *An Introduction to the Study of Mammals, Living and Extinct.* (London : Adam and Charles Black.)

EVANS, A. H. (1899). Birds. *The Cambridge Natural History*, vol. IX. (London : Macmillan and Co., Limited.)

GADOW, H. (1901). Amphibia and Reptiles. *Ibid.* vol. VIII.

GÜNTHER, A. C. (1880). *An Introduction to the Study of Fishes.* (Edinburgh : Adam and Charles Black.)

CLAUS, C. (1887). *Lehrbuch der Zoologie.* (Marburg and Leipzig : Elwert'sche Buchhandlung.)

On several occasions, when in doubt, recourse was had to the British Museum Catalogue.

GENERAL SUMMARY OF THE RESULTS OF 16,000 PRECIPITIN-TESTS CONDUCTED BY G. H. F. NUTTALL.

I. *Tests with Anti-Primate Sera.*

These tests were conducted by means of antisera for man (825 tests), chimpanzee (47 tests), ourang (81 tests), *Cercopithecus* (733 tests). Maximum reactions were only obtained with bloods of Primates. The degrees of reaction obtained indicate a close relationship between the *Hominidae* and *Simiidae*, a more distant relationship with the *Cercopithecidae*, the bloods of *Cebidae* and *Hapalidae* giving still smaller reactions than the last, when we consider the results obtained with the first three antisera. The tests with antiserum for *Cercopithecus* gave the largest reactions with bloods of *Cercopithecidae*, next with those of *Hominidae* and *Simiidae*, but slight reactions with those of *Cebidae* and *Hapalidae*. All four antisera failed to produce reactions with the two bloods of *Lemuridae* tested, except when sufficiently powerful to also produce reactions with other mammalian bloods. From this we may conclude that the Lemurs properly belong to an Order separate from the other Primates.

II. *Tests with Anti-Insectivore Sera.*

These tests were conducted with but a single antiserum, 383 bloods being examined. Maximum reactions were only obtained with the blood of *Erinaceus*, whilst slight reactions were only obtained with a very few (3·5 %) of all the other bloods examined, although the antiserum was very powerful.

III. *Tests with Anti-Carnivore Sera.*

Four antisera were used in this group: anti-cat (785 tests), anti-hyaena (378 tests), anti-dog (777 tests), anti-seal (358 tests). The results in all four cases agreed in showing a preponderance of larger reactions amongst the bloods of Carnivora, as distinguished from other Mammalia, maximum reactions usually taking place amongst more closely related forms in the sense of descriptive zoology.

IV. *Tests with Anti-Ungulate Sera:* (a) Artiodactyla.

1. Anti-pig serum (818 tests) only gave maximum reactions with bloods of *Suidae*, moderate reactions with those of *Pecora* and *Tylopoda*. Of the bloods of the Artiodactyla, as a whole, 66 % reacted in some degree. Owing to the power of the antisera used a "mammalian reaction" was frequently observed, 51 % of the non-ungulate bloods giving slight reactions. All of the three cetacean bloods examined gave a moderate or slight reaction.

2. Anti-llama serum (363 tests) only gave maximum reactions with its homologous blood, moderate reactions with the bloods of an allied species and that of *Camelus*. The percentage of slight reactions was high only amongst Ungulata.

3. A considerable number of antisera were prepared for *Pecora*, and tested on a large number of bloods:

Antiserum for	No. of tests	Result showed
Mexican-deer	749	Close connection between Cervidae and Bovidae, next to other Artiodactyla, Cetacea, Equidae
Reindeer	69	The same
Hog-deer	699	Close connection between Cervidae and Bovidae, next to other Artiodactyla, Cetacea
Antelope	686	The same
Ox	790	Close connection between Cervidae and Bovidae, next to Cetacea, Equidae, Suidae
Sheep	701	Close connection between Cervidae and Bovidae, next came other Artiodactyla, Equidae, Cetacea

(b) Perissodactyla.

4. Tests with Anti-equine sera.

These tests were conducted with three antisera for the bloods of the horse (790 tests), donkey (94 tests), and zebra (94 tests). The antisera only gave maximum reactions with equine bloods, slight reactions only with bloods of some *Artiodactyla* and *Cetacea*. (Time limit.)

V. *Tests with Anti-Cetacean Serum.*

Anti-whale serum, tested on 94 bloods, gave a maximum reaction only with cetacean bloods, slight reactions with those of *Pecora* and *Suina*, and some faint cloudings with those of other Mammalia.

VI. *Tests with Anti-Marsupial Serum.*

A weak anti-wallaby serum, tested on 691 bloods, showed a close relationship amongst *Marsupialia*, exclusive of the carnivorous Thylacine. Only a few other bloods gave slight reactions.

VII. *Tests with Anti-Avian Sera.*

1. Tests were made by means of antisera for the blood of the fowl and ostrich upon 792 and 649 bloods respectively. They demonstrated a similarity in the blood constitution of all birds, which was in sharp contrast to what had been observed with mammalian bloods when acted upon by anti-mammalian sera. Differences in the degree of reaction were observed, but did not permit of drawing any conclusions. Slight or faint reactions were observed with some reptilian bloods; only a very few mammalian bloods (0% and 0.6% respectively) showed faint reactions.

2. Tests made with antisera for egg-white of the fowl (789 tests) and emu (630 tests) gave maximum reactions with the egg-white of birds, moderate reactions with avian bloods, and distinct but slight reactions with reptilian bloods, notably those of *Crocodilia* and *Chelonia*.

VIII. *Tests with Anti-Reptilian Sera.*

Two antisera were used, the one for the blood of *Chelonia* (666 tests), the other for *Alligator* (468 tests). The first produced maximum reactions with chelonian bloods, lesser reactions with crocodilian, and slight reactions with avian egg-whites. The anti-alligator serum produced maximum reactions with *Crocodilia*, followed by *Chelonia*, avian egg-whites, and least with avian bloods. (See also Section VIII.)

IX. *Tests with Anti-Amphibian Serum.*

Anti-frog serum, tested on 551 bloods, only gave reactions with the bloods of *Ranidae*.

X. *Tests with Anti-Crustacean Serum.*

Anti-lobster serum, tested on 450 bloods, only gave reactions with the bloods of *Decapoda*.

Key to Signs used in the Tables.

In the following tables, the animals are placed in their zoological order, the blood samples tested being numbered continuously in the first column, in the second column in the order in which they were tested. As far as possible the scientific and common names of each animal have been given, this being followed by its habitat. The collector's name is placed in brackets. Where the bloods from one source are numerous, the name is represented by an inital, thus (Z) denotes that the blood was collected at the Zoological Society's Gardens, London; (R) stands for de Rothschild; (N) for Nuttall; (F) for Foster; (L. C.) for Dr Louis Cobbett; (G.-S.) for Dr Graham-Smith; (Lane) and (Mitchell) are two of our laboratory attendants. Where the habitat of the animal or the collector's name is omitted refer to the preceding samples for particulars.

Where not otherwise stated, the bloods were all collected and dried on filter-paper. Fluid sera are indicated by the abbreviations "fl." or "fl. in cap." (cap. standing for capillary tubes). Some sera were dried on glass, and the dry scales kept bottled and used for tests; such are described as "scales." In some cases the fluid serum had been filtered (fl. ser. filt.), in others it is stated that the blood sample was collected from a wound or cut, and again from what disease the animal died, etc.

The names of the antisera used are given at the top of the columns on the right. The antisera obtained from rabbits are named according to the animal's blood or serum with which the rabbit had been treated. These antisera are also ordered zoologically from left to right in a manner corresponding to the whole series of bloods.

The order of the antisera at the top of the page will serve therefore as an index to the order of the series of bloods tested. The columns giving the reactions have been grouped in blocks in their order of classification, so as to facilitate a comprehension of the reactions taking place, for instance amongst the Ungulata, as an order. As long as Mammalian bloods are under consideration, the anti-mammalian sera are thus grouped; the order to which the antiserum belongs being given at the foot of the page. When we come to the test upon avian bloods on the other hand, all columns referring to the tests with anti-mammalian sera are included in one "block."

By reading from left to right the reactions given by any one blood with the different antisera will be gathered, whereas if the columns are read from top to bottom, the reactions given by each antiserum with the different bloods in succession will be gathered.

The reactions which took place between each antiserum and each blood are included in the space formed by the intersection of the lines which include the name of the animal whose blood was tested (on the left), and those which include the name of the antiserum (above). The degrees of reaction are indicated by the following signs :—

+	denotes great reaction		**.** denotes	no reaction
+	„ marked clouding		/ „	blood insoluble
×	„ medium clouding		\ „	blood gave clouded solution
*	„ faint clouding			

The time when the reaction was noted is stated below the above signs in minutes, anywhere from 1 to 240 (4 hrs.). In a number of my earlier tests, the observations were not extended beyond 30 minutes, and the deposits taking place after 16—24 hours were not studied. This was not the case however in most of the tests. The amounts of deposit noted on the following day are indicated by the letters beneath the sign for the early or immediate reaction, noted above, as also beneath the number of minutes, as follows:—

D	denotes	very large deposit
D	„	moderate deposit
d	„	small deposit
tr	„	trace of deposit
c	„	cloudiness persisting after 16—24 hrs.

In the earlier tests where a reading of these deposits was omitted, their significance not being appreciated at the time, the probable amount of deposit which would have been noted is indicated by the above *letters reversed*.

To make the matter clearer by a few examples, I state the meaning of the following signs.

$\left.\begin{array}{l}+\\30\\\mathbf{tr}\end{array}\right\}$ means marked clouding in 30 minutes, trace of deposit after 16—24 hrs.

$\left.\begin{array}{l}\cdot\\\mathbf{D}\end{array}\right\}$ (no time being noted) denotes marked clouding in 5—15 minutes, moderate deposit after 16—24 hours

$\cdot$ no reaction, usually up to 5 hrs, and no deposit subsequently

From a medico-legal standpoint, no reactions other than those of the first class should be accepted, control tests being naturally made with other antisera of corresponding power for their homologous bloods.

Class MAMMALIA
1. Order PRIMATES

1. Suborder ANTHROPOIDEA
Fam. **Hominidae** — European

	No.	Species	Description	Man	Chimpanzee	Ourang	Monkey	Hedgehog	Cat	Hyaena	Dog	Seal
1	14.	*Homo sapiens.*	Woman, autopsy, scales Cambridge (N) 15. iii. 01	+ ɑ		· tr			·		·	
2	34.	,, ,,	Man, autopsy (N) 2. iii. 01	+ ɑ		* 120 d	·		·	·	·	·
3	35.	,, ,,	putrified, dried in scales (N) 15. iii. 01	+		·			·		·	
4	36.	,, ,,	W. M., cut, fl. in cap. (N) 13. iv. 01	+ ɑ							·	
5	37.	,, ,,	Dr D., cut, fl. in cap. (N) 23. iv. 01	+							·	
6	38.	,, ,,	Dr N., cut, fl. in cap. (N) 21. iv. 01	+ ɑ							·	
7	39.	,, ,,	W. M., cut, fl. in cap.	+ ɑ							·	
8	408.	,, ,,	Serum from Blister, Dr N.'s foot (N) 20. iv. 01	+ ɑ							·	
9	409.	,, ,,	Serum from Burn-blister Dr N. (N) 23. iv. 01	× 15 d		* 120 d	* 180	·		* 180 tr		
10	410.	,, ,,	Dr D., cut (N) 23. iv. 01	× 15 d		* 120 d	* 180	·		* 180 tr		
11	413.	,, ,,	W., cut (N) 17. vii. 01	× 15 D		* 120 d	* 180 tr	·		* 180 tr		
12	415.	,, ,,	Ascitic Fluid from autopsy (N) 23. x. 01	+ D		* 10 d	* 180 tr	·		* 180		
13	416.	,, ,,	W. M., cut (N) 28. x. 01	× 15 D	+ ɑ	* 120 d	* 45 d	·		* 180 tr		
14	417.	,, ,,	B. C., cut (N) 29. x. 01	+ 15 D		* 120 d	* 120 d	·		· tr ?		
15	418.	,, ,,	A. C., cut (N) 4. xi. 01	+ 15 D		* 120 d	* 180 tr	·		·		
16	420.	,, ,,	A. C., cut (N) 13. xi. 01	+ 15 D		* 120 d	* 180 tr	·		·		
17	422.	,, ,,	Dr N., cut (N) 10. i. 02	+ D			·	·		·	·	·
18	449.	,, ,,	A., cut (N) 18. i. 02	+ D		* 120 d	* 45 tr	·	·	* 45	·	× 40 tr
19	475.	,, ,,	W. M., cut (N) 30. i. 02	+ D		* 120 d	* 40	·	* ·	* 45	·	* 40 tr
20	575.	,, ,,	A. C., cut (N) 18. ii. 02	+ 30 D		* 120 d	* 45	·	·	* 45	* 75	* 40 tr
21	576.	,, ,,	Dr N., cut (N) 18. ii. 02	+ 30 D					* 120		·	
22	609.	,, ,,	B. C., cut (N) 4. iii. 02	+ D		* 120 d	* 200 tr	·	·	* ? 85	·	× 85
23	610.	,, ,,	C. S., cut (N) 10. iii. 02	+ D'		* 240 tr	* 200 tr	·	·	× 20	·	× 85
24	702.	,, ,,	Dr N., cut (N) 8. v. 02	+ D		/	* 200 tr ?	·		* 85 tr	·	* 85 tr

Primates — Insectivora — Carnivora

10 Bloods 10—16 were tested 8 Nov. 1902 with the 9 antisera noted.

Mammalia									Marsupialia	Aves				Reptilia		Amphibia	Crustacea
Pig	Llama	Hog-deer	Mexican-deer	Antelope	Sheep	Ox	Horse	Wallaby	(Wallaby)	Fowl	Ostrich	Fowl-egg	Emu-egg	Turtle	Alligator	Frog	Lobster
•		•	•	•	•	•	•		•	•		•	•	•			
tr •	•	+ tr	* 30 d	+ d	•	•	•		•	•		•	•	•			
×		\	•	\	•	•	•		•	\	\	\	\	\	\	\	\
tr		•	tr	•	•	•	•		•	•		•	•	•			
				•	•	•	•		•	•		•	•	•			
					•	•	•		•	•		•	•	•			
		•		•			•		•	•		•	•	•			
					•		•			•		•	•				
* 180 tr					•		•			•		•	•				
* 180 tr					•		• tr			•		•	•				
* 180 tr					•		• tr			•		•	•				
* 180 tr					•		•			•		•	•				
* 180 tr					•		• tr			•		•	•				
* 180 tr					•		• tr			•		•	•				
* 180 tr					•		•			•		•	•				
×	•	•		•	•	*				•	•	•	•	•	•	•	•
+ D	•	•	*	•	* 120	* 120	tr ?		•	•	•	•	•	•	•	•	•
* d	* ? 300	tr ?		tr	•				•	•	•	•	•	•	•	•	•
* 15 D	•		* ? tr	* ? 60	•	?			•	•	•	•	•	•	•	•	•
		•	* 20 tr	* ? 60	•		•		•	•	•	•	•	•	•	•	
•	•	•	•	•	•	•	•			•	•	•	•	•		•	
										•		•	•	•		•	
* 60	•	•	•	•	•	•	•		•	•				•			

Ungulata   Marsupialia

12 Solution less diluted than in tests 10 to 16.   **18, 20** Anti-pig serum very powerful.

Class MAMMALIA
1. *Order* PRIMATES

ANTISERA FOR.........

	No.	Ref.	Species / Source	Man	Chimpanzee	Ourang	Monkey	Hedgehog	Cat	Hyaena	Dog	Seal
1. *Suborder* ANTHROPOIDEA	25	778.	*Homo sapiens.* G. N., cut (N) 12. vi. 02	+ 15 D		* 120 d	·	·	·	·	·	·
Fam. Hominidae European	26	779.	,, ,, E. G., cut (Gardner) 13. v. 02	+ 15 D		* 240 tr	·	·	·	·	·	·
	27	785.	,, ,, C. C., wound (N) 16. v. 02	+ 30 tr		/	·	·	·	tr ?	·	·
	28	786.	,, ,, B. C., cut (N) 31. v. 02	+ 30 d		* 240 tr	* 30 tr	·	·	tr ?	·	·
Mongolian	29	327.	,, ,, Chinaman, beri-beri London (Daniels) 14. xii. 01	+ ɑ		* 240 tr	* 40		·	* 15 tr ?	·	·
	30	380.	,, ,, Chinaman, beri-beri Shanghai, China (Stanley) ca. 27. xi. 01	+ ɑ		* 240 tr	* 40		·	· tr ?	·	·
E. Indian	31	381.	,, ,, Punjaub Sikh Shanghai, China (Stanley) 20. x. 01	+ ɑ		* 240 tr	·		·	tr	·	·
	32	328.	,, ,, E. Indian, Punjaub London (Daniels) 13. xii. 01	+ ɑ		* 240 tr	* 40		·	* 15 tr ?	·	·
Negro	33	819.	,, ,, Negro Lagos, Africa (Strachan) 16. ii. 02	+ 30 d		* ? tr	× 30 tr ?	·	·	× 15 tr ?	* 15	* 35
	34	820.	,, ,, 16. ii. 02	+ 30 d		* ? tr	× 30	·	·	× 15 tr ?	* 15	* tr
	35	821.	,, ,, 16. ii. 02	+ 30 d		* ? tr	× 30	·	·	× 15 tr ?	* 15	· 35 tr
Fam. Simiidae	36	254.	*Simia satyrus* L. Ourang-outang Borneo (Z) 30. x. 01	+ D	+ 45 D	+ D	* 41 d	* 320	* 90	× 35		·
	37	297.	,, ,, (Grünbaum) 01, *fl.* & dry	+ D		* d	·	·	·	* ? 15 tr	· tr ?	
	38	906.	,, ,, (Z) 14. x. 02, *fl.*	+ D		+ i D	* 180 tr	·		* 180 tr		
	39	237.	*Anthropopithecus troglodytes.* Chimpanzee W. Africa (Z) 16. x. 01	+ D	+ 45 D	+ D	* 15 d	* 320	·	* 35 tr	·	* ? 45
	40	298.	*Anthropopithecus troglodytes* (Grünbaum) xi. 01, *fl.* and dry	+ D		* d	+ 20	·	·	* ? 15 tr	·	·
	41	329.	*Anthropopithecus troglodytes* (Grünbaum) 7. xii. 01	+ ɑ	+ 45 D	× 15 D	* 40 d	·	·	× 15 tr ?	·	·
	42	496.	*Anthropopithecus troglodytes* (Grünbaum) 12. ii. 02	+ D		+ 5 D	· tr	·	·	+ 45 tr	d	* 40
	43	299.	*Gorilla savagei.* Gorilla Equatorial Africa (Grünbaum) xi. 01	+ ɑ		· tr	* 40	·	·	* ? 80	·	·

Primates			Insectivora	Carnivora		

42 Anti-pig very powerful.

Mammalia									Aves				Reptilia		Amphibia	Crustacea
Pig	Llama	Hog-deer	Mexican-deer	Antelope	Sheep	Ox	Horse	Wallaby	Fowl	Ostrich	Fowl-egg	Emu-egg	Turtle	Alligator	Frog	Lobster
*60 tr	•	tr?	tr	tr	tr	tr	tr	•	•		•		•			
*60 tr	•	•	tr	tr	tr	tr	tr	•	•		•		•			
*40	•	•	tr	•	•	•	•		•	•	•		•	•		
*? 40	•	•	*240 tr	•	•	•	•	•	•		•		•			
×	•	•	•	•	•	•	•		•	•	•	•	•	•	•	•
•	•	•	•	*?	•	•	•		•	•	•	•	•	•	•	•
•	•	/	•	/	•	•	•		•	/	•	/	•	•	•	•
×	•	•	*65	•	•	•	•		•	•	•	•	•	•	•	•
×40	•	•	*50 tr			•	•	*240 tr	•	•	•					
×40	•	•	*45 tr			•	•	*240 tr?	•	•	•					
×40	•	•	*45 tr			•	•	*240 tr?		,	•					
+d	•	•	*90	•	•	•	•	•	•	•	•	•	•	•	•	•
•	•	•	•	•	•	•	•	•	•	•	•	•	•	•	•	•
• tr	•	•	•	•	•	•	*35 tr		•	•	•					
*d	•	•	*60	•	•	•	•	•	•	•	•	•	•	•	•	•
•	•	•	*60	•	•	•	•	•	•	•	•	•	•	•	•	•
×	•	•	•	•	•	•	•	•	•	•	•	•	•	•	•	•
+D	*300	d	*	×	*120 d	•	•	×	•	•	•	•	•	•		
•	•	•	•	•	•	•	•		•	•	•	•	•	•	•	

Ungulata   Marsupialia

Class MAMMALIA
1. *Order* PRIMATES

ANTISERA FOR.........

			Mammalia								
			Man	Chimpanzee	Ourang	Monkey	Hedgehog	Cat	Hyaena	Dog	Seal
1. *Suborder* ANTHROPOIDEA	44	364. *Cynocephalus mormon.* Mandrill W. Africa (Z) 10. xii. 01, *fl.*	+ D	·	* 15 d	× 40 tr	·	* 90	* 15		
***Fam.* Cercopithecidae**	45	501. *Cynocephalus sphinx* Linn. Guinea Baboon W. Africa (Z) 14. ii. 02, *fl.*	* d		+ 1 d	* 45 d	·	* 90	* 45 tr	d	·
	46	703. *Cynocephalus babuin* Desm. Baboon Africa (Lühe, Königsberg Z) 10.v.02, *fl.*	+ d	· 45 tr	* 15 d	× 45 d	·	* 90 tr	* 85 tr	·	* 85
	47	287. *Cynocephalus porcarius* Bodd. Chacma Baboon S. Africa (Z) 18. xi. 01	+ b	45 tr	* 15 d	+ 40 d	·	·	· tr	·	·
	48	784. *Cynocephalus porcarius* (Z) 4. vii. 02, d. intussusception, *fl.*	+ 30 D		+ 1 D	× 30 tr			tr		
	49	789. *Gelada Baboon,* d. of tuberculosis ? (New York Zoo., Langmann) 5. v. 02	·		* 120 d	× 30	·	·	tr	·	·
	50	98. *Macacus cynomolgus.* Macaque Monkey India (Rogers) 1. vii. 01	+ b		· d	+ 60				·	
	51	224. *Macacus cynomolgus* (Z) 21. viii. 01	+ b	·	* 15 d	* 10 tr	·	·	* 35 tr		·
	52	255. *Macacus ocreatus.* Ashy-black Macaque India (Z) 30. x. 01	+ D	*	× 15 d	* 10 tr	? 320	·	× 35 tr		·
	53	99. *Macacus assamiensis* Bhootan (Rogers) 1. vii. 01	+ b		· tr	+ 60	*? 45		*? 90		
	54	59. *Macacus rhesus.* Rhesus Monkey India (Z) 11. v. 01	+ b	+ b	+ 5 d	+ 60 d	·	·	·	·	
	55	97. *Macacus rhesus.* Rhesus Monkey (Rogers) 25. vi. 01	+ b	+ b	+ 5 D	+ 60	·		·	·	·
	56	124. *Macacus rhesus.* Rhesus Monkey (Z) 19. vii. 01	+ b	*	+ 5 d	+ 60 d	·	·	* 30 tr	·	·
	57	907. *Macacus rhesus.* Rhesus Monkey d. in Engl. (L. C.) 02, tuberculosis, septicemia; *fl.* ser.	+ 1 d			* 30 tr			* 180 tr		
	58	908. *Macacus rhesus.* Rhesus Monkey d. in Engl. (L. C.) 16. x. 02, dysentery; *fl.* ser.	+ 1 D			* 30 tr		·	* 180 tr		
	59	790. *Bonneted Macaque Monkey,* d. tuberculosis ? (New York Zoo., Langmann) ca. 14. vi. 02	·		· d	·	·	·	* ?	·	·
	60	118. *Cercopithecus melogenys.* Black-cheeked Monkey W. Africa (Z) 22. vii. 01	+ b	+ b	* 5 d	+ 60 d	·	·	× 30 d	·	·
	61	918. *Cercopithecus spec. ?* Monkey Brit. Centr. Africa (Dodds) p. 23. viii. 02	+ 180 tr			* 180 tr	·				
	62	58. *Cercopithecus lalandii* Is. Geoffr. Vervet Monkey S. Africa (Z) 15. v. 01.	+ b	·	* 120 d	60 d	·	·	·	·	·
			Primates				**Insectivora**	**Carnivora**			

44 Note time of reaction with anti-Mexican-deer.

45, 51, 52, 56, 60 Anti-pig serum very powerful.

48 Note cause of death.

49 Anti-human powerful; note negative results with Nos. 59 and 75. Filter-paper not mine, may be the cause of negative result? A second test with anti-monkey gave · tr? Also possibly over-exposed to sun.

Mammalia									Aves				Reptilia		Amphibia	Crustacea
Pig	Llama	Hog-deer	Mexican-deer	Antelope	Sheep	Ox	Horse	Wallaby	Fowl	Ostrich	Fowl-egg	Emu-egg	Turtle	Alligator	Frog	Lobster
+	·	·	+ 90 D	·	·	·	·	* 90	·	·	·	·	·	·	·	·
+ D	+ 45 tr	·	+ 90 d	·	* 120 d	·	* 90	* 75	·	·	·		·	·	·	·
* 60	·	·	* 90 d		·	·	* 90 d	* 75 tr	·	·	·		·		·	
·	·	·	·	·	·	·	·	·	·	·	·		·	·	·	
* 40	tr ?	·	cl			tr ?	tr	·		·	·					
·	·	·	·		·	tr ?	·	·		·						
·				·	·	·	·		·		·				·	
+ D	·	·	* 60	tr	·	*	·	·	·	·			·	·	·	
+ D	·	·	* 60	·	·	·	·	·	·	·	·		·	·	·	
·	·	·	·	·	·	·	·	·	·	·	·		·	·	·	
·	·	·	* 60 tr	·	·	·	·	·	·	·	·					
·				·	·	·	·	·	·	·					·	
· + D	·	·	* 60 tr	·	·	·	·	·	·	·	·	·	·	·	·	·
* 180 tr	·	·	·	·	·	·	·	·	·	·	·	·	·	·	·	·
· tr							* 35 tr	·	·	·	·	·			·	
·	·	·	·	·	·	·	·	·	·	·	·		·	·	·	
+ D	·	·	tr	* 30	·	·	·	·	·	·	·	·	·	·	·	·
·				·	·	·	·		·	·	·				·	·
·	·	·	·	·	·	·	·	·	·	·	·		·	·	·	·

Ungulata   Marsupialia

58 Great reaction with × man, doubtless due to dysentery causing blood concentration.

59 See note to 49.

61 Reaction retarded because of little going into solution.

Class **MAMMALIA**
1. *Order* **PRIMATES**

ANTISERA FOR........

				Man	Chimpanzee	Ourang	Monkey	Hedgehog	Cat	Hyaena	Dog	Seal
1. *Suborder* ANTHROPOIDEA	63	53.	*Cercopithecus campbelli* Waterh. Campbell's Monkey W. Africa (Z) 25. iv. 01	+ b	*	15 d	+ 60 d	•	•	* 15 tr?	•	•
Fam. Cercopithecidae	64	75.	*Cercopithecus diana.* Diana Monkey W. Africa (Z) 25. v. 01	+ b	*	× 5 d	+ 60 d	*? 320	•	* 30 tr	•	•
	65	606.	*Cercopithecus (Papio) hamadryas* (L.), juv. "Mantelpavian," hab. Arabia & Abyssinia (Lühe) 14. ii. 02	+ D	* 45 tr	* 15 d	+ 45 D	•	•	* 85	•	•
	66	286.	*Cercopithecus mona.* Mona Monkey W. Africa (Z) 20. xi. 02	+ b	* 45 d	* 120 d	+ 60 tr	•	•	tr	•	•
	67	515.	*Cercopithecus pygerythrus* F. Cuv. Mozambique Monkey E. Africa (Z) 14. ii. 02	+ D	• 45 d	* 15 d	× 45 D	•	•	* 45 tr	•	tr
	68	742.	*Cercopithecus petaurista* Schreb. Lesser White-nosed Monkey W. Africa (Z) 10. vi. 02 healthy, *fl.*	15 D	* 45 d		× 45 D	•	•	* 85 tr	•	•
	69	52.	*Cercopithecus patas.* Patas Monkey W. Africa (Z) 24. iv. 01	+ b	*	× 5 d	+ 60 d	•	•	•	•	•
	70	238.	*Cercopithecus patas* (Z) 15. x. 01	+ b	*	× 5 d	* 10	•	•	× 35 tr	•	•
	71	111.	*Cercopithecus callitrichus* Is. Geoffr. Green Monkey W. Africa (Z) 31. vii. 01	+ b	* 45 tr	* 5 d	+ 60 tr	*? 320	•	× 30 tr	•	•
	72	895.	*Cercopithecus callitrichus* (Z) 13. ix. 02, d. of pneumonia	+ 30 tr	•	* 5 d	+ 30 tr	•	•	• tr	•	•
	73	270.	*Cercopithecus callitrichus* (Z) 12. xi. 01	+ b	*	* 5 d	* 10 tr	•	•	tr?	•	•
	74	382.	"Small Chinese Monkey" Canton, China (Stanley, Shanghai) 4. xi. 01	+ b	•	* 5 d	40	•	•	* ? 15	•	•
	75	788.	Monkey, d. of enteritis ? (New York Zoo., Langmann) 30. iv. 02	•	•	• tr	30 tr	•	•	• tr	•	•
	76	271.	*Semnopithecus melalophos.* Simpae Monkey Sumatra (Z) 10. xi. 01	+ b	•	• tr	* 10 d	•	•	* 80 tr	•	•
	77	747.	*Semnopithecus rubicundus* Sarawak, Borneo (Hose) posted 10. iv. 02	+ 60 tr	•	• tr	* 45 tr	•	•	* ? 300 tr	•	tr
	78	100.	*Semnopithecus entellus* Sangur, India (Rogers) 1. vii. 01	+ b	•	• tr	+ 60	•	•	•	•	•
	79	146.	*Semnopithecus entellus* India (Copenhagen Zoo., Schierbeck) 23. ix. 01	+ b	•	* 15 d	* 60 tr	•	•	* 30 tr	•	•
Fam. Cebidae	80	88.	*Mycetes seniculus.* Red Howler Colombia (Z) 24. vi. 01	×	? 45	* 15 d	× 100 tr	*? 320	•	* 30 tr	•	•
	81	246.	*Uacaria rubiconda.* Red-faced Ouakari Upper Amazons (Z) 21. ix. 01	×	*	• d	+ 20 tr	•	•	* 25 tr	•	•
	82	288.	*Nyctipithecus trivirgatus.* Three-banded Douroucouli Guiana (Z) 20. xi. 01	+ b	•	* 15 d	? 60	•	•	* 15 tr	•	•
				Primates				Insectivora	Carnivora			

67, 70 Anti-pig very powerful.   **75** See note to No. 49.   **79, 81** Anti-pig powerful.

Pig	Llama	Hog-deer	Mexican-deer	Antelope	Sheep	Ox	Horse	Wallaby		Fowl	Ostrich	Fowl-egg	Emu-egg		Turtle	Alligator		Frog		Lobster
·	·	·	* 60 tr	·	·	·	·	·		·	·	·	·		·	·		·		·
·	·	·	* 100 tr	·	·	·	·	·		·	·	·	·		·	·		·		·
·	·	·	tr	·	·	·	·	tr ?		·	·	·	·		·	·		·		·
·	·	·	·	·	·	·	·			·	·	·	·		·	·		·		·
+D	· tr	·	*?	*	*d	d	·	·		·	·	·	·		·	·		·		·
·	*85 tr	*30 tr	*90 tr	*30 tr	*? 60	*? 60 tr	tr	·		·	·	·	·		·	·		·		·
·	·	·	*60 tr	·	·	·	·	·		·	·	·	·		·	·		·		·
+D	·	·	*60	·	·	·	·	·		·	·	·	·		·	·		·		·
*	·	·	*100 tr	*30	·	·	·	·		·	·	·	·		·	·		·		·
*30	·	·	·	·	·	*? 30	*? 30	·		·	·	·	·		·	·		·		·
·	·	·	·	·	·	·	·	·		·	*	·	·		·	·		·		·
·	·	·	·	·	·	·	·	·		·	·	·	·		·	·		·		·
·	·	·	·	·	·	·	·	·		·	·	·	·		·	·		·		·
·	·	·	·	·	·	·	·	·		·	·	·	·		·	·		·		·
· tr	·	·	+30	·	·	tr	·	·		·	·	·	·		·	·		·		·
·	·	·	·	·	·	·	·			·	·	·	·		·	·				
+d cl	·	*90 tr?	*60	·	·	×	·	·		·	·	·	·		·	·		·		·
*	·	·	· tr?	*?	·	·	·	·		·	·	·	·		·	·		·		·
+d	· tr	·	·	·	·	·	·	·		·	tr	·	·		·	·		·		·
·	·	·	*90	·	·	·	·	·		·	·	·	·		·	·		·		·

| | Ungulata | | | | | | | Marsupialia | | | | | | | | | | | | |

80 Gave · reaction with weak anti-human, which acted on Cercopithecidae.

15—2

Class MAMMALIA
1. Order PRIMATES

ANTISERA FOR.........

			Man	Chimpanzee	Ourang	Monkey	Hedgehog	Cat	Hyaena	Dog	Seal
1. Suborder ANTHROPOIDEA	**83**	480. *Chrysothrix sciurea* L. Squirrel Monkey Guiana (Z) 1. ii. 02	* D	. 45 tr	. d	. tr	.	/	/	.	/
Fam. Cebidae	**84**	584. *Chrysothrix sciurea* Pará, Brazil (Hagmann) 14. ii. 02	.	.	.	.	.	.	* 85	.	* 85
	85	697. *Ateles vellerosus.* Long-eared Spider-Monkey Brazil (Z) 11. iv. 02 (organs healthy) *fl.*	× d	* ? 45	.	* ? 45	.	.	. tr?	* tr?	.
	86	580. *Ateles ater* F. Cuv. Black-faced Spider-Monkey E. Peru (Z) 28. ii. 02	.	. 45 tr?	* 15 tr	? 45 tr?	.	.	. 45	.	.
	87	431. *Ateles geoffroyi* Kuhl. Black-handed Spider-Monkey C. America (Z) 28. xii. 01	× D	.	* 15 tr	* 260	.	.	* ? 15	.	* 55
	88	513. *Ateles geoffroyi* (Z) 11. ii. 02, *fl.*	.	. 45 tr	. tr	? 45 tr	.	.	.	.	. tr
	89	894. *Lagothrix humboldti* Geoffr. Humboldt's Lagothrix Upper Amazon (Z) 23. ix. 02, d. constipation	* 30 tr	.	× 5 d	* 30 tr	.	.	. tr?	.	.
	90	285. *Lagothrix humboldti* (Z) 20. xi. 01	+ b	.	* 15 tr	+ 60 tr	.	.	+ 60 tr	.	.
	91	119. *Cebus albifrons* Geoffr. White-fronted Capuchin S. America (Z) 25. vii. 01	*	.	. tr	* 60 tr	.	.	.	.	.
	92	481. *Cebus albifrons* (Z) 3. ii. 02	× D	.	. 15 tr	.	.	× tr	* 45	.	.
Fam. Hapalidae	**93**	226. *Hapale pygmoea.* Pigmy Marmoset Upper Amazon (Z) 23. ix. 01	.	* tr	.	.	.	.	. tr	.	.
	94	225. *Midas oedipus* L. Pinche Monkey Colombia (Z) 16. ix. 01	×	.	. tr	.	.	.	* 35 tr	.	.
	95	514. *Midas oedipus* (Z) 21. ii. 02	.	* 45 tr	. tr	* 45 tr	.	.	. tr	.	.
	96	296. *Hapale jacchus.* Marmoset (Brazenor) 11. xi. 01	*	*	.	.	.	.	.	.	.
2. Suborder LEMUROIDEA	**97**	227. *Lemur rufifrons.* Red-fronted Lemur Madagascar (Z) 26. ix. 01	.	.	.	.	* ? 320	.	× 35 tr	.	.
	98	241. *Lemur xanthomystax.* Yellow-cheeked Lemur Madagascar (Z) 23. x. 01	.	.	. tr	.	.	.	* 35	.	.
				Primates			Insectivora		Carnivora		

85 Note time of reaction with anti-Mexican-deer.

83, 86, 88, 91, 92, 93, 94, 95 Anti-pig powerful.

Pig	Llama	Hog-deer	Mexican-deer	Antelope	Sheep	Ox	Horse	Wallaby	Fowl	Ostrich	Fowl-egg	Emu-egg	Turtle	Alligator	Frog	Lobster
*D	/	tr	·	tr	·	*120	·	·	·	·	·	·	·	·	·	·
*180 d	·	·	·	·	·	·	·	·	·	·	·	·	·	·	·	·
·	·	·	×90 D	·	tr?	·	·	·	·	·	·	·	·	·	·	·
·D	·	·	·	·	·	·	·	·	·	·	·	·	·	·	·	·
×D	·	·	·d	·	·	·	·	tr	·	·	·	·	·	·	·	·
×D	·	·	·tr	·	·	·	·	·	·	·	·	·	·	·	·	·
·	·	·	·	·	·	·	*?30	·	·	·	·	·	·	·	·	·
·	tr?	·	*60	·	·	·	·	·	·	·	·	·	·	·	·	·
+D	·	·	·tr	·	·	·	·	·	·	·	·	·	·	·	·	·
*D	·	tr	·	tr	·	120	·	·	·	·	·	·	·	·	·	·
*D	·	·	·	·	·	×	·	·	·	·	·	·	·	·	·	·
*d	·	·	·	·	·	×	·	·	·	·	·	·	·	·	·	·
*D	·	·	·	·	·	d	·	·	·	·	·	·	·	·	·	·
·	·	·	·	·	·	·	·	·	·	·	·	·	·	·	·	·
+D	*?240 tr?	·	*60	tr?	·	×	·	·	·	·	·	·	·	·	·	·
*d	·	·	*60	·	·	·	·	·	·	·	·	·	·	·	·	·

Ungulata   Marsu-pialia

97, 98 Anti-pig very powerful. Very powerful anti-human had effect, but also on other mammalia.

Class MAMMALIA
2. Order CHIROPTERA
ANTISERA FOR.........

					Man	Monkey	Hedgehog	Cat	Hyaena	Dog	Seal
1. Suborder MEGACHIROPTERA		**99**	234.	*Pteropus medius.* Indian Fruit-Bat India (Z) 21. ix. 01	·	·	*? 320	·	*? 35	·	·
Fam. Pteropodidae		**100**	236.	*Pteropus medius* India (Z) 19. x. 01	·	·	·	·	* 35	·	·
		101	849.	*Rousettus amplexicaudatus* Geoff. Fruit-Bat Ceylon (Rothschild) 24. i. 02.	* 240	·	·	·	* ? 15 tr	* 240	·
2. Suborder MICROCHIROPTERA		**102**	259.	*Rhinolophus ferrum-equinum* England (Farren) x. 01	·	·	·	·	·	·	·
Section Vespertilionina		**103**	856.	*Rhinolophus minor* Temm. Okenawa, Japan (R) 15. iii. 02							
Fam. Rhinolophidae		**104**	809.	*Hipposideros diadema* Ceylon (R) 23. i. 02	·	·	·	·	·	·	·
Fam. Vespertilionidae		**105**	64.	*Plecotus auritus* L. Long-eared Bat Ireland (N) 7. iv. 01	·	·	* ? 320	·	tr	·	·
		106	394.	*Plecotus auritus* England (N) 12. xi. 01	·	·	·	·	* ? 90	·	·
		107	759.	*Plecotus auritus* (Brazenor) 5. v. 02	·	tr	·	·	tr	·	tr
		108	127.	*Scotophilus pipistrellus* England (R) 7. ix. 01	·	·	* ? 320	·	· tr	×	tr
		109	442.	*Vespertilio dorianus* Dob. Paraguay (F) 9. xii. 01	·	·	·	·	·	·	·
		110	441.	*Myotis nigricans* Wied. Paraguay (F) 11. xii. 01	·	＼	＼	·	＼	·	＼
		111	923.	*Myotis ruber* Paraguay (F) 21. ii. 02	* 180	·	·	·	·		
		112	928.	*Lasiurus borealis bonariensis* Paraguay (F) 1. v. 02	· tr	·	·	·	·		
Fam. Emballonuridae		**113**	445.	*Molossus rufus* Geoff. Paraguay (F) 16. xii. 01	·	· tr	·	·	* 45 tr	·	* 40
		114	374.	*Molossus cerastis* Thos., nov. spec. Paraguay (F) 2. xii. 01	·	·	·	·	* 15 tr?	·	·
		115	446.	*Molossus fosteri* Thos., nov. spec. Paraguay (F) 5. xii. 01	·	·	·	·	· tr	·	* 40
		116	373.	*Molossus temmincki* Lund. Paraguay (F) 27. xi. 01	·	·	·	·	* 95 tr?	·	·
		117	440.	*Molossus temmincki* (F) 17. xii. 01	d	·	·	·	·	·	tr?
		118	921.	*Molossus glaucinus* (?) Paraguay (F) 6. iii. 02	* ? 180	·	·	·	·		
		119	656.	*Nyctinomus laticaudatus* Geoff. Paraguay (F) 14. ii. 02	·	·	* 290	·	·	·	·
					Primates		Insectivora	Carnivora			

99, 100, 102 Anti-pig very powerful.

Mammalia									Aves				Reptilia		Amphibia	Crustacea
Pig	Llama	Hog-deer	Mexican-deer	Antelope	Sheep	Ox	Horse	Wallaby	Fowl	Ostrich	Fowl-egg	Emu-egg	Turtle	Alligator	Frog	Lobster
*d	*? 240	·	·	·	·	·	·	·	·	·	·	·	·	·	·	·
*d	·	·	·	·	·	·	·	·	·	·	·	·				
*? 240	·					*? 240	·		·		·					
*d		·	·	·	·	·	·	·	·	·	·	·	·	·	·	·
·	·		·			·	·	·			·			·		
·	·	·		·			·	·	·	·	·	·				·
·	·	·	·			·		·	·	·	·	·	·		·	
tr	tr?	·	·	·	·	·	·	·	·		·	·	tr?			·
·	·	/	·	/	·	·	·	·	·	/	·	/	·	·	·	·
*30 d	·	·	·	·	·	·	·	·	·	·	·	·	·		·	·
*? 60	·	\	·	·	·	·	·	·	·	·	·	·	·		·	
*25					·	·			·		·					
·					·	·			·							
*60 d	·	·	*	·	·		tr?	·	·	·	·	·	·		·	
·	·	·	*45	·	·	·	·	·	·	·	·	·	·	·	·	·
*30 d	·	·	*	·	·	·	·	·	·	·	·	·	·		·	·
·	·		*65	·	·	·	·	·	·	·	·	·	·		·	·
*30	tr?		*	·	·	·	·	·	·	·	·	·	·		·	·
·						·	·	·	·	·	·	·	·		·	
·	tr?	·	·	·			·	·	·	·	·	·	·	·	·	·

| | | Ungulata | | | | | | Marsu-pialia | | | | | | | | |

Class MAMMALIA
2. Order CHIROPTERA

ANTISERA FOR.........

			Man	Monkey	Hedgehog	Cat	Hyaena	Dog	Seal
2. Suborder MICROCHIROPTERA	**120**	375. *Hemiderma brevicauda* Wied. Paraguay (F) 25. xi. 01	.	.	.	.	* ? 15 tr?	.	.
Section Vespertilionina	**121**	657. *Glossophaga soricina* Pall. Paraguay (F) 26. xii. 01	.	.	.	.	.	.	* ? 85
Fam. Phyllostomatidae	**122**	377. *Sturnira lilium* Geoff. Paraguay (F) 27. xi. 01	.	.	.	.	.	.	.
	123	655. *Artibeus literatus* Licht. Paraguay (F) 8. ii. 02	.	.	.	.	.	.	.
	124	376. *Desmodus rotundus* Geoff. Paraguay (F) 25. xi. 01	.	.	.	.	* 15 tr	.	.
	125	443. *Pygoderma bilabiatum* Wagn. Paraguay (F) 19. xii. 01	.	.	.	.	.	.	* 40
	126	372. *Vampyrops lineatus* Geoff. Paraguay (F) 27. xi. 01	.	.	.	.	tr	.	.
	127	444. *Vampyrops lineatus* 5. xii. 01	.	.	.	* ?	.	.	.
Fam. ?	**128**	190. *Chiroptera* spec. ? "Short-eared Bat" England (Garrood) 22. viii. 01	.	.	tr?	.	tr	.	.

3. Order INSECTIVORA

			Man	Monkey	Hedgehog	Cat	Hyaena	Dog	Seal
Fam. Erinaceidae	**129**	133. *Erinaceus europaeus.* Hedgehog England (Garrood) 9. viii. 01	.		+ 30 D	.	* 30 tr	×	
	130	134. *Erinaceus europaeus* Denmark (Schierbeck) ix. 01	.	.	.				
	131	232. *Erinaceus europaeus* England (Z) 30. ix. 01	.	.	+ 30 d		* 35 tr	.	
	132	294. *Erinaceus europaeus* (Bird) 19. xi. 01	.	.	+ 45 d		* ? 45 tr	.	
	133	783. *Erinaceus europaeus* (N) 23. vi. 02, *fl.*	.	.	+ 5 D		tr	.	
	134	846. *Erinaceus europaeus* (R) 11. i. 02	.	.	+ 5 D		tr?	* 240	tr
	135	900. *Erinaceus europaeus* (4 sera, mixed) England (N) 19. viii. 02, *fl.* (used for treatment)			+ 3 D		d	.	.
Fam. Soricidae	**136**	857. *Crocidura coerulea.* Shrew Okenawa, Japan (R) 9. iii. 02	.	.		.	.	.	.
	137	858. *Crocidura coerulea* (R) 9. iii. 02				.	.	.	.
Fam. Talpidae	**138**	80. *Talpa europea* L. Mole England (N) 10. vi. 01	.			.	.	.	.
	139	460. *Talpa europea* (Bird) 19. xii. 01	.		tr?		.	.	.

Primates   Insecti-vora   Carnivora

128 Anti-pig very powerful, as also in tests of most of the preceding.

Pig	Llama	Hog-deer	Mexican-deer	Antelope	Sheep	Ox	Horse	Wallaby		Fowl	Ostrich	Fowl-egg	Emu-egg		Turtle	Alligator		Frog		Lobster
•	•	•	*45	•	•	•	•	•		•	•	•	•		•	•		•		•
•	•	•	•	•	•	•	•	•		•	•	•	•		•	•		•		•
•	•	•	*45	•	•	•	•	•		•	•	•	•		•	•		•		•
•	•	•	•	•	•	•	•	•		•	•	•	•		•	•		•		•
•	•	•	*45	•	•	•	•	•		•	•	•	•		•	•		•		•
*30 d	•	•	*	•	•	•	•	•		•	•	•	•		•	•		•		•
•	•	•	•	•	•	•	•	•		•	•	•	•		•	•		•		•
*30 d	•	•	*	•	•	•	•	•		•	•	•	•		•	•		•		•
×D	tr ?	•	•	•	•	•	•	•		•	•	•	•		•	•		•		•
	•							•												
*d	•	•	•	•				•		•	•	•	•		•	•		•		•
×d	*? 240	•	•	•	•	•	•	•		•	•	•	•		•	•		•		•
•	•	•	•	•	•	•	•	•		•	•	•	•		•	•		•		•
*40	tr ?	•	*45 tr	•			tr ?	tr ?		•	•	•	•		•	•		•		•
•	tr ?					•	•			•			•			•				
	tr																			
•	•					•	•			•		•			•	•				
•	•		•	•	•	•	•			•		•			•	•				
•	•	•	•	•	•	•	•	•		•	•	•	•		•	•		•		•
*60	•	•	*tr ?	•	•	•	tr ?	tr ?		•	•	•	•		•	•		•		•

              Ungulata   Marsupialia

130, 131   Anti-pig powerful.

Class **MAMMALIA**

3. *Order* **INSECTIVORA**

ANTISERA FOR.........

					Mammalia						
					Man	Monkey	Hedgehog	Cat	Hyaena	Dog	Seal
Fam. Talpidae	**140**	701.	*Talpa europea* Scotland (W. Evans) 10. v. 02		* ? 60	·	·	·	* 85 tr	·	·
	141	782.	*Talpa europea.* From 2 specimens England (N) 23. vi. 02, *fl.*		·	·	·	·	* 85 tr	·	* 85
Fam. Centetidae	**142**	247.	*Centetes ecaudatus* Schreb. Tenrec Madagascar (Z) 27. ix. 01		*	·	* 320	·	240 tr	·	·
	143	433.	*Centetes ecaudatus* 3. i. 02, *fl.*		d	* ? 260	·	·	* 15	·	·

4. *Order* **CARNIVORA**

					Man	Monkey	Hedgehog	Cat	Hyaena	Dog	Seal
1. *Suborder* **CARNIVORA VERA**	**144**	103.	*Felis tigris* L. Tiger India (Z) 13. viii. 01, *fl.*		·	·	·	× 75 D	+ 30 b	·	·
1. *Section* **Aeluroidea**	**145**	738.	*Felis tigris* Amur (Lühe, Königsberg Z.) 21. v.02, *fl.*		× 120 tr	·	·	* 30 tr	+ 85 D	* 120 tr	· tr
Fam. Felidae	**146**	300.	*Felis caracal* Schreb. Caracal S. Abyssinia (Z) ca. 15. xi. 01, *fl.*		×	·	·	+ 90 d	+ 5 d	·	·
	147	137.	*Felis* spec. Leopard Agra, India (Hankin) 12. vii. 01		·	·	tr	·	+ 30 d	·	× 30
	148	556.	*Felis pardus.* Indian Panther Bombay, India (Phipson) 14. i. 02		·	·	·	* 120 tr?	* 45 D	× 45 tr	· d
	149	509.	*Felis pardalis.* Ocelot Brazil (Hagmann) 30. i. 02		·	·	·	* d	+ 45 d	·	* 40 tr
	150	798.	*Felis pardalis* (Goeldi) 5. v. 02		·	tr?	·	* 45 tr	+ 5 D	·	·
	151	797.	*Felis onca.* Jaguar Brazil (Goeldi) 4. v. 02		·	tr?	·	* 45 tr	+ 5 D	·	·
	152	588.	*Felis* spec. California Lynx California (Robison) 30. xii. 01		* tr	·	·	× 120 tr	+ 85 D	* tr	× 85
	153	16.	*Felis domesticus.* Domestic Cat; scales England (N) 29. iii. 01		·	·	·	* 30 tr	+ 15 d	·	tr?
	154	17.	*Felis domesticus* 31. iii. 01			·	·	* 60 tr	+ 15 d	·	·
	155	29.	*Felis domesticus* 29. iii. 01		·	·	·	·	* 5 d	·	·
	156	30.	*Felis domesticus* 4. iv. 01		·		·	* 90	* 5 d	·	·
	157	31.	*Felis domesticus* 12. iii. 01		·	·	·	* 30	* 5 d	·	·
					Primates		Insecti-vora	Carnivora			

142 Anti-pig powerful.   **147** Went badly into solution, which explains · reaction with *weak* anti-cat.

Mammalia									Aves				Reptilia		Amphibia	Crustacea
Pig	Llama	Hog-deer	Mexican-deer	Antelope	Sheep	Ox	Horse	Wallaby	Fowl	Ostrich	Fowl-egg	Emu-egg	Turtle	Alligator	Frog	Lobster
* 60	•	•	•	•	•	•	•	•	•		•	•	•		•	
•	* ? 300	•	* 45 tr	•				tr?	•	•	•	•	•	•		•
+ D	•	•	•	•	•	•	•	•	•	tr	•	•	•	•	•	•
•	•	•	× 90	•	•	•	•	•	•	•	•	•	•	•	•	•
•	•	•	* 90 d	•	•	•	•	* 75	•	•	•	•	•	•	•	•
* 60	tr	•	* 30	•	•	tr	•	•	•	•	•	•	•	•	•	•
•	•	•	* 90 d	•	•	•	•	× 90	•	•	•	•	•	•	•	•
•	tr	•	•	•	•	•	•	•	•	•	•	•	•	•	•	•
* cl	•	* ? 75	tr	•	* 180 tr?	* ? 180	•	•	•	•	•	•	•	•	•	•
* d	•	•	•	•	•	•	•	•	•	•	•	•	•	•	•	•
× 40	•	•	* 45 tr	•	•	•	•	•	•	•	•	•	•	•	•	•
* 40	•	•	* ? 45	•	•	•	•	•	•	•	•	•	•	•	•	•
•	•	•	•	•	•	•	•	•	•	•	•	•	•	•	•	•
× 30 d	•	•	•	•	•	•	•	•	•	•	•	•	•	•	•	•
•	•	•	•	•	•	•	•	•	•	•	•	•	•	•	•	•
* ? tr	•	•	•	•	•	•	•	•	•	•	•	•	•	•	•	•
* tr	•	•	•	•	•	•	•	•	•	•	•	•	•	•	•	•
× tr	•	•	•	•	•	•	•	tr	•	•	•	•	•	•	•	•

Ungulata (Pig–Horse)   Marsupialia (Wallaby)

155 Dried over a year and seven months when tested, dissolved with difficulty.

Class MAMMALIA
4. Order CARNIVORA

ANTISERA FOR........

			Species / locality / date	Man	Monkey	Hedgehog	Cat	Hyaena	Dog	Seal
1. Suborder **CARNIVORA VERA** **1. Section** **Aeluroidea** **Fam.** **Felidae**	**158**	32.	Felis domesticus 4. iv. 01. Scales	·	·		*90		·	
	159	135.	Felis domesticus Seeland (Schierbeck) ix. 01	×	·	*320	×60 tr	+30 d	×	×30 tr
	160	317.	Felis domesticus England (R) 2. xii. 01	×	·	·	*5	*15 tr	·	· tr
	161	395.	Felis domesticus (N) 28. x. 01	×	*? 260	·	×30 tr	×15 d	*tr	·
	162	739.	Felis domesticus 8. v. 02, fl.	+120 d	·	·	+30 d	+85 D	+120 tr	·
Fam. **Viverridae**	**163**	493.	Viverra civetta Schreb. African Civet Cat Africa (Z) 8. ii. 02	·	·	·	*	+45	*	*75
	164	569.	"Civet Cat" or "Skunk" Orange River, S. Africa (Parkinson) 10. xii. 01	·	·	·	·	*120 tr	·	*40 tr
	165	362.	Genetta tigrina. Blotched Cenet Africa (Z) 9. xii. 01	·	·	*45	·	+15	·	·
	166	612.	Genetta spec. nr Lake Nyassa, C. Africa (Dodds) p. 8. ii. 02	·	·	·	·	*85	·	·
	167	572.	Paradoxurus grayi. Himalayan Paradoxure India (Calcutta Zoo., Rogers) received 21. ii. 02	*? 90	·	·	tr	×45 d	*30	×40
	168	748.	Paradoxurus hermaphroditus. "Munsang" Sarawak, Borneo (Hose) posted 10. iv. 02	*60 tr	tr	·	·	· tr	·	·
	169	86.	Paradoxurus niger Desm. Common Paradoxure Java (Z) 20. vi. 01	·	·	·	·	·	·	·
	170	517.	Paradoxurus niger India (Z) 13. ii. 02	*d	·	·	×	+45 tr	*	·
	172	84.	Herpestes griseus Geoffr. Grey Ichneumon India (Z) 15. vi. 01	·	·	·	·	×30 tr	·	·
	173	89.	Herpestes pulverulentes Wagner. Dusty Ichneumon S. Africa (Z) 24. vi. 01	·	·	·	·	*30 tr	·	·
	174	159.	Herpestes pulverulentes 6. viii. 01	·	·	·	·	*35 d	·	·
	175	105.	Cynictis penicillata. Levaillant's Cynictis S. Africa (Z) 13. viii. 01	·	·	·	·	·	·	·
	176	60.	Suricata tetradactyla Schreb. Suricate S. Africa (Z) 6. v. 01	·	·	·	·	*15 d	·	20 tr
	177	102.	Suricata tetradactyla 10. viii. 01	·	·	*? 320	·	+30	·	·
	178	905.	Suricata tetradactyla 17. x. 02 pleurisy, fl.	*180	·	·		+10 tr		
Fam. **Proteleidae**	**179**	494.	Proteles cristatus Sparrm. Aard Wolf S. Africa (Z) 10. ii. 02	·	·	·	·	+45 D	·	*75

Primates	Insectivora	Carnivora			

159, 160 Anti-human and anti-pig powerful.

	Mammalia									Aves				Reptilia		Amphibia	Crustacea
Pig	Llama	Hog-deer	Mexican-deer	Antelope	Sheep	Ox	Horse	Wallaby		Fowl	Ostrich	Fowl-egg	Emu-egg	Turtle	Alligator	Frog	Lobster
·	·	·	·	·	·	·	·	·		·	·	·	·	·	·	·	·
d	·	*	·	·	·	·	·	·		·	·	·	·	·	·	·	·
×D	·	tr	·	tr ?	·	·	·	·		·	·	·	·	·	·	·	·
×	·	·	·	·	·	·	·	·		·	·	·	·	·	·	·	·
·	·	·	×265	·	·	·	·	·		·	·	·	·	·	·	·	·
×60	* ? 300 tr ?	·	* 30 tr	110	110	* 120 tr	* 30	·		·	·	·	·	·	·	·	·
d *	* 300 tr	tr	tr	·	·	·	·	*?		·	·	·	·	·	·	·	·
×	·	·	·	·	·	·	·	·		·	·	·	·	·	·	·	·
·	·	·	·	·	·	·	·	·		·	·	·	·	·	·	·	·
* 180 cl	tr	* 75 tr	* 120 tr	60	·	* 180	·	·		·	·	·	·	·	·	·	·
tr	·	× 30	·	·	·	tr	00	·		·	·	·	·	·	·	·	·
*	·	·	·	·	·	·	·	·		·	·	·	·	·	·	·	·
×D	·	×d	*	*	·	*d tr ?	·	·		·	·	·	·	·	·	·	·
*	·	·	* 60 cl	·	·	·	·	·		·	·	·	·	·	·	·	·
*	·	·	·	·	·	·	·	·		·	·	·	·	·	·	·	·
*d	·	·	·	·	·	·	·	·		·	·	·	·	·	·	·	·
·	·	·	tr	·	·	·	·	·		·	·	·	·	·	·	·	·
·	·	·	·	·	·	·	·	·		·	·	·	·	·	·	·	·
*	·	·	* ? 100	·	·	·	·	·		·	·	·	·	·	·	·	·
* 180 tr	·	·	·	·	·	·	·	·		·	·	·	·	·	·	·	·
*d	·	tr	·	·	·	·	·	·		·	·	·	·	·	·	·	·

Ungulata   Marsupialia

170 Anti-pig powerful.

237

Class MAMMALIA
4. Order CARNIVORA

ANTISERA FOR.........

	No.		Species / locality	Man	Monkey	Hedgehog	Cat	Hyaena	Dog	Seal
1. Suborder **CARNIVORA VERA**	180	284.	*Hyaena striata* Zimm. Striped Hyaena. Africa (Flower, Cairo) 13. ix. 01	·	·	·	* 80	+ 15 D	·	·
1. Section **Aeluroidea** *Fam.* **Hyaenidae**	181	855.	*Hyaena striata* N. Africa (Z) 31. vii. 02 diseased, *fl.*	·	·	·	+ 240 cl	+ 2 D	* 30	* 20 tr
2. Section **Cynoidea** *Fam.* **Canidae**	182	695.	*Canis familiaris.* Dog. England (N) 27. iv. 02, filtered; *fl.*	+ d	·	·	/	× 85 tr cl	+ D	tr
	183	15.	*Canis familiaris.* Terrier Dog; scales. 6. iii. 01	·	·	·	·	+ 15 d	+ D	tr ?
	184	780.	*Canis familiaris.* Collie Dog; aged. (Z) 19. vi. 02, *fl.*	·	·	·	* 240 tr	* 300 tr	+ 15 D	·
	185	397.	*Canis familiaris.* Mongrel. (N) 2. xii. 01	+ b	·	·	·	× 15 d	+ D	* 55
	186	330.	*Canis familiaris.* 10. xii. 01	×	·	·	·	× 15 tr ?	+ D	·
	187	727.	*Canis.* Pariah Dog. Bhadarwa, Kashmir (Donald) 13. ii. 02	·	·	·	·	· tr ?	* 5 tr	·
	188	289.	*Canis lupus.* Common Wolf. Europe (Z) 18. xi. 01	×	·	·	·	* 80	×	·
	189	890.	*Canis lupus.* Chitral, India (Fulton) 28. v. 02	·	·	·	·	· tr	·	× 15 tr
	190	794.	*Canis* spec. Wolf; young. ? (New York Zoo., Langmann) posted 14. vi. 02	* 30	* 240 tr ?	·	cl	× 5 d	+ 15 D	* 35 tr ?
	191	891.	*Canis lupus* ?. Wolf, 1 month old. Chitral, India (Fulton) about v. 02	·	·	·	·	·	·	·
	192	589.	*Canis latrans.* Coyote. California (Robison) 4. ii. 02	tr	·	·	* 240 tr	* 290 tr	+ D	·
	193	272.	*Canis anthus.* Jackal. N. Africa (Z) 3. xi. 01	*	·	·	·	× 80 tr	+ D	·
	194	228.	*Canis aureus.* Common Jackal. India (Z) 10. ix. 01	* ?	·	* 320	·	+ 35 d	+ D	+ 30
	195	359.	*Canis aureus.* (Z) 3. xii. 01	·	·	* ? 45	·	+ 15	+ D	·
	196	718.	*Canis aureus.* Bhadarwa, Kashmir (Donald) 11. ii. 02	·	·	·	·	tr	tr	·
	197	85.	*Canis mesomelas* Schreb. Black-backed Jackal. S. Africa (Z) 18. vi. 01	·	·	·	·	+ 30 d	+ D	* 30
	198	283.	*Canis vulpes* L. Common Fox. England (Farren) 19. xi. 01	·	·	·	·	·	+ D	·
	199	385.	*Canis vulpes.* xi. 01	·	·	·	·	× 15 d	×	* 55
				Primates	Insectivora		Carnivora			

180 Sample 11 months old on paper when tested with anti-hyaena.

181 Fluid serum 3 months old when tested by anti-hyaena, etc. This serum used for treating the rabbit which yielded the anti-serum.

Mammalia									Aves				Reptilia		Amphibia	Crustacea
Pig	Llama	Hog-deer	Mexican-deer	Antelope	Sheep	Ox	Horse	Wallaby	Fowl	Ostrich	Fowl-egg	Emu-egg	Turtle	Alligator	Frog	Lobster
·	·	·	·	·	·	·	·	·	·	·	·	·	·	·	·	·
*30	tr ?				·				·						·	
/	·	·	/	*?/tr?	*tr	/	·	/								
*30 tr	·	·	·	·	·	tr	·		·				·			
×40	·	·	×45 tr					· tr	·	·	·					
·			*265	·	·	·	·	·					·	·	·	·
×	·	·	·	·	·	·	·		·	·	*?	·	·	·	·	·
·	·	·	·	·	·	·	·		·	·	·	·	·	·	·	·
·	·	·	·	·	·	·	·	·	·	·	·	·	·	·	·	·
*40		·	*240 tr	·	tr	·				*30	·	·			·	·
·	·	·	·	·	·	·	·	·	·	·	·	·	·	·	·	·
·	·	·	·	·	·	·	·	·	·	·	·	·	·	·	·	·
×d	*240 tr	·	·	·	·	×	·		·	·	·	·	·	·	·	·
×	·	·	·	·	·	·	·		·	·	·	·	·	·	·	·
·	·	·	·	·	·	·	·	·	·	·	·	·	·	·	·	·
·	*? 240	·	*120 tr?		·	·	·	·	·	·	·	·	·	·	·	·
·																·
·	·	·	·	·	·	·	·	·	·	·	·	·	·	·	·	·

Ungulata   Marsupialia

191 Foam-test doubtful, no reactions.

Class MAMMALIA
4. Order CARNIVORA

ANTISERA FOR.........

			Mammalia						
			Man	Monkey	Hedgehog	Cat	Hyaena	Dog	Seal
1. Suborder CARNIVORA VERA									
2. Section Cynoidea	**200**	386. *Canis vulpes* xii. 01	*	·	/	·	/	×	/
Fam. Canidae	**201**	675. *Canis vulpes* Mecklenburg (Kuse) 16. xii. 02	+60 d	·		·	×85 d	+D	*85 tr
	202	498. *Canis pallidus.* Pale Fennec Fox Soudan (Z) 4. ii. 02	·	·		·	×45 tr	*d	×40 tr
	203	242. *Canis cerdo.* Fennec Fox N. Africa (Z) 23. x. 01	*	·		·	+35 d	+b	×30
	204	654. *Canis azarae* Paraquay (F) 26. i. 02	·	·		·	·tr	·	*85 tr
	205	256. *Canis procyonides* Japan (Z) 28. x. 01	·	·		*?320	*35 d	+b	*45
	205a	934. *Canis pallipes* Sykes. Indian Wolf India (Z) 23. x. 02, *fl. ser.*	*15 tr	·	·		+5 d		·
3. Section Arctoidea									
Fam. Ursidae	**206**	76. *Ursus americanus.* Black Bear N. America (Z) 28. v. 01	·	·		·	·	·	·
	207	435. *Ursus arctos* Linn. Brown Bear N. Europe (Z) 14. i. 02, *fl.*	*d	·		*?260	+55 d	+d	*55 tr?
	208	367. *Ursus tibetanus* F. Cuv. Himalayan Bear E. Asia (Z) 27. xii. 01, *fl.*	·	·		*90 tr	+15 d	*d	·
	209	885. *Ursus tibetanus* Bhadarwa, Kashmir State (Donald) 4. vi. 02	·	·		·	*15 tr		×15 d
Fam. Procyonidae	**210**	634. *Ælurus fulgens* F. Cuv. Panda Nepal (Z) 23. iii. 02, d. of bronchitis	·	*?200		·	+85 d	·	×85 tr
	211	366. *Procyon lotor* Linn. Raccoon N. America (Z) 27. xii. 01, *fl.*	·	·		×90		*d	
	212	508. *Nasua socialis.* Coati Brazil (Hagmann) 29. i. 02	·	·		·	*45 tr	*	·tr
	213	273. *Nasua rufa.* Ring-tailed Coati S. America (Z) 5. xi. 01	·	·		·	+80 tr	·	·
	214	360. *Nasua rufa* 6. xii. 01	·	·		·	+90	·	·
	215	781. *Nasua rufa* 17. vi. 02, killed fighting	·	·		·	*300 tr	*60	tr
	216	577. *Nasua spec.?* "Coati mundi" Brazil (Tuckett) recd. 1. iii. 02	·	·		*120	×45 tr cl	?75	*40 tr
Fam. Mustelidae	**217**	114. *Lutra vulgaris* Erxl. Common Otter England (Z) 30. vii. 01	*	\	·	\	×30 tr	·	·
			Primates	Insectivora		Carnivora			

205 *a* This blood tested separately 29. xi. 02. Note degrees of reaction and time differences. Anti-man much more powerful than anti-monkey.

Pig	Llama	Hog-deer	Mexican-deer	Antelope	Sheep	Ox	Horse	Wallaby		Fowl	Ostrich	Fowl-egg	Emu-egg		Turtle	Alligator		Frog		Lobster
·	/	·	·	·	·	·	·	·		·	·	·	·		·	·		·		·
·	·	·	·	·	* ? 60	* 60	·	·		·	·	·	·		·	·		·		·
× d	·	tr	* ?	·	d	·	·	×		·	·	·	·		·	·		·		·
× d	·	·	·	·	·	·	·	·		·	·	·	·		·	·		·		·
·	·	·	·	·	·	·	·	·		* 60 tr ?	·	·	·		·	·		·		·
* d	·	·	·	·	·	·	·	·		·	·	·	·		·	·		·		·
× 15 d	·	·	·	·	·	·	·	·		·	·	·	·		·	·		·		·
·	·	·	* 100 d	·	·	·	·	·		·	·	·	·		·	·		·		·
×	·	·	* 90	·	·	·	·	·		·	·	·	·		·	·		·		·
×	·	·	× 90 d	·	·	·	·	* ?		·	·	·	·		·	·		·		·
·	·	·	·	·	·	·	·	·		·	·	·	·		·	·		·		·
·	* ? 85	·	·	·	·	·	·	·		·	·	·	·		·	·		·		·
×	·	·	× 90 tr	·	·	·	·	* 75		·	·	·	·		·	·		·		·
·	× d	·	·	·	·	·	·	·		·	·	·	·		·	·		·		·
·	·	·	* 60	·	·	·	·	·		·	·	·	·		·	·		·		·
×	·	·	·	·	·	·	·	·		·	·	·	* ?		·	·		·		·
·	tr ?	·	* 45 tr	·	·	·	·	tr		·	·	·	·		·	·		·		·
× D	·	* ? 75	* 20 tr	* ? 60	* ? 180	* ?	·	·		·	·	·	·		·	·		·		·
× D	·	·	\	·	·	·	·	\		·	·	·	·		·	·		·		·

| Ungulata | | | | | | | | Marsupialia | | | | | | | | | | | | |

Class MAMMALIA
4. Order CARNIVORA
ANTISERA FOR........

1. *Suborder* **CARNIVORA VERA**

3. *Section* **Arctoidea**

Fam. **Mustelidae**

No.	Cat.	Species	Man	Monkey	Hedgehog	Cat	Hyaena	Dog	Seal
218	257.	*Lutra vulgaris* (Farren) x. 01	·		? 320	·	* 35 tr	·	·
219	499.	*Lutra vulgaris* Ireland (Scharff) recd. 12. ii. 02	·	· tr ?	·	* 75 tr	* 75 tr	+ D	·
220	579.	*Lutra vulgaris* England (Z) 4. iii. 02, *fl.*	·	\	\	· tr	\	·	\
221	704.	*Meles taxus.* Badger England (Cropper) 10. v. 02	* 60	·	·	* 30	× 85 tr	·	· tr
222	758.	*Meles taxus* (Brazenor) 25. iv. 02	* 60 tr	·	·	·	× 85 tr	·	· tr
223	245.	*Ictonyx zorilla* Thunb. Cape Zorilla S. Africa (Z) 24. viii. 01	·	·	* 320	·	* 35 tr	·	× 30
224	361.	*Ictonyx zorilla* 30. xi. 01	·	·	\	·	\		\
225	795.	"Sand Badger" China (New York Zoo., Langmann) p. 14. vi. 02	·	·	tr ?	·	+ 5 d	·	· tr
226	22.	*Mustela putorius.* Pole-cat England (N) 25. iv. 01	·	·	·	tr ?	· 20 d	·	· 20 d
227	113.	*Mustela putorius* Germany (Kuse) 7. viii. 01	·	· /	·	· /	× 350 tr	·	× 30 d
228	396.	*Mustela putorius* England (Z) 16. v. 01	·	·	* ? 45	·	× 15 tr	* tr	·
229	630.	*Mustela putorius.* Ferret (Brazenor) 4. ii. 02	·	·	·	·	· tr	·	* 85
230	274.	*Mustela martes.* Pine Marten England (Z) 10. xi. 01	·	·	\ ·	·	\ * 80	·	\
231	516.	*Mustela foina* Erxl. Beech Marten Russia (Z) 18. ii. 02	* d	·	* 60	*	× 45 tr	× tr	× 40 d
232	315.	*Mustela erminea.* Stoat England (R) 9. xi. 01	·	·	·	·	tr ?		
233	354.	*Mustela erminea* (Brazenor) 9. xii. 01	·	·	·	·	+ 15 d	·	* 55 tr
234	355.	*Mustela erminea* Canada (Hedley) 30. xi. 01	·	·	\	·	\	·	\
235	629.	*Mustela vulgaris.* Common Weasel England (Brazenor) 11. i. 02	·	· tr	·	·	· tr	·	* 85 tr
236	847.	*Mustela vulgaris* (R) 14. iv. 02	·	·	·	·	* 15 d	* 240	* 35 d
237	899.	*Mustela* spec. ? Weasel Bhadarwa, Kashmir State (Donald) 4. iv. 02	·	·	·	·	* tr ?	·	·

Primates | Insectivora | Carnivora

222, 229, 235 · with anti-reindeer.

Pig	Llama	Hog-deer	Mexican-deer	Antelope	Sheep	Ox	Horse	Wallaby	Fowl	Ostrich	Fowl-egg	Emu-egg	Turtle	Alligator	Frog	Lobster
*d	•	•	•	•				•	•	•	•	•	•	•	•	•
*cl	•	•	•	•	d	•	.	tr	•	•	•	•	•	•		
•	\	•	•	•	•	•	•	•								
*60	•	•	*30	•	•	•	•	•	•	•	•	•	•	•		•
tr	tr ?	*60 tr		tr ?		tr	•	•	•	•	•	•	•			•
*d	tr	•						•	•	tr	•		•			•
×	\	•				•		•	•	•	•		•			
•	•	•	•	•	tr ?			•	•	•	•	•	•			
tr	•	•	•					•			•				•	•
•	•	•	/	•	•	•	•	/	•	•	•	•	•	•	•	•
•	•	•	•	•	•	•	•	•	•	•	•	•	•		•	•
•	*? 300 \ •	•	•	•	•	•	•	•	•	•	•	•	•	•	•	•
×D tr	*45 tr	×d	*	×	*240 d	•	•	*?	•	•	•	•	•	•	•	•
•	•	•	•	•	•	•	•		•	•	•	•	•	•	•	•
×	•	•	•	•	•	•			•	•	•	•	•	•	•	•
•	\	•	•	•	•	•	•		•	•	•	•	•	•	•	•
•	•	•	•	•	•	•		•	•	•	•	•	•		•	
•	•	•	•	•	•	•	•		•	•	•	•	•			
•	•					•			•	•	•	•	•			

| | Ungulata | | Marsu-pialia | | | | | | | | | | | | | |

233 *? 40 with anti-reindeer.

Class MAMMALIA
4. Order CARNIVORA
ANTISERA FOR........

				Man	Monkey	Hedgehog	Cat	Hyaena	Dog	Seal
								Mammalia		
2. *Suborder* PINNIPEDIA *Fam.* Otariidae	**238**	853.	*Otaria californiana.* California Sea-lion N. Pacific Ocean (Z) 24. vii. 02, marasmus, *fl.*	·	·	· tr ?	·	· 20 D	* 240	* 20 d
	239	910.	*Otaria californiana* (Washington D. C., Zoo.) 2. x. 01, pneumonia	* 180	·	·	·	× 10 tr		
	240	791.	*Otaria* spec. Sea-lion, d. of pneumonia ? (New York Zoo., Langmann) 26. iv. 02	·	* 240 tr	·	·	× 5 tr	·	* 35 tr
Fam. Phocidae	**241**	500.	*Halichoerus grypus.* Gray Seal Ireland (Scharff) 14. xi. 01	·	·	·	·	* 110 tr	·	+ 40 tr
	242	854.	*Phoca vitulina* L. Common Seal, d. young British Seas (Z) 31. vii. 02, *fl.*, no dis.	·	·	· tr ?	·	· 20 D	* 240	* 20 d

5. Order RODENTIA

				Man	Monkey	Hedgehog	Cat	Hyaena	Dog	Seal
Fam. Sciuridae	**243**	24.	*Sciurus vulgaris.* Common Squirrel England (Lane) 22. iv. 01	·	· tr ?	·	·	·	·	· tr
	244	130.	*Sciurus vulgaris* Germany (N) 2. ix. 01	×	·	·	·	· tr ?	·	·
	245	131.	*Sciurus vulgaris* 3. ix. 01	·	·	·	·	·	·	·
	246	132.	*Sciurus vulgaris* (Schlieffen) 16. ix. 01	·	·	·	·	·	·	·
	247	140.	*Sciurus vulgaris* (Kuse) 7. viii. 01	·	·	* ? 320	·	* ? tr	·	·
	248	258.	*Sciurus vulgaris* England (Farren) vii. 01	·	·	·	·	· tr ?	·	·
	249	356.	*Sciurus vulgaris* (Brazenor) 25. xi. 01	·	·	·	·	* ? tr ?	·	· tr ?
	250	808.	*Sciurus tristriatus.* Squirrel Kandy, Ceylon (R) 24. i. 02	·	·	·	·	·	* ? 60	·
	251	749.	*Sciurus lowii.* Squirrel Sarawak, Borneo (Hose) p. 10. iv. 02	* 40 tr	·	·	·	·	·	·
	252	571.	*Sciurus giganteus.* Large Indian Squirrel Calcutta, India (Rogers) 14. xi. 01	·	·	·	·	·	·	·
	253	557.	*Sciurus palmarum.* Palm Squirrel Ahmedabad, W. India (Mason) 27. i. 02	* 75	·	·	·	·	·	·
	254	715.	*Pteromys inornatus.* Red Flying Squirrel Bhadarwa, Kashmir (Donald) 2. iii. 02	* 60	·	·	·	* 85 tr	·	·
	255	292.	*Sciuropterus fimbriatus.* Gray Flying Squirrel Chitral, India (Z) 18. xi. 01	×	·	·	·	* ? 15 tr	·	·
	256	231.	*Cynomys ludovicianus* Wagn. Prairie Marmot N. America (Z) 13. ix. 01	·	·	·	·	* 35 tr	·	·

Primates   Insectivora   Carnivora

242 This serum used for treating the rabbit which yielded the anti-seal serum, which was feeble.

Pig	Llama	Hog-deer	Mexican-deer	Antelope	Sheep	Ox	Horse	Wallaby	Fowl	Ostrich	Fowl-egg	Emu-egg	Turtle	Alligator	Frog		Lobster
*240	tr ?			•					•		•						
*180 tr •		•	•	•		•	•	•	•	•	•						
*120 tr	tr	tr		•	d	•	•	•	•	•	•	•	•		•		
*? 240	tr ?			•					•		•						
tr ?	•	•	•	•	×	×	•	•	•	•	•	•	•				
*	•	•	tr	•		•	•	•	•	•	•	•	•		•		•
*	•	•	•	•		•		•	•	•	•	•	•		•		•
* tr	•	•	•	•			•	•	•	•	•	•	•		•		•
*	*? tr	•	•	•		•	•	•	•	•	•	•	•		•		•
•	•	•	•	•		•	•	•	•	•	•	•	•		•		•
•	tr ?	•	•	•		•	•	•	•	•	•	•	•				
•	•	•	•	•		•	•	•	•	•	•	•	•				
tr	•	•	•	•		tr	•	•	•				•				
*	•	•	•	*20		•	•	•	•	•	•	•	•				
*60	tr	•	*? 120 tr ?	•		•	•	•	•	•	•	•	•		•		
•	•	•	*60	•		•	•	•	•	•	•	•	•	•	•		•
* tr	•	•	•	•			•	•	•	•	•	•	•	•		•	•

Ungulata   Marsupialia

Class MAMMALIA
5. *Order* RODENTIA

ANTISERA FOR.........

			Man	Monkey	Hedgehog	Cat	Hyaena	Dog	Seal
Fam. Sciuridae	**257**	518. *Cynomys ludovicianus* Ord. Prairie Marmot N. America (Z) 17. ii. 02	*	·	·	*	× 45 tr	×	·
Section **Myomorpha**									
Fam. Myoxidae	**258**	340. *Myoxus avellanarius* L. Dormouse England (R) 14. xi. 01	·	·	tr ?	·	tr ?	·	tr ?
Fam. Muridae	**259**	104. *Gerbillus shawi* Algeria (Z) 10. viii. 01	·					·	
	260	719. *Gerbillus indicus.* Indian Gerbille Bombay Presidency (Phipson) 11. ii. 02	* tr	·	·	·	tr ?	·	
	261	293. *Arvicola pratensis.* Red Field Vole England (Bird) 8. xi. 01	·	·	·	·	· d	·	
	262	260. *Hypudaeus glareolus* Schr. Bank Vole England (R) 3. xi. 01	·	·	/	·	/	·	/
	263	295. *Hypudaeus glareolus* (R) 5. xi. 01	·	·	·	·	* ? 15 tr ?	·	tr ?
	264	636. *Mus minutus (messorius).* Harvest Mouse England (Bird) 17. ii. 02	·	·	·	·	·	·	
	265	473. *Mus minutus (messorius)* 27. i. 02	·	·	·	·	·	·	* ? 40
	266	337. *Mus sylvaticus.* Long-tailed Field Mouse England (Bird) 25. xi. 01	·	·	·	·	* ? 15 tr	·	·
	267	357. *Mus sylvaticus* (R) 4. xii. 01	·	·	·	·	* ? 55 tr ?	·	·
	268	387. *Mus musculus* L. Common House Mouse England (N) 25. xi. 01	·	·	·	·	·	·	·
	269	388. *Mus musculus* 1. i. 02	·	·	·	·	·	·	·
	270	392. *Mus musculus* 28. x. 01	·	·	·	·	·	·	·
	271	393. *Mus musculus* 29. x. 01	·	·	·	·	·	·	·
	272	423. *Mus musculus* (Mitchell) ix. 02	·	·	·	·	·	·	·
	273	472. *Mus musculus* (N) 27. i. 02	·	·	tr ?	·	·	·	·
	274	859. *Mus spec.* (of M. musculus group). Mouse Okenawa, Japan (R) 9. iii. 02							
	275	95. *Mus decumanus* Pall. House Rat England (N) 18. vii. 01	*	·	·	·	·	·	·
	276	316. *Mus decumanus.* Rat England (R) 30. xi. 01	·	·	·	·	* ? 15 tr	·	·
	277	61. *Mus rattus.* Black Rat Ireland (N) 4. v. 01	·	·	·	·	·	·	·

Primates	Insecti- vora	Carnivora

274 Foam test doubtful, no reactions.

	Mammalia										Aves				Reptilia		Amphibia	Crustacea
Pig	Llama	Hog-deer	Mexican-deer	Antelope	Sheep	Ox	Horse	Wallaby		Fowl	Ostrich	Fowl-egg	Emu-egg		Turtle	Alligator	Frog	Lobster
× D	·	* tr	*	×	· tr	· tr	·	·		·	·	·	·		·	·	·	
	tr ?	·	·	·	·	·	·			·	·	·	·		·	·	·	·
·			·	·	·	·				·	·	·	·		·	·	·	
×	·	*	·	·	·	·	·	·		·	·	·	·		·	·	·	·
·	tr ?	\	·	\	·	·	·	·		·	\	·	\		·	·	·	·
*	/	·	·	·	·	·	·	·		·	·	·	·		·	·	·	·
·	·	·	·	·	·	·	·	·		·	·	·	·		·	·	·	·
·	·	·	·	·	·	·	·	·		·	·	·	·		·	·	·	·
* d	* ? 300	tr ?	·	·	·	·	·	·		·	·	·	·		·	·	·	·
·	·	·	·	·	·	·	·	·		·	·	·	·		·	·	·	·
*	·	·	·	·	·	·	·	·		·	·	·	·		·	·	·	·
·	·	·	·	·	·	·	·	·		·	·	·	·		·	·	·	·
·	·	·	·	·	·	·	·	·		·	·	·	·		·	·	·	·
·	·	·	* ? 265	·	·	·	·	·		·	·	·	·		·	·	·	·
·	·	·	* ? 265	·	·	·	·	·		·	·	·	·		·	·	·	·
·	* ? 300	·	·	·	·	·	·	·		·	·	·	·		·	·	·	·
·	· tr ?	·	·	·	·	·	·	·		·	·	·	·		·	·	·	·
*	·	·	·	·	·	·	·	·		·	·	·	·		·	·	·	·
*	·	·	·	·	·	·	·	·		·	·	·	·		·	·	·	·
·	·	·	·	·	·	·	·	·		·	·	·	·		·	·	·	·

| Ungulata | Marsu-
pialia |

Class MAMMALIA
5. Order RODENTIA
ANTISERA FOR.........

				Man	Monkey	Hedgehog	Cat	Hyaena	Dog	Seal
Section **Myomorpha** *Fam.* **Muridae**	**278**	48.	*Mus rattus.* Black Rat, albino England (N) ca. 12. iv. 01	•					•	
	279	705.	*Nesocia bengalensis.* Indian Mole Rat Khandesh, Deccan (Millard) 27. iii. 02	*60 tr	•		•	•	•	
	280	519.	*Cricetomys gambianus* Waterh. Gambian Pouched Rat W. Africa (Z) 19. ii. 02	•	•		•	•	•	
	281	581.	*Cricetomys gambianus* (Z) 22. ii. 02	•	•	•	•	•	•	
	282	924.	*Oryzomys angonya.* Rat Paraguay (Foster) 12. ii. 02	•				•		
Fam. **Spalacidae**	**283**	115.	*Myoscalops argenteo-cinereus.* Silvery Mole E. Africa (Z) 29. vii. 01	•	•	•	•	× 30 tr	•	
Section **Hystricomorpha** *Fam.* **Octodontidae**	**284**	633.	*Capromys pilorides* (Say). Fournier's Capromys Cuba (Z) 26. iii. 02	•	•	•	•	* ? 85 tr	•	
Fam. **Hystricidae**	**285**	621.	*Synetheres (Cercolabes) prehensilis.* Tree Porcupine Pará, Brazil (Hagmann) 4. iii. 02	* 45 d	•	•	•	•	* 45 d	× 85
	286	622.	*Synetheres (Cercolabes) prehensilis* (Hagmann) 3. iii. 02	* 45 d	•	•	•	•	•	•
	287	290.	*Atherura africana.* African Brush-tailed Porcupine W. Africa (Z) 22. xi. 01	•	•	•	•	•	* 80 tr	•
Fam. **Dasyproctidae**	**288**	436.	*Dasyprocta cristata* (Desm.). Agouti W. Indies (Z) 6. i. 02, *fl.*	•	•	\	× 90	\	•	\
	289	230.	*Dasyprocta cristata* 20. viii. 01	* ?	•	•	•	* 35 tr	•	* 45
	290	801.	*Dasyprocta agouti.* Agouti Trinidad, W. Indies (Tulloch) 3. v. 02	•	•	•	•	• tr	•	•
	291	457.	*Caelogenys paca* L. Spotted Cavy S. America (Z) 24. i. 02	•	•	•	•	•	•	•
Fam. **Caviidae**	**292**	922.	*Cavia aperea* Paraguay (Foster)	* ? 180	•	•		* 180		
	293	11.	*Cavia cobaya* Schreb. Guinea-pig, scales England (N) 11. iii. 01	•	•			•	•	
	294	18.	*Cavia cobaya, fl.* pleural exud. diphtheria iv. 01	•						
	295	19.	*Cavia cobaya* 11. iii. 01	•	•		•	•	•	
	296	27.	*Cavia cobaya, fl.* ser. decomp., sealed 2. iii. 01					•		
				Primates	**Insecti- vora**		**Carnivora**			

284 • with anti-reindeer.

Pig	Llama	Hog-deer	Mexican-deer	Antelope	Sheep	Ox	Horse	Wallaby	Fowl	Ostrich	Fowl-egg	Emu-egg	Turtle	Alligator	Frog	Lobster
			•		•	•	•									
•	tr ?	•	•	•	•	•	•	•	•		•		•		•	
•	•	tr	*	•	•	•	•	•	•	•	•	•	•		•	
* 180 •	•	•	•	•	•	•	•		•		•		•		•	
							•									
d	•			•	•	•	•	•	•	•	•	•	•	•	•	•
•	•	•	•	•	•	•	•	•	•	•	•	•	•		•	
•	tr	•	•	•	•	•	•	•	•	•	•	•	•		•	
•	•	•	•	•	•	•	•	•	•	•	•	•	•	•	•	•
•	\	•	* 90 d	•	•	•	* 90	× 90 tr	•	•	•	•	•	•	•	•
* d	•	•	•	•	•	•	•	•	•	•	•	•	•	•	•	•
* 40	•	•	tr	•	•	•	•	•	•	•	•		•		•	•
* 30 d	•	•	•	•	•	•	•	•	•	•	•		•	•	•	•
* 180					•		* 180		•	•	•		•			
•		•	•		•	•	•	•	•	•	•		•		•	
			•		•	•	•	•								
•	•	•	•	•	•	•	•	•	•	•	•	•	•		•	
		•	tr	•				•				•	•			

Ungulata   Marsu-pialia

Class MAMMALIA
5. Order RODENTIA

ANTISERA FOR.........

Section / Fam.	No.	No.	Species / locality	Man	Monkey	Hedgehog	Cat	Hyæna	Dog	Seal
Section **Hystricomorpha** *Fam.* **Caviidae**	297	737.	*Dolichotis patachonica* Shaw. Patagonian Cavy Patagonia (b. at Z) 29. iv. 02 (*fl.* coagulated)	/	·	·	·	tr ?	/	* ? 290
	298	50.	*Hydrochoerus capybara.* Capybara S. America (Z) 24. iv. 01	·	·	·	·	·	·	·
	299	626.	*Hydrochoerus capybara* Pará, Brazil (Hagmann) 3. iii. 02	* 45 d	·	·	·	·	·	·
Fam. **Leporidae**	300	447.	*Lepus brasiliensis.* Hare Paraguay (Foster) 19. xii. 01	·	·	·	·	·	·	·
	301	136.	*Lepus europaeus.* Common Hare Seeland (Schierbeck) ix. 01	·	·	·	·	tr ?	·	·
	302	339.	*Lepus europaeus* England (N) 26. xi. 01	·	·	·	·	tr ?	·	·
	303	164.	*Lepus europaeus* Germany (Kuse) 26. vii. 01	·	·	·	·	tr ?	·	·
	304	128.	*Lepus variabilis.* Scotch Hare Scotland (R) 30. viii. 01	·	·	·	·	tr	*	·
	305	33.	*Lepus cuniculus.* Tame Rabbit England (N) 22. iii. 01	·	·	·	·	·	·	·
	306	390.	*Lepus cuniculus.* Wild Rabbit Ireland (N) 4. v. 01	·	·	·	·	·	·	·
	307	391.	*Lepus cuniculus* 4. v. 01	·	·	·	·	·	·	·
	308	848.	*Lepus cuniculus* England (R) 14. xii. 02	·	·	·	·	·	·	tr ?
	309	587.	*Lepus cuniculus.* Tame Rabbit Brazil (Hagmann) 16. ii. 02	·	·	·	·	·	·	·
	310	565.	"Long-eared Gray Hare" Orange River, S. Africa (Parkinson) 17. i. 02	·	·	·	·	·	* 75	·
	311	562.	"Bushy-tailed Red Hare" Orange River, S. Africa (Parkinson) 22. xii. 01	·	·	·	·	·	* 75	·
				Primates		Insectivora	Carnivora			

| | Mammalia | | | | | | | | | Aves | | | | Reptilia | | Amphibia | Crustacea |
Pig	Llama	Hog-deer	Mexican-deer	Antelope	Sheep	Ox	Horse	Wallaby		Fowl	Ostrich	Fowl-egg	Emu-egg		Turtle	Alligator	Frog	Lobster
·	*?300	/	·	·	/	/	·	·		/	·	·	/		·		·	·
·	·	·	·	·	·	·	·	·		·	·	·	·		·		·	·
·	·	·	·	·	·	·	·	·		·	·	·	·		·		·	·
·	·	·	·	·	·	·	·	·		·	·	·	·		·	·	·	·
·	·	·	·	·	·	·	·	·		·	·	·	·		·	·	·	·
·	·	·	·	·	·	·	·	·		·	·	·	·		·	·	·	·
·	·	·	·	·	·	·	·	·		·	·	·	·		·	·	·	·
·	·	·	·	·	·	·	·	·		·	·	·	·		·	·	·	·
·	·	·	·	·	·	·	·	·		·	·	·	·		·	·	·	·
·	·	·	·	·	·	·	·	·		·	·	·	·		·	·	·	·
·	·	·	·	·	·	·	·	·		·	·	·	·		·	·	·	·
·	·	·	·	·	·	·	·	·		·	·	·	·		·	·	·	·
·	·	·	·	·	·	·	·	·		·	·	·	·		·	·	·	·
×d	·	·	·	·	·	*?180		·		·	·	·	·		·	·		·

Ungulata   Marsupialia

251

Class MAMMALIA
6. Order UNGULATA
ANTISERA FOR.........

			Man	Monkey	Hedgehog	Cat	Hyaena	Dog	Seal
1. Suborder ARTIODACTYLA	**312**	49. *Sus scrofa domestica.* Domestic Pig England (N) 24. iv. 01	•	•	•	•	* 15	•	
Group **Suina**	**313**	696. *Sus scrofa domestica.* (N) 29. iv. 02 (*fl.* and dr. in sc. filtered)	× d	•	•	•	× 85 tr	tr ?	•
Fam. **Suidae**	**314**	744. *Sus scrofa.* Wild Boar Mecklenburg (von Oertzen) 25. v. 02	* 60 tr	•	•	•	* ? 85 tr	•	•
	315	897. *Sus* spec. Wild Boar Singapore (W. Kerr) 2. viii. 02	•	•	•	•	* 15		* 15
Fam. **Dicotylidae**	**316**	600. *Dicotyles tajuca* L. Collared peccary America (Lühe) 18. iii. 02	* tr	•	•	•	× 85 tr	•	•
Group **Tylopoda**	**317**	151. *Camelus dromedarius* Linn. Dromedary Cairo Abattoir, Egypt (Littlewood) 9. ix. 01	*	•	•	•	* 30 tr	•	* ? 45
Fam. **Camelidae**	**318**	379. *Auchenia glama* Linn. Llama S. America (Nat. Zool. Park, D. C., Salmon) ix. 01	×	•	•	•	* 90	•	•
	319	892. *Auchenia huanacos* Molina. Huanaco Bolivia (Z) 23. ix. 02, d. of rupture of pulmonary artery; *fl. ser.*	•	•	tr ?	•	* 15 tr	•	•
	320	932. *Auchenia huanacos* b. in England, d. young (Z) 21. x. 02, *fl. ser.*	•	•	•	•	• tr ?	•	•
Group **Tragulina** *Fam.* **Tragulidae**	**321**	77. *Tragulus meminna* Erxl. Chevrotain India (Z) 17. v. 01	•	•	•	•	* 240 tr	•	•
Group **Pecora** (True Ruminants)	**322**	889. *Moschus moschiferus.* Musk Deer Bhadarwa, Kashmir State (Donald) 30. v. 02	•	•	•	•	• tr	•	•
Fam. **Cervidae**	**323**	152. *Cervulus vaginalis.* Barking Deer S. Sylhet, India (Dalgetty) 23. viii. 01	•	•	•	•	• tr	•	•
	324	623. *Cervus rufus* Brazil (Hagmann) 6. iii. 02	* 45 d	•	•	•	* 45	•	•
	325	229. *Cervus aristotelis.* Sambur Deer b. in Zoo., London (Z) 16. ix. 01	* ?	•	•	•	* 35 tr	•	•
	326	502. *Cervus porcinus* Zimm. Hog Deer India (Z) 15. ii. 02, *fl. ser.*	•	•	•	* 90 tr	* 45 tr	•	•
	327	291. *Cervus sika.* Japanese Deer b. in Zoo., London (Z) 15. xi. 01	*	•	•	•	* 15 tr	•	•

| Primates | Insectivora | Carnivora |

316 Anti-pig serum weak, had no effect on over 20 other bloods except that of Cervus axis which gave $\frac{*}{120}$ reaction much later.

Pig	Llama	Hog-deer	Mexican-deer	Reindeer	Antelope	Sheep	Ox	Horse	Donkey	Zebra	Wallaby	Fowl	Ostrich	Fowl-egg	Emu-egg	Turtle	Alligator	Frog	Lobster
+	·	* 30	+ 60 d		* 60	·	·	·	·	* 230	·	·	·	·	·	·	·	·	·
+ D	* 85 tr	* tr	× 90 D	* 15 tr	* tr	d	·	× 90 tr	* 30	* 230 tr	·	·	·	·	·	·	·	·	·
+ D	·	* 60 tr	× 30 d		? 30 tr	·	tr	· tr	* 230	* 230 tr	·	·	·	·	·	·	·	·	·
+ D	·						+ 30 cl	·	* 30	* 230 tr	·	·	·	·	·	·	·	·	·
* 15 tr	× 85 tr	·	× 20 tr		* tr	* tr	·	·	* 30	·	·	·	·	·	·	·	·	·	·
+ D	+ 40 d	* 90 tr	* 60		* 90 tr?	·	·	·	·	·	·	·	·	·	·	·	·	·	·
·	+ 20 d	·	* 30		·	·	·	·	·	·	·	·	·	·	·	·	·	·	·
* 30	+ 20 D												·	·	·	·	·	·	·
ℳ 5 cl	+ D												·	·	·	·	·	·	·
·	·	·	* 30 tr		* 30 tr	·	·	·	·	·	·	·	·	·	·	·	·	·	·
·	·	·				·	·	·	·	·	·	·	·	·	·	·	·	·	·
·	·	**+ D**	* 80		* tr	·	·	·	·	·	·	·	·	·	·	·	·	·	·
·	·	× 45 d	+ 15 d		× 30 d	* 45 d	* 30 d	·	·	tr	·	·	·	·	·	·	·	·	·
+ D	* ? 24 tr	+ d	+ 30 d		* d	+	+	·	·	·	·	·	·	·	·	·	·	·	·
+ D	· tr	× 15 d	**+ D**	* 15 d	* 15 d	+ D	+ D	+ 90 tr	·	·	* 75 tr	·	·	·	·	·	·	·	·
·	·	+ d	+ 30 d		* tr	+	**+ D**	·	·	·	·	·	·	·	·	·	·	·	·

| Ungulata | | | | | | | | | | | Marsupialia |

253

Class MAMMALIA
6. Order UNGULATA
ANTISERA FOR.........

Taxon	No.	Ref.	Species / Locality	Man	Monkey	Hedgehog	Cat	Hyaena	Dog	Seal
1. Suborder ARTIODACTYLA *Group* **Pecora** (True Ruminants) *Fam.* **Cervidae**	328	63.	*Cervus axis* Erxl. Axis Deer India (Z) 13. v. 01	•		•	•	•	•	•
	329	601.	*Cervus axis* Erxl. neonat. Axis Deer India (Lühe) 6. iii. 02	•		•	•	•	•	•
	330	743.	*Cervus axis* d. 1 day old b. in Zoo., London (Z), 10. vi. 02, *fl. ser.*	•	•	•	•	*? 85 tr	•	•
	331	658.	*Cervus dama.* Common Fallow Deer Mecklenburg (G. v. Oertzen) 10. ii. 02	•	•	•	•	* 85 tr	•	•
	332	746.	*Cervus dama* England (Brazenor) 22. iv. 02	•	tr	•	•	* 290 tr	•	•
	333	154.	*Cervus elephas* L. Red Deer Germany (Schlieffen) 15. ix. 01	•	•	•	•	• tr	•	•
	334	745.	*Cervus elephas.* Red Deer, calf Germany (von Oertzen) 10. iv. 02	* 60 tr	•	•	•	* 290 tr	•	•
	335	129.	*Cervus capreolus* L. Roebuck Germany (N) 27. viii. 01	×	•	•	* 320	× 35 tr	•	•
	336	163.	*Cervus capreolus* (Kuse) 26. viii. 01	•	•	•	/	tr ?	•	•
	337	483.	*Rangifer tarandus.* Reindeer N. Europe (Lühe, Konigsberg Zoo.) 4. ii. 02, *fl.*	tr ?	•	•	*? 90	* 45	•	•
	338	631.	*Cariacus mexicanus* H. Smith. Mexican Deer Mexico (Z) 9.iv.02, *fl.ser.* (d. of phthisis and pleurisy)	•	•	•	•	* 85 tr	•	*? 290
Fam. **Bovidae**	339	25.	*Connochaetes gnu.* White-tailed Gnu, clot dried in scales S. Africa (d. in England, N) 12. iv. 01	•	•	•	•	•	•	•
	340	486.	*Cobus unctuosus* Laurill. Sing-Sing Antelope (Z) 8. ii. 02 also *fl.*	•	•	•	•	* 45	•	•
	341	62.	*Gazella arabica.* Gazelle Arabia (Z) 13. v. 01	•	•	•	•	•	•	•
	342	109.	*Gazella subgutterosa.* Persian Gazelle Central Asia (Z) 31. vii. 01	•	•	•	*? 320	•	•	•
	343	792.	"American Antelope" ? (New York Zoo., Langmann) p. 14. vi. 02	* 30	* 240	•	•	× 15 tr	*? 60	•
	344	567.	"Deuker Bok" Orange River, S. Africa (Parkinson) 11. xii. 01	* 75	•	•	•	•	•	•
	345	568.	"Stein Bok" Orange River, S. Africa (Parkinson) 20. xii. 01	•	•	•	•	•	•	•
	346	239.	*Tragelaphus sylvaticus.* Bosch-Bok S. Africa (Z) 15. x. 01	×	•	•	•	* 35	•	•
	347	510.	*Æpyceros melampus.* Palla Brit. Centr. Africa (Dodds) 1. i. 02	•	•	•	•	*? 120	*	*? 40

Primates | Insectivora | Carnivora

329 Anti-pig serum weak.

Pig	Llama	Hog-deer	Mexican-deer	Reindeer	Antelope	Sheep	Ox	Horse	Donkey	Zebra	Wallaby	Fowl	Ostrich	Fowl-egg	Emu-egg	Turtle	Alligator	Frog	Lobster
·	·	*tr	+60 D		+D	+	+	·	·	·	·	·	·	·	·	·	·	·	·
*120	tr	×d	*d		*tr	·	·	·	·		·	·	·	·	·	·	·	·	
*60 tr	*85 tr	+30 D	+30 D		+30 D	+60	tr	tr			·	·	·	·		·	·	·	
·	tr	*30 D	+45 d	tr	*45 d	+30 D	*tr	·	·			·	·	*60	·	·	·	·	·
*60 tr	·	+60 D	tr		+30 d	*60	*60 tr	·				·	·	·		·	·	·	
+D	·	+D	×60 tr		+d	+	+D	·			·	·	·	·	·	·	·	·	·
*60 tr	*300 tr	+60 D	+30 d		30 D	+60	×60	tr	·		·	·	·	·		·	·	·	·
+D	·	+30 d	+60 D		+30 d	+	+D	·	·	*230	·	·	·	·	·	·	·	·	·
·	·	*d	/		·	+	+D	·	·	·	/	·	·	·	·	·	·	·	·
+D	*300	+15 D	+15 D	15 d	+D	+D	+D	×90 tr	·	·	*? 70	·	·	·	·	·	·	·	·
·	*85	+D	+D	*d	+D	+D	+30 d	·	*30	*40	·	·	·	·	·	·	·	·	·
·	·	*tr	*tr		*d	·	+D	·			·	·	·	·		·	·	·	
tr		tr	tr		d	·	D	·			·	·	·	·		·	·	·	
+D	*300	×d	+D	*tr	+D	+D	+D	·	·	*40 tr	*75	·	·	·	·	·	·	·	
·	·	*tr	+60 d		+D	+	+D	·	·	*240	·	·	·	·	·	·	·	·	·
*	·	*30 tr	+60 d		+30 d	·	·	·	·		·	·	·	·	·	·	·	·	·
*40	·	*30	×45 tr			·	*75 tr	·	·		·	·	·	·	·	·	·	·	·
+D	·	+30 d	+20 d		+20	+D	*	·	·	tr	·	·	·	·	·	·	·	·	·
+d	·	+30 d	+20 d		+20	+D	+	·	·		·	·	·	·	·	·	·	·	·
+D	tr	*tr	×30 tr cl		+d	+d	+	·	·		·	·	·	·	·	·	·	·	·
*cl	tr ?	·			*180	*360					·	·	·	·	·	·	·	·	·

Ungulata   Marsupialia

Class MAMMALIA
6. Order UNGULATA

ANTISERA FOR.........

1. Suborder ARTIODACTYLA

Group **Pecora** (True Ruminants)

Fam. **Bovidae**

No.		Species / locality	Man	Monkey	Hedgehog	Cat	Hyaena	Dog	Seal
348	243.	Capra megaceros. Markhoor N. E. India (Z) 23. x. 01, *fl. ser.*	*	·	· tr	·	× 35 tr	·	·
349	282.	Capra jemlaica. Tahr (Wild Goat) Chamba State, India (Ainsworth) 21. x. 01	*	·	·	·	* 80 tr	·	·
350	149.	Capra hircus Linn. Egyptian Goat Cairo Abattoir (Littlewood) ix. 01	×	·	* 320	·	* 30 tr	·	? 45
351	596.	Capra hircus. Common Goat England (Z) 3. iii. 02, *fl. ser.*	*	·	* 85	·	* 85 tr	·	·
352	158.	Capra hircus (Z) 8. viii. 01	·	·	* 320	·	+ 35 tr	·	·
353	542.	Capra falconeri. Markhoor Chitral, India (Leslie) 29. i. 02	* 30 tr	·	·	·	* 45	·	·
354	7.	Ovis aries. Domestic Sheep, dr. in scales England (N) 28. ii. 01	×	·	·	·	tr	·	·
355	8.	Ovis aries 19. iii. 01	·	·	·	·	·	·	·
356	9.	Ovis aries Slightly decomposed	·	tr ?	·	·	·	·	·
357	10.	Ovis aries	·	tr ?	·	·	·	·	·
358	45.	Ovis aries, defibr. bl. 19. ii. 01	·	·	·	·	·	·	·
359	96.	Ovis aries. St Kilda Sheep (Lane) 11. vii. 01	·	·	·	·	* ? 30 tr	·	·
360	726.	Ovis spec. ? Domestic Sheep Bhadarwa, Kashmir (Donald) 13. ii. 02	* 60 tr	·	·	·	· tr ?	tr ?	·
361	150.	Ovis aries L. Bedouin Breed Cairo Abattoir (Littlewood) ix. 01	*	·	* 320	·	* 30	·	* ? 45
362	604.	Ovis musimon Schreb. Moufflon Sardinia and Corsica (Lühe) 26. ii. 02	* ?	·	·	·	* ? 85	·	·
363	541.	Ovis vignei typica. Oorial Chitral, India (Sweet) 1. ii. 02	·	·	·	·	·	·	·
364	83.	Ovis burrhel. Burrhel Sheep Himalayas (Z) 31. v. 01	·	·	·	·	* 30 tr	·	·
365	145.	Ovis burrhel b. in Zoo., London (Z) ca. 8. viii. 01	·	·	·	·	* ? 320	·	·
366	595.	Ovis burrhel (Z) 10. iii. 02, *fl. ser.*	·	·	·	·	× 85 tr	·	·
367	693.	Ovis tragelaphus Desm. Aoudad, young N. Africa (Lühe, Königsberg Zoo.) 26. iii. 02	* tr	·	·	·	× 85 tr	·	* ? 290

Primates	Insectivora	Carnivora

Pig	Llama	Hog-deer	Mexican-deer	Reindeer	Antelope	Sheep	Ox	Horse	Donkey	Zebra	Wallaby	Fowl	Ostrich	Fowl-egg	Emu-egg	Turtle	Alligator	Frog	Lobster
+ D	* ? 240 tr	+ d	+ 90 D	* 15 d	+ d	+ b	+	·			* 90	·	·	·	·	·	·	·	·
·	·	* d	+ 30 tr cl		+ d	+ b	+	·	·	·	·	·	·	·	·	·	·	·	·
+ D	* 240 tr	+ D	+ 60 tr cl		+ D	+ Œ	+	·	·	·	·	·	·	·	·	·	·	·	·
·	· tr	+ D	+ D	* 15 tr	+ d	+ D	+ D	·	* 30	* 40	·	·	·	·	·	·	·	·	·
+ D		+ d	+ 60 tr		+ D	+ Œ	+	·			·	·	·	·	·	·	·	·	·
+ D	* 120 tr	+ 30 d	+ 20 d		+ 20 d	+ D	+ D	·			·	·	·	·	·	·	·	·	·
* 30 tr	· tr	+ d	+ d		+ d	+ Œ	+	·			·	·	·	·	·	·	·	·	·
					+ Œ	+	·	·	·		·	·	·	·	·	·	·	·	·
+ 30 d		+ D	+ D		+ D	+ D	+ d	·			·	·	·	·	·	·	·	·	·
+ 30 tr		* d	+ D		+ D	+ D	+ d	·			·	·	·	·	·	·	·	·	·
·		* 30 tr	* 100		* 30 d	+ Œ	+	·	·	·	·	·	·	·	·	·	·	·	·
*	·	* 30 tr	+ 60 tr		+ 30 d	+ Œ	+	·			·	·	·	·	·	·	·	·	·
·	·	* tr	* 30 tr		+ d	+ D	× tr	·			·	·	·	·	·	·	·	·	·
+ D	* 240 tr	+ D	+ 60 tr cl		+ D	+ Œ	+	·	·	·	·	·	·	·	·	·	·	·	·
·	· tr	* d	· d		* D	+ D	* tr	·	·	·	·	·	·	·	·	·	·	·	·
+ D	·	+ 30 d	* 20 d		* 20 d	+ D	+ D	·	·	·	·	·	·	·	·	·	·	·	·
·	* ? 240	× tr	+ 60 d		+ d	+ Œ	×	·			·	·	·	·	·	·	·	·	·
* d	* 30	× tr	* 60		* tr	+	+	·			·	·	·	·	·	·	·	·	·
	* 300 tr	× 15 d	+ D	* 15 d	× 15 D		* ? 90	* 30	* 40	·		·	·	·	·	·	·	·	·
· tr	·	* d	* 30		× d	+ D	·	· 90		* 230	·	·	·	·	·	·	·	·	·

Ungulata   Marsupialia

Mammalia

Class MAMMALIA

6. *Order* UNGULATA

	No.		ANTISERA FOR.........	Man	Monkey	Hedgehog	Cat	Hyaena	Dog	Seal
1. *Suborder* **ARTIODACTYLA** *Group* **Pecora** (True Ruminants) *Fam.* **Bovidae**	**368**	4.	*Bos taurus* L. Ox, filtered ser., scales England (N) 19. ii. 01	×	•	• tr ?	•	* ? 15	•	• tr ?
	369	6.	*Bos taurus* 23. ii. 01	•	•	•	•	•	•	•
	370	42.	*Bos taurus* 20. ii. 01	•					•	
	371	147.	*Bos taurus*. Egyptian Cow Cairo Abattoir (Littlewood) ix. 01	×	•	* ? 320 tr	•	* ? 30 tr	•	•
	372	148.	*Bos bubalis*. Water Buffalo Egypt, Cairo Abattoir (Littlewood) ix. 01	×	•	* ? 320	•	* ? 30 tr	•	•
	373	383.	*Bos* spec. ? Buffalo (agricultural) Shanghai, China (Stanley) ca. 27. xi. 01	*	•	•	•	* ? 15 tr	•	•
	374	717.	*Bos gaurus* Evans. Indian Bison Kandesh, India (Simcox) 17. iii. 02	× 60 tr	•	•	•	× 85 tr	* 5 tr	* 290
2. *Suborder* **PERISSODACTYLA** *Fam.* **Tapiridae**	**375**	796.	*Tapirus americanus*. Tapir Brazil (Goeldi) 20. v. 02	•	• tr ?	•	•	• tr	•	•
Fam. **Equidae**	**376**	1.	*Equus caballus*. Horse, antidiphther. ser. dr. in sc. bottled (L. C.) 1895	•	•		•		•	
	377	2.	*Equus caballus*. Antidiphther. ser. dr. in sc. bottled (L. C.) 1897	•	•		•		•	
	378	594.	*Equus caballus*. Antidiphther. ser. fl. England (B. W. and Co.) 4. i. 1897	•	•		•		•	
	379	3.	*Equus caballus*. Normal filtered ser. sc. bottled (N) 4. iv. 01	×	•	•	•	•	•	•
	380	694.	*Equus caballus* England (N) 20. iv. 02, filtered ser., *fl.*	* tr	•	•	* ? 90	* 85 tr	* ? tr ?	* ? 85
	381	929.	*Equus asinus* L. Donkey England (N) 3. xi. 02, *fl. ser.* filt.	* 180 tr	• tr	•		* 180 tr		
	382	930.	*Equus asinus* (McFadyean) 6. xi. 02, *fl. ser.*	•	•	•		•		
	383	903.	*Equus grevyi* Günther. Grevy's Zebra, young Abyssinia (Z) 13. x. 02, *fl. ser.*	* 15 d	•	•		* 10 tr		
3. *Suborder* **HYRACOIDEA** *Fam.* **Hyracidae**	**384**	563.	*Hyrax* spec. "Rock Rabbit" Orange River, S. Africa (Parkinson) 24. xii. 01	* 75	•	•	•	• tr ?	•	•
				Primates		Insectivora	Carnivora			

376, 377 Foam-test doubtful, not included in small tables.

Pig	Llama	Hog-deer	Mexican-deer	Reindeer	Antelope	Sheep	Ox	Horse	Donkey	Zebra	Wallaby	Fowl	Ostrich	Fowl-egg	Emu-egg	Turtle	Alligator	Frog	Lobster
*30 d	·	**+** d	**+** D		**+** D	+ d	**+**	·			·	·		·	·	·			
· tr	·	tr	tr		* d	+ tr	+ d	·			·	·		·	·	·			
·						+ d	**+**	·				·		·	·	·		·	·
+ D	*240 tr	**+** D	+ D		**+** D	+ d	**+**	·	*230	*40 d	·	·		·	·	·		·	·
+ D	*240 tr	**+** D	+ D		**+** D	+ d	**+**	·	*230	*40 d	·	·		·	·	·		·	·
·	·	× d	+ 30 d		+ d	·	**+**	·			·	·		·	·	·		·	·
*60 tr	*300 tr	* tr	+ 30 d		+ d	× d	+ d	·	*? 230	*40	·	·		·	·	·		·	·
*40	·	·	*45 tr		·	·	·	·	*? 240	*40	·	·		·	·	·			
·	·	·	·	·	·	·	·	·		·		·	·	·	·	·			
		* tr	·	·	* tr	·	·	·			·	·	·	·	·	·			
· tr	tr?	· tr?	·		· tr	* tr	·	+ cl			·	·	·	·	·	·		·	·
· tr	·	·	·		·	·	·	**+ D**			·	·	·	·	·	·		·	·
*30	*85 tr		*90 D	*15	? tr?	* d	+ 30 d	**+ D**	**+** 5 **D**	**+** 10 **D**	·	·	·	·	·	·			
× 25 d							tr?	**+** 1 **D**	**+** 5 **D**	**+** 10 **D**	·	·	·	·	·	·			
× 10							·	**+** 1 **D**	**+** 5 **D**	**+** 10 **D**	·	·	·	·	·	·			
· tr						tr		**+** 1 **D**	**+** 5 **D**	**+** 10 **D**	·	·	·	·	·	·			
*	·	·	·		*60	·	·	·			·	·	·	·	·	·		·	

| | | | Ungulata | | | | | | | Marsupialia | | | | | | | | | |

373 Anti-sheep serum weak.

Class **MAMMALIA**

7. *Order* **CETACEA**

ANTISERA FOR.........

8. *Order* **EDENTATA**

			Man	Monkey	Hedgehog	Cat	Hyaena	Dog	Seal
3. *Suborder* Hyracoidea	**385**	698. *Balaenoptera rostrata.* Rorqual, also *fl.* Killed near Bergen, Norway (Brunchorst) 30. iv. 02	* tr	.	.	.	. tr	* ? tr?	* ? 85
***Fam.* Hyracidae**	**385** *a*	935. *Balaenoptera rostrata* Killed near Tromsoe, Norway (Torup) p. 6. xi. 02, *fl. ser.*	* 30 tr	.	.		*		
	386	485. *Phocaena communis.* Porpoise N. Coast of France (von Oertzen, Havre) ca. 2. ii. 02, *fl.*	* tr	.	.	.	* 45	.	.
***Fam.* Bradypodidae**	**387**	160. *Bradypus tridactylus.* Three-toed Sloth British Guiana (Z) 5. viii. 01	.	.	* ? 320	.	tr	.	.
	388	505. *Bradypus marmoratus.* Sloth Brazil (Hagmann) 29. i. 02	.	.	.	*	* ? 45 tr	.	.
***Fam.* Myrmecophagidae**	**389**	233. *Myrmecophaga jubata.* Great Ant-eater S. America (Z) 25. ix. 01	.	.	.	.	* 35 tr	.	.
	390	799. *Myrmecophaga jubata* Brazil (Goeldi) 5. v. 02	.	.	.	.	tr tr	.	.
	391	507. *Tamandua tetradactyla.* Tamandua Ant-eater Brazil (Hagmann) 29. i. 02	.	.	.	.	* 45 tr	.	.
	392	586. *Tamandua tetradactyla* 16. ii. 02	.	.	.	.	* 85 tr	.	* 290
	393	625. *Tamandua tetradactyla* 3. iii. 02	d	.	.	.	.	.	* 85
***Fam.* Dasypodidae**	**394**	235. *Dasypus villosus.* Hairy Armadillo La Plata (Z) 21. viii. 01	.	.	.	.	* ? 35	.	.
	395	371. *Dasypus sexcinctus* L. Armadillo, male Paraguay (Foster) 30. xi. 01	.	.	\	.	\	.	\
	396	506. *Dasypus setosus.* Armadillo Brazil (Hagmann) 29. i. 02	.	.	.	.	.	*	.
	397	110. *Tatusia peba.* Peba Armadillo S. America (Z) 1. viii. 01	.	.	.	.	× 30 tr	.	.
	398	585. *Tatusia novemcincta* (young). Peba Armadillo Brazil (Hagmann) 15. ii. 02	.	.	.	.	.	.	.
***Fam.* Orycteropodidae**	**399**	561. *Orycteropus capensis.* "Ant Bear" or "Aard Vaark" Orange River, S. Africa (Parkinson) 22. i. 02	.	.	.	.	.	.	.
			Primates		Insectivora		Carnivora		

385 * tr with anti-reindeer.

Mammalia										Aves				Reptilia		Amphibia	Crustacea
Pig	Llama	Hog-deer	Mexican-deer	Antelope	Sheep	Ox	Horse	Donkey	Wallaby	Fowl	Ostrich	Fowl-egg	Emu-egg	Turtle	Alligator	Frog	Lobster
* 15 d	•	tr	× 90 **D**	* ? tr	60 D	* 30 D	* 90 tr ?	•	•	•	•		•				
+ 5 d						60 tr	•	•		•	•	•					
* D	•	*	•	tr	•	* 120 d	•		•	•	•	•		•	•	•	•
* d	•	*	* 60	•	•	•	•		•	•	•			•	•	•	•
* d	•	•	* ?	•	•	•	•		•	•	•			•	•	•	•
* d	? 240	•	* 60 tr	•	•	•	•		•	•	•			•	•	•	•
″ 40	•	•	″ 45 tr	•	•	•	•		•	•	•						
* cl	•	•	•	•	•	•	•		•	•	•	•		•		•	•
* 180 cl	•	•	•	•	•	•	•		•	•	•	•	•	•		•	
•	•	•	•	•	•	•	•		•	•	•	•		•		•	•
* d	tr	\	•	\	•	•	•		•	•	\		\	•	•	•	•
•	\	•	•	•	•	•	•		•	•	•	•	•	•	•	•	•
* d	•	•	•	•	•	•	•		•	•	•	•	•	•	•	•	
*	•	•	tr	•	•	•	•		•	•	•			•	•	•	•
* 180	•	•	•	•	•	•	•		•	•	•	•	•	•		•	
* ? cl	•	•	•	•	•	•	•		•	•	•	•	•	•	•		

Ungulata   Marsupialia

Class MAMMALIA
9. Order MARSUPIALIA
ANTISERA FOR………

			Man	Monkey	Hedgehog	Cat	Hyaena	Dog	Seal
Suborder POLYPROTODONTIA	**400**	800. *Didelphys marsupialis.* Opossum. Trinidad, W. Indies (Tulloch) 20.iv.02	·	·	·	·	·	·	·
Fam. Didelphyidae	**401**	448. *Didelphys marsupialis* Paraguay (Foster) 19. xii. 01	·	·	·	·	·	·	·
	402	927. *Didelphys marsupialis* 5. iv. 02	·	·	·	·	·	·	·
	403	432. *Didelphys lanigera.* Woolly Opossum. Colombia (Z) 2. i. 02	d	·	·	·	*15	·	·
	404	926. *Didelphys cinerea* Paraguay (Foster) 7. iii. 02	·	·	·	·	·	·	·
	405	923. *Metachirus nudicaudatus.* Opossum. Paraguay (Foster) 9. iv. 02	*? 180	·	·	·	·	·	·
Fam. Dasyuridae	**406**	482. *Thylacinus cynocephalus.* Thylacine. Tasmania (Z) 1. ii. 02, *fl.*	·	·	·	*? 90	·	·	·
Fam. Peramelidae	**407**	116. *Perameles* spec. inc. Bandicoot. Australia (Z) 29. vii. 01	·	·	·	·	tr	·	·
Suborder DIPROTODONTIA	**408**	240. *Trichosurus vulpecula* Kerr. Vulpine Phalanger. Australia (Z) 18. x. 01	·	·	·	·	· tr	·	·
Fam. Phalangeridae	**409**	458. *Trichosurus vulpecula* 22. i. 02	·	·	·	·	·	·	·
	410	896. *Petaurus sciureus* Shaw. Squirrel-like Phalanger. Australia (born at Z.) recd. 26. ix. 02	·	·	·	·	·	·	·
Fam. Macropodidae	**411**	51. *Hypsiprymnus rufescens* Gray. Rufous Rat-Kangaroo. N. S. Wales (Z) 24. iv. 01	·	·	·	·	·	·	·
	412	459. *Onychogale frenata* Gould. Bridled Wallaby. Australia (Z) 23. i. 02	·	·	·	·	·	·	·
	413	497. *Onychogale frenata* 6. ii. 02	·	·	·	·	·	*	·
	414	488. *Onychogale unguifera* Gould. Nail-tailed Wallaby. N. W. Australia (Z) 3. ii. 02, *fl.*	·	·	·	·	tr	·	tr
	415	503. *Onychogale unguifera* 13. ii. 02	·	·	*? 60	·	·	·	·
	416	611. *Onychogale unguifera* 22. iii. 02, *fl. ser.* filt., sealed	·	·	·	·	·	·	·
	417	487. *Petrogale xanthopus* Gray. Yellow-footed Rock-Kangaroo. Australia (Z) 10. ii. 02, *fl.*	·	·	·	·	·	·	·
			Primates		Insectivora	Carnivora			

402, 404, 405 Anti-wallaby serum very weak.

262

| | Mammalia | | | | | | | | | Aves | | | | Reptilia | | Amphibia | Crustacea |
Pig	Llama	Hog-deer	Mexican-deer	Antelope	Sheep	Ox	Horse	Wallaby	Fowl	Ostrich	Fowl-egg	Emu-egg	Turtle	Alligator	Frog	Lobster
*40	·	·	*45 tr		·	·	·	·		·	·	·	·	·	·	·
*30	·	·	·		·	·	·	*	·	·	·		·	·	·	·
·					·	·		·	·	·	·		·	·		
×	·	·	·	·		·	·	·	·	·	·		·	·	·	·
·					·	·		·	·	·	·			·		
·					·	·		·	·	·	·			·		
*60 d		*? 90			·	·	*? 90	×75			·	·	·	·	·	·
d	·	·	·	·	·	·	·	·	·	·	·	·	·	·	·	·
*d	·	·	·	·	·	·	·	·	·	·	·		·	·	·	·
*30 d	·	·	·	·	·	·	·	·	·	·	·	·	·	·	·	·
*30	·		·		·	*? 30		·	·	·		·			·	
·	·	·	·	tr	·	·	·	+ 60 tr cl	·	·	·	·	·	·	·	·
×d	·	·	·	·	·	·	·	+ 75 D	·	·	·	·	·	·	·	·
*d	·	tr	·	·	·	·	·	×D	·	·	·	·	·	·	·	·
*d	*300	·	·	·	·	·	·	+ 75 D	·	·	·	·	·	·	·	·
*d	·	·	·	·	·	·	·	+ 75 D	·	·	·	·	·	·	·	·
·	·	·	·	·	·	·	·	+ 75 d cl	·	·	·	·	·	·	·	·
*d	·	·	·		·	·	·	+ 75 D	·	·	·	·	·	·	·	·

| Ungulata | | | | | | | | Marsupialia |

			Mammalia						
Class **MAMMALIA** 9. *Order* **MARSUPIALIA** ANTISERA FOR.........			Man	Monkey	Hedgehog	Cat	Hyaena	Dog	Seal
Suborder **DIPROTODONTIA**	418	74. *Petrogale penicillata.* Brush-tailed Kangaroo N. S. Wales (Z) 28. v. 01	•	* 100	* ? 320	•	* 30 tr	•	•
Fam. **Macropodidae**	419	484. *Petrogale penicillata* Australia (Lühe, Königsberg Zoo.) 4. ii. 02, *fl.*	tr ?	•	•	•	•	•	•
	420 421§	573. *Macropus giganteus.* Great Kangaroo Australia (Calcutta Zoo., Rogers) 1. ix. 01	•	•	•	•	•	•	•
	422	275. *Macropus giganteus* (Z) 2. xi. 01	•	•	* ? 85	•	* 80 tr	•	•
	423	676. *Macropus ruficollis.* Scrub Wallaby Oberon, N. S. Wales (Cashman) 30. xii. 01	•	•	* ? 290	•	•	•	* 85 tr ?
	424	117. *Macropus bennetti* Waterh. Bennett's Wallaby Tasmania (Z) 12. vii. 01	•	* 100 tr ?	•	•	•	•	•
	425	512. *Macropus bennetti* 22. ii. 02, *fl.*	•	•	•	•	•	•	tr ?
	426	605. *Macropus bennetti* (Lühe) 13. ii. 02	•	•	•	•	* ? 85	•	•
		10. *Order* **MONOTREMATA**							
	427	699. *Ornithorhynchus paradoxus* Blumenb. Duck-billed Platypus N. S. Wales (Palmer, Sidney) 29. iii. 02	tr ?	•	•	•	•	tr ?	•
			Primates	Insectivora		Carnivora			

§ The omitted No. 421 is replaced by 385a in the series.

	Mammalia										Aves					Reptilia		Amphibia	Crustacea
Pig	Llama	Hog-deer	Mexican-deer	Antelope	Sheep	Ox	Horse	Wallaby		Fowl	Ostrich	Fowl-egg	Emu-egg		Turtle	Alligator		Frog	Lobster
•	•	•	•	•	•	•	•	+60 d		•	•	•	•		•	•		•	•
*60 d	•	•	•	•	•	•	•	+75 d		•	•	•	•		•	•		•	•
*? 180	•	•	tr?	•	•	•	•	* tr		•	•	•	•		•	•		•	•
•	•	•	*60	•	•	•	•	+45		•	•	•	•		•	•		•	•
•	•	•	•	•	•	•	•	+15		•	•	•	•		•	•		•	•
*d	•	•	tr?	•	•	•	•	+60 d		•	•	•	•		•	•		•	•
×d	•	•	•	•	•	•	•	+75 D		•	•	•	•		•	•			•
•	•	•	•	•	•	•	•	+15 d		•	•	•	•		•	•		•	•
•	•	•	•	•	•	•	•	•		•	•	•	•		•				•

| | | | Ungulata | | | | | Marsu-pialia |

Class AVES

I. *Division* RATITAE

II. *Division* CARINATAE

1. *Order* COLYMBIFORMES

2. *Order* PROCELLARIFORMES

		ANTISERA FOR.........	Man	Monkey	Cat	Dog	Pig	Hog-deer
Fam. Struthionidae	**428**	740. *Struthio molybdophanes* Reichenow. Ostrich — Somali-land (Z) 7. vi. 02, *fl.*	•	•	•	•	•	•
	429	489. *Struthio molybdophanes* 20. xii. 01, *fl.*	•	•	•	•	•	•
	430	120. *Struthio australis.* S. African Ostrich — S. Africa (Z) 8. vii. 01	•	•	•	•	•	•
	431	602. *Struthio* spec. Ostrich — Africa (Lühe) 21. iii. 02, *fl.*	•	•	•	•	•	•
Fam. Rheidae	**432**	504. *Rhea americana.* Rhea — Brazil (Hagmann) 29. i. 02	•	•	*	•	•	•
	433	342. *Rhea americana* — S. America (Z) 25. xi. 01	•	•	•	•	•	•
	434	305. *Rhea americana* (R) 11. xi. 01	•	•	•	•	•	•
Fam. Casuaridae	**435**	277. *Casuarius bicarunculatus.* Cassowary — Arroo Islands (Z) 14. xi. 01	•	•	•	•	•	•
	436	304. *Dromaeus novae-hollandiae.* Emu — Australia ? (R) 11. xi. 01	•	•	•	•	•	•
Fam. Colymbidae	**437**	650. *Colymbus septentrionalis.* Red-throated Diver — England (B) 27. ii. 02	•	•	•	•	•	•
Fam. Podicipedidae	**438**	700. *Podiceps* spec. Diver, Ger. "Haubentaucher" — Mecklenburg (von Oertzen) 23. iv. 02	• tr ?	•	•	*? tr ?	•	•
	439	253. *Podiceps......* Lesser Grebe — England (Farren) x. 01	•	•	•	•	•	＼
	440	311. *Tachybaptes fluviatilis.* Little Grebe — England (B) 21. xi. 01	•	•	•	•	•	•
	441	644. *Tachybaptes fluviatilis* 21. xii. 01	•	•	•	•	•	•
	442	645. *Podiceps auritus.* Eared Grebe — England (B) 3. ii. 02	•	•	•	•	•	•
	443	221. *Podiceps cristatus.* Crested Grebe — Europe (Z) 26. ix. 01	•	•	•	•	•	＼
	444	301. *Podiceps cristatus* 18. xi. 01	•	•	•	•	•	＼
Fam. Procellariidae	**445**	842. *Fulmarus glacialis.* Fulmar Petrel — England (Clarke) 1. ii. 02	•	•	•	*? 240	•	

431 • with antisera for seal, hyaena, hedgehog, llama.

Mammalia						Aves				Reptilia		Amphibia	Crustacea
Mexican-deer	Antelope	Sheep	Ox	Horse	Kangaroo	Fowl	Ostrich	Fowl-egg	Emu-egg	Turtle	Alligator	Frog	Lobster
•	tr ?	•	tr	•	•	+ 45 d		* 60 tr		tr	•		
•	•	•	•	•	•	+ **D**	+ **D**	* 60	•	•	•	•	•
•	•	•	•	•	•	+	+ **D**	+ d	* 30 d	•	•	•	•
•	•	•	•	•	•	*	+ **D**		*	* ?	•	•	•
•	•	•	•	•	•	* d	* d	•	•	•	•	•	•
•	•	•	•	•	•	+	+ d	•	d	•	•	•	•
•	•	•	•	•	•	+ D	+ d	•	* tr	•	•	•	•
•	•	•	•	•	•	+ d	+ d	•	* tr	•	•	•	•
	•	•	•	•	•	* d	× tr	+	* tr	•	•	•	•
•	•	•	•	•	•	* tr	* 30	•	•	•		•	
•	•	•	•	•	•	+ d	* d	•	tr ?	•		•	
•	\	•	•	•	•	* d	\	•	\	•	•	•	•
•	•	•	•	•	•	* d	* tr	* ?	tr ?	•	•	•	•
•	•	•	•	•	•	*	•	•	•	•	•	•	•
•	•	•	•	•	•	*	* 60	•	•	•	•	•	•
•	\	•	•	•	•	* ? d	\	•	\	•	tr ?	•	•
•	\	•	•	•		+ D	* ?	+	\	•	•	•	•
			•			• tr	•						

Class AVES

II. *Division* CARINATAE

3. *Order* CICONIIFORMES

ANTISERA FOR........

				Man	Monkey	Cat	Dog	Pig	Hog-deer
1. *Suborder* **Steganopodes** *Fam.* Phalacrocoracidae	**446**	646.	*Phalacrocorax graculus.* Shag England (B) 10. ii. 02	•	•	•	•	•	•
Fam. **Fregatidae**	**447**	733.	*Fregata aquila.* Frigate, or Man-of-War Bird Trinidad, W. Indies (Tulloch) 2. iv. 02	/	•	•	/	•	/
Fam. **Pelecanidae**	**448**	511.	*Pelecanus onocrotalus* L. White Pelican N. Africa (Z) 18. ii. 02 *fl.*	•	tr ?	•	•	•	•
	449	466.	*Pelecanus* spec. Pelican Trinidad (Tulloch) 1. i. 02	•	•	•	•	•	•
2. *Suborder* **Ardeae**	**450**	155.	*Ardea candidissima.* Snowy Egret Brazil (Z) 6. viii. 01	•	•	•	•	•	•
Fam. **Ardeidae**	**451**	608.	*Ardea purpurea* L. Purple Heron S. Europe (Lühe) 27. ii. 02	•	•	•	•	•	•
	452	183.	*Ardea cinerea.* Common Heron Germany (Kuse) 17. ix. 01	•	•	•	•	/	/
	453	251.	*Ardea cinerea* England (Farren) viii. 01	•	•	•	•	•	/
	454	648.	*Ardea cinerea* (B) 2. ii. 02	•	•	•	•	•	•
	455	710.	*Ardea cinerea* Kashmir State, India (Donald) 16. iv. 02	/	/	/	/	/	•
	456	368.	*Nycticorax violaceus* Linn. Violaceous Night Heron S. America (Z) 18 xii. 01	•	•	•	•	•	•
	457	566.	" Large Black and White Heron " Orange R., S. Africa (Parkinson) 3. i. 02	•	•	•	•	•	•
	458	453.	*Botaurus stellaris.* Bittern England (Clarke) 18. i. 02	•	•	•	•	•	•
3. *Suborder* **Ciconiae** *Fam.* **Ciconiidae**	**459**	126.	*Ciconia nigra* Bechst. Black Stork Europe (Z) 27. vii. 01	•	•	•	•	•	•
	460	529.	*Ciconia alba.* White Stork India (Inglis) 18. i. 02	•	•	•	•	•	•
	461	852.	*Dissura maquari* Bodd. Maquari Stork S. America (Z) 18. vii. 02 *fl.*, killed, blind	•	•	•	•	•	•
	462	527.	*Dissura episcopus.* White-necked Stork India (Inglis) 3. ii. 02	•	•	•	•	•	•
4. *Suborder* **Phoenicopteri** *Fam.* **Phoenicopteridae**	**463**	851.	*Phoenicopterus ruber* Linn. Ruddy Flamingo N. America (Z) 22. vii. 02, *fl.* no dis.	•	•	•	•	•	•

448, 451 · with 4 antisera for seal (tr ?), hyaena, hedgehog, llama.

| Mammalia | | | | | | Aves | | | | Reptilia | | Amphibia | Crustacea |
Mexican-deer	Antelope	Sheep	Ox	Horse	Kangaroo	Fowl	Ostrich	Fowl-egg	Emu-egg	Turtle	Alligator	Frog	Lobster
•	•	•	•	•	•	*	* 60	•	•			•	•
•	/	/	/	•	•	/		•	/	•			
tr? •	•	•	•	•	•	+ D •	+ D * d	•	d •			•	•
•	•	•	•	•	•	* d * D	* *? 90 tr	•	•			•	•
•	/	/	/	•	•	/	/	/	/	/	/	/	/
•	/	•	•	•	•	•	/	•	/	/			
/	•	/	/	/	/	/	•	/	•	/			
•	•	•	•	•	•	+ D	* tr	+	•	•	•	•	•
•	•	•	•	•	•	* d	* d	•	•	•	•	•	•
•	•	•	•	•	•	* d	* D	•	tr?	•	•	•	•
•	•	•	•	•	•	+ b	+ 30 D	+ D	* 30 d	•	+	•	•
•	•	•	•	•	•	* D	* d	•	* 210 d	•	•	•	•
•	•	•	•	•	•	+ 35 d	* 240 tr	•	* 210	•		•	•
•	•	•	•	•	•	* d	* d						
•	•	•	•	•	•	+ 35 d	tr?						

447, 452, 455 But slightly or not soluble. **454** Insoluble.

Class AVES
II. _Division_ CARINATAE
4. _Order_ ANSERIFORMES
ANTISERA FOR.........

			Man	Monkey	Cat	Dog	Pig	Hog-deer
464	220.	_Mergus albellus_ L. Smew Europe (Z) 5. ix. 01	•	•	•	•	•	•
465	651.	_Mergus albellus_ England (B) 10. ii. 02	•	•	•	•	•	
466	261.	_Mergus serrator._ Red-breasted Merganser England (R) 5. xi. 01	•	•	•	•	•	•
467	312.	_Mergus serrator_ (B) 12. xi. 01	•	•	•	•	•	•
468	672.	_Clangula glaucion._ Golden-Eye Orkney (B) iii. 02	•	•	•	•	•	•
469	461.	_Clangula glaucion_ England (Bird) 6. xii. 01		•	•	•	•	•
470	347.	_Clangula glaucion_ (B) 30. xi. 01	•	•	•	•	•	•
471	341.	_Clangula glaucion_ (R) 25. xi. 01	•		•	•	•	•
472	549.	_Netta rufina._ Red-crested Pochard Darbhanga, India (Inglis) 18. i. 02	•		•	•	•	•
473	263.	_Nyroca ferina._ Pochard England (R) 5. xi. 01	•		•	•	•	•
474	262.	_Fuligula cristata._ Tufted Duck England (R) 5. xi. 01	•		•	•	•	•
475	649.	_Fuligula cristata_ (B) 8. i. 02	•		•	•	•	•
476	276.	_Dendrocygna arcuata._ Wandering Tree-Duck E. Indies (Z) 12. xi. 01	•		•	•	•	\
477	590.	_Anas boscas._ Domesticated Duck England (N) 7. iii. 02, _fl. ser._	•	•	•	•	•	•
478	659.	_Anas boscas_ Mecklenburg (G. von Oertzen) 19. ii. 02	•		•	•	•	•
479	404.	_Anas boscas_ England (N) 16. v. 01	•	•	•	•	•	•
480	405.	_Anas boscas_ 19. x. 01	•	•	•	•	•	•
481	917.	_Anas boscas_ (Cobbett) v. 1900 (_fl. ser._ sealed over 2 years)	•	•			* ? 25	
482	141.	_Anas boscas_? Wild Duck Germany (Kuse) 8. viii. 01	•	•	•	•	•	•
483	831.	_Anas boscas_ England (R) 21. ii. 02	•		•	•		
484	833.	_Mareca penelope._ Widgeon England (Clarke) 25. i. 02	•	•	•	•		
485	161.	_Tadorna tadornoides._ Sheldrake S. Australia (Z) 9. viii. 01	•	•	•	•	•	tr?
486	548.	_Casarca rutila._ Brahminy Duck Tirhut, India (Inglis) 17. i. 02	•	•	•	•	•	
487	712.	_Nettion crecca._ Green-winged Teal Kashmir (Donald) 19. iii. 02	•		•		•	
488	881.	_Nettion crecca_ Okenawa, Japan (R) 14. iii. 02						

4. _Suborder_
Phoenicopteri
Fam.
Anatidae
Subfam.
Merginae

Subfam.
Fuligulinae

Subfam.
Anatina

488 Insoluble.

	Mammalia						Aves				Reptilia		Amphibia	Crustacea
Mexican-deer	Antelope	Sheep	Ox	Horse	Kangaroo		Fowl	Ostrich	Fowl-egg	Emu-egg	Turtle	Alligator	Frog	Lobster
•	•	•	•	•	•		*d	* tr	•	•	•	•	•	•
•	•	•	•	•	•		*60 tr	*60 tr	•	•	•	•	•	•
•	•	•	•	•	•		+d	* tr	*	•	•	•	•	•
•	•	•	•	•	•		*d	* tr	•	•	•	•	•	•
•	•	•	•	•	•		+60 d	+60 d	•	•	•	•	•	•
•	•	•	•	•	•		+D	+D	60	•	•	•	•	•
•	•	•	•	•	•		+Ↄ	*? tr?	•	tr?	•	•	•	•
•	•	•	•	•	•		+Ↄ	+ tr	•	•	•	•	•	•
•	•	•	•	•	•		**+D**	*d	•	*? 240	•	•	•	•
•	•	•	•	•	•		**+D**	* tr	*	•	•	•	•	•
•	•	•	•	•	•		+d	* tr	*	•	•	•	•	•
•	•	•	•	•	•		+d	* 60	•	•	•	•	•	•
•	\	•	•	•	•		+d	\	*	\	•	•	•	•
•	•	•	•	•	•		**+D**	**+D**	•	tr	*? 90	*? 90	•	•
•	•	•	•	•	•		+60 D	+60 D	*60	•	•	•	•	•
*? 265	•	•	•	•	•		+Ↄ	* tr?	•	•	•	•	•	•
•	•	•	•	•	•		+Ↄ	•	•	•	•	•	•	•
•	•	*40 tr	•	•	•		+10 d	× 15 tr	•	•	•	•	•	•
•	•	•	•	•	•		*d	*	d	•	•	•	•	•
•	•	•	•	•	•		*15	•	d	•	•	•	•	•
•	•	•	•	•	•		+15 Ↄ	•	•	•	•	•	•	•
•	•	•	•	•	•		+D	+d	*	•	•	•	•	•
•	•	•	•	•	•		**+D**	*d	•	•	•	•	•	•

II. *Division* CARINATAE
4. *Order* ANSERIFORMES
ANTISERA FOR........

			Mammalia					
			Man	Monkey	Cat	Dog	Pig	Hog-deer
4. Suborder Phoenicopteri	**489**	832. *Nettion crecca.* Green-winged Teal England (R) 12. ii. 02	•	•	•	•		
Fam. **Anatidae**	**490**	187. *Nettion crecca* Germany (Schlieffen) ix. 01	•	•	•	•	•	•
Subfam. **Anatina**	**491**	677. *Anser spec.* Wild Goose Mecklenburg (Kuse) 2. x. 01	•	•	•	•	•	•
Subfam. **Anserinae**	**492**	671. *Anser spec.* Upland Goose England (B) 4. iv. 02	•	•		•	•	•
Subfam. **Plectropterinae**	**493**	73. *Aex sponsa.* Summer Duck N. America (Z) 25. v. 01	•	•	•	•	•	•
	494	219. *Rhodonessa caryophyllacea.* Pink-headed Duck India (Z) 14. viii. 01	•	•	•	•	•	\
	495	492. *Cairina moschata.* Muscovy Duck Trinidad, W.I. (Tulloch) 29. i. 02	•	•	•	•	•	tr ?
Subfam. **Cygninae**	**496**	23. *Cygnus olor* Gm. Mute Swan England (Lane) 24. iv. 01	•	•	•	•	tr ?	•
	497	309. *Cygnus olor* (B) 8. xi. 01	•	•	•	•	*	•
	498	121. *Cygnus olor* (Z) 25. vi. 01	•	•	•	•	•	•

5. *Order* FALCONIFORMES

			Man	Monkey	Cat	Dog	Pig	Hog-deer
1. Suborder Cathartae	**499**	495. *Cathartes atratus* Bartr. Black Vulture S. America (Z) 10. ii. 02	•	•	•	•	•	tr ?
Fam. **Cathartidae**	**500**	520. *Cathartes atratus* 19. ii. 02	•	•	•	•	•	•
	501	438. *Cathartes atratus* ? " Corbeau " Trinidad, B.W.Indies(Tulloch) 30.xii.01	•	•	•	•	•	
	502	369. *Sarcorhamphus gryphus* L. Condor Vulture S. America (Z) 27. xii. 01, *fl. ser.*	•	•	•	•	•	
2. Suborder Accipitres	**503**	707. *Gyps fulvescens.* Vulture Kashmir (Donald) 22. iii. 02	•		•	•	•	
Fam. **Vulturidae**	**504**	188. *Vultur monachus* ? Vulture S. Sylhet, India (Dalgetty) 12. ix. 01	•	•	•	•	•	
Fam. **Falconidae**	**505**	888. *Circus macrurus.* Pale Harrier Kashmir State (Donald) 17. iv. 02	•	•	•		•	
	506	559. *Milvus govinda.* Pariah Kite Bombay, India (Mason) 3. ii. 02	•	•	•	•	•	
	507	716. *Milvus govinda* Kashmir (Donald) 2. iii. 02	•	•	•	•	•	
	508	93. *Accipiter nisus.* Sparrow Hawk Ireland (Dillon) 7. v. 01	•	•	•	•	•	
	509	193. *Accipiter nisus* England (Beauford) 9. viii. 01	•	•	•	•	•	

505 Note source and p. 63 in text.

	Mammalia						Aves				Reptilia		Amphibia	Crustacea
Mexican-deer	Antelope	Sheep	Ox	Horse	Wallaby		Fowl	Ostrich	Fowl-egg	Emu-egg	Turtle	Alligator	Frog	Lobster
					•		*15	•	•					
•	•	•	•	•	•		*d	tr	•	•	•	•	•	•
•	•	•	•	•	•		+D	+60 D		•	•		•	
•	•	•	•	•	•		*d	tr						
•	•	•	•	•	•		+α	*tr	•	tr?	•	•		•
•	\	•	•	•	•		+d	\	d	\	•	tr?		•
•	•	•	•	•	•		+D	*d	*		•	•		
	•	*	*	•	•		+D		•	•				
•	•	•	•	•	•		*d	*	*d					•
•	•	•	•	•	•		+α	*d	+					
•	•	•	•	•	•		*D	*d	•	•	•	•	•	
•	•	•	•	•	•		+D	*D	•	*d	•	•		•
•	•	•	•	•	•		+D	*d	•	•	•	•	•	
•	•	•	•	•	•		+α	+d	•	•				•
•	•	•	•	•	•		*60		•	•	•	•	•	
•	•	•	•	•	•		+D	tr	•	•	•	•	•	
			•	•	•									
•	•	•	•	•	•		d	*D	•	•	•		•	
•	•	•	•	•	•		*5		•	•	•			
•	•	•	•	•	•		*	*30	•	•	•	•	•	•
•	•	•	•	•	•		*d	*tr	•	•	•	•		•

507 Also • with 4 antisera for seal, hedgehog, hyaena, llama.

Class AVES
II. Division CARINATAE
5. Order FALCONIFORMES
ANTISERA FOR.........

			Man	Monkey	Cat	Dog	Pig	Hog-deer
2. Suborder								
Accipitres								
Fam.	**510**	249. *Accipiter nisus.* Sparrow Hawk (Farren) viii. 01		•	•		•	•
Falconidae	**511**	464. *Accipiter nisus* (Bird) 27. xii. 01	•		•	•		•
	512	828. *Accipiter nisus* (R) 2. vi. 02	•	•	•	•		
	513	709. *Accipiter virgatus.* Sparrow Hawk Kashmir (Donald) 11. iv. 02	•	•	•	•		•
	514	162. *Astur palumbarius.* Goshawk Germany (Kuse) 27. viii. 01	•	•	•	•		•
	515	213. *Astur palumbarius* England (Z) 14. viii. 01	•	•	•	•	•	•
	516	79. *Falco peregrinus.* Peregrine Falcon England (Lane) vi. 01	•	•	•	•	•	•
	517	530. *Falco peregrinus.* India (Inglis) 17. i. 02	•	•	•	•		•
	518	91. *Falco tinnunculus.* Kestrel Ireland (Dillon) 16. v. 01	•	•	•	•	•	•
	519	180. *Falco tinnunculus* Denmark (Schierbeck) ix. 01	•	•	•	•		•
	520	185. *Falco tinnunculus* Germany (Schlieffen) ix. 01	•	•	•	•		•
	521	400. *Falco tinnunculus* England (Lane) vi. 01	•	•	•	•	•	•
	522	775. *Falco aesalon.* Merlin Orkney, N.B. (B) 21. iv. 02	•	•	•	•	•	•
	523	252. *Tinnunculus alaudarius.* Kestrel Hawk England (Farren) vi. 01	•	•	•	•		•
	524	830. *Tinnunculus alaudarius* (R) 5. vi. 02	•	•	•	•	•	•
	525	706. *Tinnunculus spec.* Cerchner's Kestrel Kashmir State (Donald) 22. iii. 02	•	•	•	•		•
	526	139. *Buteo vulgaris* L. Buzzard Germany (Kuse) 9. viii. 01	•	•	•	•		tr ?
	527	176. *Buteo vulgaris* (N) 27. viii. 01	•	•	•	•		•
	528	177. *Buteo vulgaris* 2. ix. 01	•		•	•	•	
	529	660. *Buteo vulgaris* N. Wales (B) 6. iv. 02	•		•	•	tr ?	
	530	829. *Buteo vulgaris* England (R) 15. ii. 02	•	•	•	•	•	•
	531	603. *Buteo lagopus* (Gmel.) "Rauchfussbussard" Germany (Lühe) 11. iii. 02	•	•	•	•	•	•
	532	714. *Buteo ferox.* Long-legged Buzzard Kashmir (Donald) 8. iii. 02	•	•	•	•		•
	533	212. *Archibuteo lagopus.* Rough-legged Buzzard Norway (Z) 14. viii. 01	•	•	•	•	•	•
	534	866. *Butastur indicus.* Hawk Okenawa, Japan (R) 14. iii. 02	•	•	•	•		•
	535	211. *Aquila chrysaetus.* Golden Eagle Norway (Z) 11. ix. 01	•	•	•	•	•	•
	536	278. *Aquila chrysaetus* Scotland (Z) 14. xi. 01	•	•	•	•		•

513 Note source; appeared to go slightly into solution.
525, 532 Note source and p. 63 in text.

Mammalia						Aves				Reptilia		Amphibia	Crustacea
Mexican-deer	Antelope	Sheep	Ox	Horse	Wallaby	Fowl	Ostrich	Fowl-egg	Emu-egg	Turtle	Alligator	Frog	Lobster
·	·				·	+ D	* tr	· cl	·	·	·	·	·
·	·	·	·	·	·	+ D	* d	·		·	·	·	·
·					·	· tr				·	·	·	·
·	·	·	·	·	·					·	·	·	·
·	·	·	·	·	·	* d	·	·	·	·	·	·	·
·	·	·	·	·	·	d	* d	·		·	· tr ?	·	·
·	·	·	·	·	·	·	* 30 d	·		·	·	·	·
·	·	·	·	·	·	+ D	* d	·	* 210 d	·	·	·	·
·	·	·	·	·	·	*	·	·	·	·	·	·	·
·	·	·	·	·	·	d	* tr	·		·	·	·	·
* ? 265	·	·	·	·	·		* tr	·		·	·	·	·
·	·	·	·	·	·	tr ?	·			·	·	·	·
·	·	·	·	·	·	*	*	·		·	·	·	·
·	·	·	·	·	·	* 15 d	·			·	·	·	·
·	·	·	·	·	·	*	*	·	·	·	·	·	·
·	·	·	·	·	·	d	tr	* ?		·	·	·	·
·	·	·	·	·	·	* d	tr	* ?		·	·	·	·
					·	tr	* 60 d	·		·	·	·	·
					·	+ 5 d	·	·		·	·	·	·
·	·	·	·		·	*	* 90	·		·	·	·	·
·	·	·		·		* d	* d	·		·	·	·	·
·	·	·	·	·	·	·				·	·	·	·
·	·	·	·	·	·	+ d	* d	·		·	·	·	·
·	·	·	·	·	·	+ d	* d	*		·	·	·	·

531 · also with anti-seal, hyaena, hedgehog, llama.
534 Foam-test practically negative.

275

Class AVES
II. Division CARINATAE
5. Order FALCONIFORMES
ANTISERA FOR.........

			Man	Monkey	Cat	Dog	Pig	Hog-deer
2. Suborder Accipitres Fam. Falconidae	**537**	531. *Haliaëtus indus.* Brahminy Kite India (Inglis) 4. ii. 02	•		•	•	•	•
	538	214. *Haliaëtus leucogaster.* White-tailed Sea Eagle Tasmania (Z) 7. x. 01	•	•	•	•		•
	539	721. *Haliaëtus leucogaster* Chatrapur, India (Fischer) 23. ii. 02						
	540	539. *Nisaetus fasciatus.* Bonelli's Eagle Deccan, India (Betham) 14. i. 02	•		•	•	•	•
	541	886. *Nisaetus prunatus.* Booted Eagle Kashmir State, India (Donald) 18. iv. 02	•	•	•	•		•
	542	69. *Spizaetus bellicosus.* Martial Hawk Eagle Brit. E. Africa (Z) 25. v. 01	•		•	•	•	•
	543	708. *Spilornis cheela.* Crested Serpent Eagle Bhadarwa, Kashmir State (Donald) 21. iii. 02	•		•	•	•	

6. Order GALLIFORMES

			Man	Monkey	Cat	Dog	Pig	Hog-deer
1. Suborder Galli Fam. Cracidae	**544**	218. *Crax globicera.* Globose Curassow Honduras (Z) 30. viii. 01	•	•	•	•	•	•
	545	389. *Meleagris gallipavo.* Domestic Turkey England (Mitchell) 24. xii. 01	•		•	•	•	•
	546	178. *Tetrao tetrix.* Blackcock Scotland (R) 28. viii. 01	•		•	•		•
	547	168. *Lagopus mutus* Leach. Ptarmigan Scotland (R) 21. viii. 01	•		•	•		•
	548	108. *Lagopus scoticus.* Grouse Scotland (Haldane) 12. viii. 01	•	•	•	•	•	•
	549	179. *Lagopus scoticus* (R) 28. viii. 01	•		•	•	•	
	550	248. *Lagopus scoticus* England (Wheeler) 3. xi. 01	•		•	•	•	
	551	528. *Coturnix communis.* Grey Quail India (Inglis) 3. ii. 02	/		•	•	•	
	552	170. *Perdix cinerea* L. Partridge Germany (Kuse) 13. viii. 01	•		•	•	•	
	553	171. *Perdix cinerea* Seeland (Schierbeck) ix. 01			•	•	•	•
	554	172. *Perdix cinerea* Germany (Schlieffen) ix. 01			•	•		•
	555	173. *Perdix cinerea* England (Leighton) 11. x. 01			•	•	•	•
	556	174. *Perdix cinerea* England (Wheeler) 20. x. 01			•			
	557	724. *Tetraogallus himalayensis.* Himalayan Snow Cock Kashmir (Donald) 27. ii. 02	/		/	•		/
	558	728. *Caccabis chukor.* "Chukor" Partridge Kashmir (Donald) 20. ii. 02	•		•	•	•	
	559	826. *Phasianus colchicus* L. Pheasant England (R) 13. i. 02	*?30	•	•	•	•	

541 No foam.   **543** Solution ?, note source.

	Mammalia						Aves				Reptilia		Amphibia	Crustacea
Mexican-deer	Antelope	Sheep	Ox	Horse	Wallaby	Fowl	Ostrich	Fowl-egg	Emu-egg	Turtle	Alligator	Frog	Lobster
•	•	•	•	•	•	+ D	* D	•	* d	•		•	
•	•	•	•	•	•	* d	* d			•		•	•
•	•	•	•	•	•	* 5 tr			tr	•			
•	•	•	•	•	•	+ D	+ D	•	tr ?	•			
			•	•	•	•	•			•	•		•
•	•	•	•	•	•	+ D	+ 30 d	•	•	•		•	
•	•	•	•	•	•	•	•						
•	•	•	•	•	•	* d	* 60 d	d	•	•	•	•	•
•	•	•	•	•	•	+ D	* d	•	•	•	•	•	•
•	•	•	•	•	•	* d	* tr	?	•	•	•	•	•
•	•	•	•	•	•	* d	* tr	•	*	•	•	•	•
•	•	•	•	•	•	*	* 30 tr	*	•	•	•	•	•
•	•	•	•	•	•	+ d	× d	?	•	•	•	•	•
•	•	•	•	•	•	+ D	* tr	•	cl	•	•	•	•
•	•	•	•	•	•	+ D	* d	•	* 210	•	•	•	•
•	•	•	•	•	•	+ D	* tr	d	•	•	•	•	•
•	•	•	•	•	•	+ D	* d	* d	•	•	•	•	•
•	•	•	•	•	•	+ D	* tr	* d	•	•	•	•	•
•	•	•	•	•	•	* D	* tr	* d	*	•	•	•	•
•	•	•	•	•	•	+ D	* tr	* d	•	•	•	•	•
•	/	•	•	•	•	•	•	•	/	•			•
•	•	•	•	•	•	* tr	* 5 tr	•		•	•	•	
					•	* 5 d	•	•					

557 Note source and text, p. 63.

277

Class AVES
II. Division CARINATAE
6. Order GALLIFORMES
ANTISERA FOR.........

			Mammalia					
			Man	Monkey	Cat	Dog	Pig	Hog-deer
1. Suborder Galli **Fam. Cracidae**	560	683. *Phasianus colchicus* L. Pheasant Mecklenburg (Kuse) 8. xi. 01	•	•	•	•*	•	•
	561	334. *Phasianus colchicus* England (Leighton) 26. xi. 01	•	•	•	*?	•	•
	562	244. *Phasianus colchicus* 23. x. 01		•	•	•	•	•
	563	181. *Phasianus colchicus* Denmark (Schierbeck) ix. 01	•	•	•	•	•	•
	564	403. *Phasianus colchicus × sinensis.* Pheasant Ireland (N) 4. v. 01	*	•	•	tr ?	•	•
	565	750. *Argusianus grayi.* Argus Pheasant Sarawak, Borneo (Hose) p. 10. iv. 02						
	566	722. *Pucrasia macrolopha.* "Koklass" Pheasant Kashmir (Donald) 21. ii. 02	•		•	•	•	•
	567	729. *Lophophorus impeyanus.* Pheasant Kashmir (Donald) 21. ii. 02	•		•	•	•	•
	568	217. *Thaumalea amherstiae.* Amherst Pheasant China (Z) 24. viii. 01	•		•	•	•	•
	569	308. *Thaumalea picta.* Golden Pheasant (B) 7. xi. 01	•		•	•	•	•
	570	751. *Euplocomus nobilis.* Fireback Pheasant Sarawak, Borneo (Hose) 10. iv. 02						
	571	731. *Euplocomus leucomelamus?* Black-crested Kalij Pheasant Kashmir (Donald) 17. ii. 02	•		•	•	•	•
	572	12. *Gallus domesticus.* Fowl, scales England (N) 4. iii. 01		•	•	•	•	•
	573	40. *Gallus domesticus.* Putrid in tube 19. x. 01		•	•		•	•
	574	182. *Gallus domesticus* Denmark (Schierbeck) ix. 01		•	•	•	•	•
	575	401. *Gallus domesticus* England (N) 19. x. 01		•	•	•	•	•

7. Order GRUIFORMES

			Man	Monkey	Cat	Dog	Pig	Hog-deer
Fam. Rallidae	576	843. *Rallus aquaticus* L. Water Rail England (R) 7. iv. 02	•		•	•	•	•
	577	67. *Crex pratensis.* Land Rail Ireland (N) 4. iv. 01	•		•	•	•	•
	578	266. *Fulica atra* L. Coot England (R) 5. xi. 01	•	•	•	•	•	•
	579	463. *Fulica atra* (Bird) 6. xii. 01	•		•	•	•	•
	580	637. *Fulica atra* 31. i. 02	•		•	•	•	•
Fam. Gruidae	581	184. *Grus communis.* Crane Germany (Schlieffen) ix. 01	•		•	•	•	•

560 · with 4 antisera for seal, hyaena, hedgehog, llama.

	Mammalia						Aves					Reptilia			Amphibia		Crustacea
Mexican-deer	Antelope	Sheep	Ox	Horse	Wallaby		Fowl	Ostrich	Fowl-egg	Emu-egg		Turtle	Alligator		Frog		Lobster
•	•	•	•	•	•		**+D**	D	•	* 180		•			•		
•	•	•	•	•	•		**+α**	+ tr	+	•		•	•				•
•	•	•	•	•	•		+ D	* tr	*	•		•	•		•		•
•	•	•	•	•	•		+ d	* tr	?	•		•	•		•		•
•	•	•	•	•	•		*	* tr		•		•	•				
•	•	•	•	•	•		+ tr		•	tr		•					
•	•	•	•	•	•		* tr		* 	* tr		•					
•	•	•	•	•	•		+ D	* d	d	•		•	•		•		•
•	•	•	•	•	•		* d	× d	•	tr ?		•	•		•		
•	•	•	•	•	•		* tr		•	•		•	•				•
•	•	•	•	•	•		**+ D**	+ 30 d	•			•	•				•
•	•	•	•	•	•		**+ D**		+ 30 d	tr					•		•
•	•	•	•	•	•		**+ D**	* d	* d	•					•		•
•	•	•	•	•	•		**+ α**	* tr ?									
				•			* 240 tr		•								
•	•	•	•	•	•		+ α	* 30 tr ?	•	•		•	•		•		•
•	•	•	•	•	•		* d	* tr	•	•					•		•
•	•	•	•	•	•		+ D	**+ D**	•	•		•			•		
							*	* 60	•								
•	•	•	•	•	•		+ D	× tr	*	*		•	•		•		•

565, 570 Certainly insoluble.

Class AVES
II. *Division* CARINATAE
8. *Order* CHARADRIIFORMES
ANTISERA FOR.........

			Mammalia					
			Man	Monkey	Cat	Dog	Pig	Hog-deer
1. *Suborder* Limicolae	**582**	167.	*Charadrius pluvialis* L. Golden Plover Scotland (R) 21. viii. 01	•	•	•	•	•	•
Fam. Charadriidae	**583**	647.	*Aegialitis hiaticola.* Ringed Plover England (B) 4. ii. 02	•		•	•	•	•
Subfam. Charadriinae	**584**	882.	*Aegialitis alexandrina* var. *dealbatus.* Snipe Okenawa, Japan (R) 15. iii. 02						
	585	535.	*Lobivanellus malabaricus.* Yellow-wattled Lapwing Ahmedabad, India (Mason) 26. i. 02	•		•	•	•	•
	586	551.	*Recurvirostra avocetta.* Avocet Tirhut, India (Inglis) 17. i. 02	•		•	•	•	•
	587	552.	*Himantopus candidus.* Stilt Tirhut, India (Inglis) 14. i. 02	•		•	•	•	•
Subfam. Tringinae	**588**	776.	*Tringa alpina.* Dunlin England (B) 19. v. 02	•		•	•	•	•
	589	845.	*Tringa canutus.* Knot England (R) 11. ii. 02	•	•·	•	•	•	•
	590	524.	*Totanus stagnalis.* Marsh Sandpiper India (Inglis) 2. ii. 02	•		•	•	•	•
	591	525.	*Totanus glareola.* Wood Sandpiper India (Inglis) 2. ii. 02	•		•	•	•	•
	592	523.	*Totanus fuscus.* Spotted Redshank India (Inglis) 1. ii. 02	•		•	•	•	•
	593	767.	*Totanus hypoleucus* Temm. Sandpiper England (B) 19. v. 02	•		•	•	•	•
	594	522.	*Limosa belgica.* Black-tailed Godwit India (Inglis) 14. i. 02	•		•	•	*\n?\n240	•
	595	674.	*Numenius arquata* L. Curlew England (B) iii. 02	•		•	•	•	•
	596	766.	*Numenius phaeopus.* European Whimbrel England (B) 19. v. 02	•		•	•	•	•
	597	836.	*Numenius phaeopus* (Clarke) 19. v. 02	•	•	•	•		
Subfam. Scolopacinae	**598**	673.	*Scolopax rusticola* L. Woodcock England (B) ii. 02			•	•	•	•
	599·	454.	*Scolopax rusticola* (Clarke) 24. xii. 01	•	•	•		•	•
	600	336.	*Scolopax rusticola* (Bird) 28. xi. 01	•	•	•	•	•	•
	601	526.	*Gallinago caelestis.* Fantail Snipe India (Inglis) 2. ii. 02	•	•	•	•	•	•
	602	837.	*Gallinago caelestis.* Full Snipe England (R) 21. ii. 02	•	•	•	•	•	•
	603	331.	*Gallinago caelestis* vel *media.* Snipe India (Dalgetty) 4. xi. 01	•	•	•	•	•	•
	604	264.	*Gallinago gallinula.* Jack Snipe England (R) 5. xi. 01	•	•	•	•	•	•

584 Reactions all negative, did not foam.

Mammalia						Aves				Reptilia		Amphibia	Crustacea
Mexican-deer	Antelope	Sheep	Ox	Horse	Wallaby	Fowl	Ostrich	Fowl-egg	Emu-egg	Turtle	Alligator	Frog	Lobster
·	·	·	·	·	·	* d	* d	·	·	·		·	·
·	·	·	·	·	·	*	* 60	·	·	·			
·	·	·	·	·	·	* 210	240	·	·	·		·	
·	·	·	·	·	·	* d	* d	·	240	·			
·	·	·	·	·	·	* d	* d	·	*	* ? 180	·	·	
·	·	·	·	·	·	45 tr	·	·	·	·		·	
·	·	·	·	·	·	* 35 tr	·	·	·				
·	·	·	·	·	·	* 210 d	* d	·	·	·		·	
·	·	·	·	·	·	* 210 d	* d	·	* 210	·		·	
·	·	·	·	·	·	× D	* D	·	* 210 d	·		·	
·	·	·	·	·	·	* 45 tr	·	·	·	·		·	
·	·	* ? 240	·	·	·	+ D	+ D	·	* 210 d	·		·	
·	·	·	·	·	·	+ d	· d	·	·	·		·	
·	·	·	·	·	·	* 45 tr	·	·	·	·		·	
·	·	·	·	·	·	× 5 d	·	·	·	·			
·	·	·	·	·	·	+ 60 d	d	·	·	·		·	
* ?	·	·	·	·	·	* 60 d	* D	·	·	·	·	·	·
·	·	·	·	·	·	+ b	* tr	*	·	·		·	·
·	·	·	·	·	·	D	* d	·	·	·		·	·
tr ?	·	·	·	·	·	180	* 30	·	·	·		·	·
·	·	·	·	·	·	* ?	*	·	·	·	·	·	·
·	·	·	·	·	·	* d	tr	·	·	·		·	·

595 · with anti-tortoise.   **599** · with anti-seal, hyaena, hedgehog, llama.

			Class **AVES** II. _Division_ **CARINATAE** 8. _Order_ **CHARADRIIFORMES** ANTISERA FOR.........	Man	Monkey	Cat	Dog	Pig	Hog-deer
1. _Suborder_ **Limicolae** _Fam._ **Charadriidae** _Subfam._ **Scolopacinae**	**605**	265.	_Gallinago gallinago._ Common Snipe England (R) 5. xi. 01	.	.	.	.	.	.
	606	730.	_Gallinago solitaria._ Solitary Snipe Kashmir (Donald) 21. ii. 02	.		.	.		.
	607	384.	_Gallinago_ spec. "Winter Snipe" Shanghai, China (Stanley) ca. 27. xi. 01	.			.		.
	608	862.	_Arenaria interpres._ Snipe Okenawa, Japan (R) 19. iii. 02						
	609	863.	_Tringoides hypoleucus._ Snipe Okenawa, Japan (R) 14. iii. 02						
	610	873.	_Ochthodromus mongolus._ Snipe Okenawa, Japan (R) 9. iii. 02						
Fam. **Chionididae**	**611**	71.	_Chionis alba_ Forst. Yellow-billed Sheath- bill England (Z) 28. v. 01	.	.	.	.	.	.
Fam. **Parridae**	**612**	614.	_Parra (africana?)_ Jaçana Lake Nyassa, C. Africa (Dodds) p. 8. ii. 02						
Fam. **Laridae**	**613**	70.	_Larus argentatus._ Herring Gull England (Z) 28. v. 01	.		.	.	.	.
	614	462.	_Larus argentatus_ (Bird) 6. xii. 01	.		.	.	.	.
	615	653.	_Larus ridibundus_ L. Black-headed Gull England (B) 24. i. 02	.		.	.	.	.
	616	834.	_Larus ridibundus_ L. Black-headed Gull England (Clarke) 15. ii. 02	.	.	.	.	.	.
	617	652.	_Larus glaucus._ Glaucous Gull England (B) 3. i. 02	.		.	.	.	.
	618	550.	_Larus brunneicephalus._ Brown-headed Gull Darbhanga, India (Inglis) 28. i. 02	.		.	.	.	.
	619	344.	_Larus canus_ L. Common Gull England (Clarke) 16. xii. 01	.	/	/	.	.	/
	620	850.	_Larus..._spec? Little Gull England (Clarke) 10. ii. 02	.	.	.			
	621	546.	_Sterna anglica._ Gull-billed Tern Darbhanga, India (Inglis) 31. i. 02	.		.			
Fam. **Alcidae**	**622**	835.	_Uria troile._ Ringed Guillemot England (Clarke) 8. ii. 02	.	.	.	.		
2. _Suborder_ **Columbae** _Fam._ **Columbidae**	**623**	825.	_Columba palumbus_ L. Wood Pigeon England (R) 1. iii. 02	.	* ? 35	.	* 15		
	624	398.	_Columba palumbus_ (Bird) 28. xi. 01	.	.	.	.	.	.
	625	399.	_Columba palumbus_ (Wheeler) ca. 22. x. 01	.	.	.	.	.	.

607 Filter paper but slightly stained, diluted blood ?.
608 Foam-test doubtful, no reactions.   **609, 610** Foam-test negative, no reactions.

	Mammalia						Aves				Reptilia		Amphibia	Crustacea
Mexican-deer	Antelope	Sheep	Ox	Horse	Wallaby		Fowl	Ostrich	Fowl-egg	Emu-egg	Turtle	Alligator	Frog	Lobster
·	·	·	·	·	·		*d	·	·	·	·	·	·	·
·	·	·	·	·	·		*tr	·	·	·	·			
·	·	·	·	·	·		·	*	·	tr ?	·		·	·
·	·	·	·	·	·		+b	*? d	·	tr ?	·	·	·	·
·	·	·	·	·	·		+b	*? d	·	tr ?	·	·	·	·
·	·	·	·	·	·		*30 d	*d	·	·	·	·	·	·
·	·	·	·	·	·		*60 tr	*60 d	·	·	·	·	·	·
·	·	·	·	·	·		*5 d	·	·	·	·		·	·
·	·	·	·	·	·		*60 tr	·	·	·	·		·	·
·	·	·	·	·	·		+D	*d	·	·	·		·	·
/	/	·	·	·	/		·	*	·	/	·	·	·	·
·	·	·	·	·	·		×5 d	·	·	·	·		·	·
·	·	·	·	·	·		*d	*d	·	*? 240	·		·	·
·	·	·	·	·	·		·	·	tr	·	·			·
·	·	·	·	·	·		*? 180 tr?	·	·	·	·			·
·	·	·	·	·	·		+D	*tr	·	·	·	·	·	·
·	·	·	·	·	·		+D	*tr	·	·	·	·	·	·

612 Foam-test doubtful, all reactions negative.   619, 622 Sample greasy.

				Mammalia					
				Man	Monkey	Cat	Dog	Pig	Hog-deer
2. *Suborder* **Columbae** *Fam.* **Columbidae**	626	332.	*Columba palumbus* L. Wood Pigeon (Leighton) 26. xi. 01	•	•	•	•	•	?
	627	313.	*Columba palumbus* (R) 13. xi. 01	•	•	•	•	•	•
	628	175.	*Columba palumbus* Germany (N) 4. ix. 01	•	•	•	•	•	•
	629	138.	*Columba palumbus* (Kuse) 10. viii. 01	•	•	•	•	•	•
	630	13.	*Columba livia.* Domesticated Pigeon (scales) England (N) 11. iii. 01	•	•	•	•	tr ?	•
	631	21.	*Columba livia* 11. iii. 01	•	•	•	•	•	•
	632	41.	*Columba livia* 9. v. 01	•	•	•	•	•	/
	633	491.	*Columba livia* Trinidad (Tulloch) 19. i. 02	•	•	•	•	•	•
	634	723.	*Columba intermedia.* Blue Rock Pigeon Kashmir (Donald) 15. ii. 02	/	•	•	/	/	•
	635	547.	*Columba eversmanni.* Eastern Stock Pigeon Tirhut, India (Inglis) 17. i. 02	•	•	•	•	•	•
	636	763.	*Columba oenas* L. Stock Dove England (B) 24. iv. 02	•	•	•	•	•	•
	637	533.	*Turtur cambayensis.* Little Brown Dove Ahmedabad, India (Mason) 26. i. 02	•	•	•	•	•	•
	638	215.	*Geopelia striata.* Barred Dove India (Z) 30. ix. 01	•	•	•	•	•	•
	639	583.	*Caloenas nicobarica* Linn. Nicobar Pigeon Indian Archipelago (Z) 26. ii. 02	•	•	•	•	•	•
	640	72.	*Caloenas nicobarica* 28. v. 01	•	•	•	•	•	•

9. *Order* **CUCULIFORMES**

				Man	Monkey	Cat	Dog	Pig	Hog-deer
1. *Suborder* **Cuculi** *Fam.* **Cuculidae**	641	112.	*Cuculus canorus* L. Cuckoo Germany (Kuse) 7. viii. 01	•	•	•	•	•	•
	642	762.	*Cuculus canorus* England (B) 23. iv. 02	•	•	•	•	•	•
Fam. **Musophagidae**	643	893.	*Turacus buffoni* Vieill. Buffon's Touracou W. Africa (Z) 17. ix. 02 d. of ascites	tr ?	•	•	•	tr ?	•
2. *Suborder* **Psittaci**	644	87.	*Ara macao* × *A. militaris.* Macaw Hybrid Bred in Italy (Z) 24. vi. 01	•	•	•	•	•	* ?
	645	122.	*Cacatua roseicapilla* Vieill. Roseate Cockatoo Australia (Z) 22. vii. 01	•	•	•	•	•	•
	646	123.	*Cacatua leadbeateri* Vig. Leadbeater's Cockatoo Australia (Z) 12. vii. 01	•	•	•	•	•	•

639, 643 • with anti-seal, hyaena, hedgehog, llama.

	Mammalia						Aves					Reptilia		Amphibia	Crustacea
Mexican-deer	Antelope	Sheep	Ox	Horse	Wallaby		Fowl	Ostrich	Fowl-egg	Emu-egg		Turtle	Alligator	Frog	Lobster
·	·	·	·	·			+ D	*	·	·		·	·	·	·
·	·	·	·	·	·		* d	tr	·	·		·	·	·	·
·	·	·	·	·	·		* d	* tr	?	·		·	·	·	·
·	·	·	·	·	·		* d	*	·	·		·	·	·	·
·	·	·	·	·			* d		·	·		·	·		·
·	·	·	·	·	·		·	·	·	·		·	·		·
·	/	·	·	·	·		·	·	/	/		·	·	·	·
·	·	·	·	·	·		* D	d	·	·		·	·		·
·	/	/	·	·	·		/	·	·	/		·	·	·	·
·	·	* ? 240	* ? 240	·	·		+ D	* D	·	* d		·	·	·	·
·	·	·	·	·	·		* 45 tr		·	·		·	·		·
·	·	·	·	·	·		* 210 d	d	·	* d		·	·	·	·
·	·	·	·	·	·		* d	* d	·	·		·	·	·	·
·	·	·	·	·	·		+ D	+ D	·	* 180		·	·	·	·
·	·	·	·	·	·		+ D	* d	·	tr ?		·	·	·	·
·	·	·	·	·	·		* 45 tr	+ tr	·	·		·	·	·	·
		·	·	·	·		+ 30 D	·	·	·		·	·	·	·
·	·	*	·	·	·		+ D	+ 30 d	·	·		·	·	·	·
tr ?	·	·	·	·	·		+ D	* d	·	·		·	·	·	·
·	·	·	·	·	·		+ D	* d	·	·		·	·	·	·

285

Class **AVES**

II. *Division* **CARINATAE**

9. *Order* **CUCULIFORMES**

ANTISERA FOR........

			Mammalia					
			Man	Monkey	Cat	Dog	Pig	Hog-deer
2. *Suborder* Psittaci	**647**	363.	*Chrysotis versicolor.* Blue-faced Amazon Parrot St Lucia W. I. (Z) 27. xi. 01	·	·	·	·	·	·
	648	346.	*Chrysotis versicolor* S. America (B) 25. xi. 01	·	·	·	·	·	
	649	814.	*Loriculus indicus.* Parrot Ceylon (R) 22. i. 02						
	650	125.	*Pœocephalus senegalus* Linn. Senegal Parrot W. Africa (Z) 24. vii. 01	·	·	·	·	·	
	651	223.	*Pœocephalus gulielmi.* Jardine's Parrot W. Africa (Z) 20. viii. 01	·	·	·	·	·	
	652	370.	*Polytelis melanura* Vig. Black-tailed Paroquet Australia (Z) 23. xii. 01	·	·	·	·	·	
	653	560.	*Palaeornis torquata.* Rose-ringed Paroquet Ahmedabad, India (Mason) 27. i. 02	·		·		·	
	654	302.	*Conurus auricapillus.* Golden-Headed Conure S. America (Z) 18. xi. 01	·	·	·	·	·	
	655	741.	*Nestor notabilis* Gould. Ka-Ka Parrot N. Zealand (Z) 2. vi. 02 killed, para- lyzed, *fl.*	·		·	·	·	
	656	156.	*Nestor notabilis* 8. viii. 01	·	·	·	·	·	

10. *Order* CORACIIFORMES

			Man	Monkey	Cat	Dog	Pig	Hog-deer
1. *Suborder* Coraciae *Fam.* **Alcedinidae**	**657**	841.	*Alcedo ispida.* Kingfisher England (Clarke) 22. i. 02	·		·	* ? 240		
	658	879.	*Alcedo ispida* var. *bengalensis.* Kingfisher Japan (R) 14. iii. 02						
	659	877.	*Alcedo ispida* var. *bengalensis* (R) 11. iii. 02						
	660	811.	*Halcyon smyrnensis fuscus.* Kingfisher Ceylon (R) 27. i. 02	·	·	·	·	·	
Fam. **Meropidae**	**661**	810.	*Merops phillipinus.* Bee Eater Ceylon (R) 27. i. 02	·	·	·	·	·	
Fam. **Bucerotidae**	**662**	216.	*Rhytodoceros undulatus.* Malayan Wrinkled Hornbill Malacca (Z) 8. x. 01	·	·	·	·	·	·
2. *Suborder* Striges *Fam.* **Strigidae**	**663**	540.	*Strix javanica.* Indian Screech Owl Bombay, India (Phipson) 23. i. 02			·	·	·	
	664	682.	*Strix flammea* L. Barn Owl Mecklenburg (Kuse) 5. x. 01	·		·	·	·	
	665	521.	*Strix flammea* India (Inglis) 28. i. 02	·		·	·	·	·

658, 659 Foam-test doubtful, no reactions.

Mammalia							Aves				Reptilia		Amphibia	Crustacea
Mexican-deer	Antelope	Sheep	Ox	Horse	Wallaby		Fowl	Ostrich	Fowl-egg	Emu-egg	Turtle	Alligator	Frog	Lobster
·	·	·	·	·	·		+ D	+ d	·	·	·	·	·	·
·	·	·	·	·	·		+ D /	+ d	*	·	·	·	·	·
·	·	·	·	·	·			* 30	tr ?	·	·	·	·	·
·	·	·	·	·	·		*	* 30 d	*	·	·	·	·	·
·	·	*	·	·	·		+ d	* tr	·	·	·	·	·	·
·	·	·	·	·	·		+ D	+ d	×	·	·	·	·	·
·	·	·	·	·	·		* 60 d	D	·	·	·	·	·	·
·	·	·	·	·	·		* d	* tr	×	·	·	·	·	·
·	tr ?	·	tr ?	·	·		+ 45 d		tr ?	·	·	·	·	·
·	·	·	·	·	·		* d	* ? tr	* tr ?	·	·	·	·	·
				·			· tr							
·	·						/	* 30	·	·				
·	·						/	* 30	·	·				
·		·	·	·			* d	* d	*	·			·	·
·	·	·	·	·			* d	+ D	·	* 240 tr ?	·		·	
·	·	·	·	·			+ 60 D	* D	·		·		·	
·	·	·	·	·			* d	* d	?	·			·	

287

Class AVES
II. Division CARINATAE
10. Order CORACIIFORMES
ANTISERA FOR.........

			Species / Locality	Man	Monkey	Cat	Dog	Pig	Hog-deer
2. Suborder **Striges** *Fam.* **Strigidae**	666	250.	*Strix flammea* L. Barn Owl England (Farren) v. 01	•	•	•	•	•	•
	667	310.	*Strix flammea* (B) 7. xi. 01	•	•	•	•	•	•
	668	345.	*Syrnium aluco* L. Tawny Owl England (Clarke) 13. xii. 01	•	•	•	•	•	•
	669	638.	*Asio brachyotus.* Long-eared Owl England (B) 20. xii. 01	•	•	•	•	•	•
	670	827.	*Asio brachyotus* (Clarke) 12. iv. 02	•		•	•	•	•
	671	532.	*Bubo bengalensis.* Rock Horned Owl Ahmedabad, India (Mason) 26. i. 02	•		•	•	•	•
	672	887.	*Bubo bengalensis* Kashmir State (Donald) 17. iv. 02						
	673	279.	*Bubo maximus.* Great Eagle Owl Europe (Z) 11. xi. 01	•	•	•	•	•	•
	674	281.	*Nyctea scandiaca.* Snowy Owl Bylott Islands. Lancaster S'd. (Z) 14. xi. 01	•	•	•	•		
3. Suborder **Pici** *Fam.* **Capitonidae**	675	720.	*Xantholaema hematocephala?* Crimson-breasted Barbet Bombay, India (Phipson) 18. ii. 02						
	676	812.	*Xantholaema rubricapilla.* Bar-bit Ceylon (R) 27. i. 02	•	•	•	•	•	•
Fam. **Rhamphastidae**	677	624.	*Rhamphastus ariel.* Toucan Brazil (Hagmann) 4. iii. 02	•	•	•	•	•	•
Fam. **Picidae**	678	639.	*Picus major* L. Great Spotted Woodpecker England (B) 4. ii. 02	•		•	•	•	•
	679	452.	*Picus major* (Clarke) 21. i. 02	•	•	•	•	•	•
	680	640.	*Gecinus viridis* Linn. Green Woodpecker England (B) 12. ii. 02	•		•	•	•	•
	681	307.	*Gecinus viridis* (B) 11. xi. 01	•		•	•	•	•
	682	678.	*Picus* spec. Woodpecker Mecklenburg (Kuse) 14. xii. 01	•		•	•	•	•
	683	761.	*Dendrocopus minor.* Lesser-spotted Woodpecker England (B) 12. v. 02	•		•	•	•	•
	684	669.	*Iynx torquilla* L. Wryneck England (B) 10. iv. 02	•		•	•	•	•

11. Order PASSERES

			Species / Locality	Man	Monkey	Cat	Dog	Pig	Hog-deer
Fam. **Alaudidae**	685	544.	*Pyrrhulauda grisea.* Ashy-crowned Finch Lark Darbhanga, India (Inglis) 4. ii. 02	•		•	•	•	•

672 Foam test doubtful, reactions all negative.
675 Foam test negative or doubtful, no reactions.

Mammalia						Aves				Reptilia		Amphibia	Crustacea
Mexican-deer	Antelope	Sheep	Ox	Horse	Wallaby	Fowl	Ostrich	Fowl-egg	Emu-egg	Turtle	Alligator	Frog	Lobster
•	•	•	•	•		*d	tr	•	•	•	•	•	•
•	•	•	•			*d	×d	* ?	•	•	•		•
•	•	•	•	•		+	*tr	•	tr ?	•	•	•	•
•	•	•	•	•		ᗡ	tr	•	•	•		•	
•						*	*60	•	•			•	
•					•	×5D	•	*15					
•	•	•	•	•	•	+D	+D	•	*210 d			•	
•	•	•	•	•	•	+d	*tr	•	•	•	•	•	•
•	•	•	•	•	•	+d	*tr	*	*?	•	•	•	•
•					•	/	*30	•					
•	•	•	•	•	•	*60 tr?	*45 d	•	•			•	
•	•	•	•	•	•	*	•	•	•			•	•
•	•	•	•	•	•	*60 D	*120 d	•	•	•	•	•	•
•	•	•	•	•	•	*	*60	•	•				•
•	•	•	•	•	•	d	*tr	*?	•			•	
•	•	•	•	•	•	*60 d	*d	•	•	•		•	
•	•	•	•	•	•	*45 tr	•	•					
•	•	•	•	•	•	+60 d	tr	•	•			•	•
•	•	•	•	•	•	*d	*d	•	•				

676 Solution weak.
677 • with anti-seal (* ? 85), hyaena, llama, hedgehog.
682 • with anti-tortoise.

Class AVES
II. *Division* CARINATAE
11. *Order* PASSERES
ANTISERA FOR.........

				Man	Monkey	Cat	Dog	Pig	Hog-deer
Fam. **Alaudidae**	**686**	534.	*Pyrrhulauda grisea.* Ashy-crowned Finch Lark Ahmedabad, India (Mason) 26. i. 02	•		•	•	•	•
	687	641.	*Alauda arvensis* L. Skylark England (B) 28. i. 02	•		•	•	•	•
Fam. **Motacillidae**	**688**	871.	*Motacilla boarula melanope.* Wagtail Okenawa, Japan (R) 13. iii. 02						
	689	349.	*Motacilla lugubris.* Pied Wagtail England (B) xi. 01	•			•		
	690	901.	*Motacilla lugubris* (R) 25. viii. 02	•	•			•	
	691	545.	*Anthus maculatus.* Indian Tree Pipit Darbhanga, India (Inglis) 31. i. 02	•		•	•	•	
	692	867.	*Anthus maculatus* Okenawa, Japan (R) 14. iii. 02						
	693	666.	*Anthus trivialis.* Tree Pipit England (B) 9. iv. 02	•		•	•		
	694	450.	*Anthus pratensis.* Meadow Pipit England (Clarke) 19. xii. 01	•			•		•
	695	451.	*Anthus obscurus.* Rock Pipit England (Clarke) 18. xii. 01	•	•	•	•	•	•
Fam. **Timeliidae**	**696**	537.	*Melacocercus terricolor.* "Sat-bhai" or Bengal Warbler Deccan, India (Phipson) 10. i. 02	•		•	•	•	
Fam. **Pycnonotidae**	**697**	538.	*Pycnonotus haemorrhous.* Common Madras Bulbul Deccan, India (Phipson) i. 02	•		•	•	•	•
	698	817.	*Kelaartia penicillata.* Yellow Bulbul Ceylon (R) 31. i. 02	•		•	•	•	•
	699	865.	*Hypsipetes squamiceps pryeri* Okenawa, Japan (R) 10. iii. 02	•	•	•	•	•	•
	700	815.	*Hypsipetes ganeesa* Ceylon (R) 31. i. 02	•	•	•	•	•	•
Fam. **Muscicapidae**	**701**	902.	*Muscicapa grisola.* Spotted Fly-catcher England (R) 22. viii. 02	•	•			•	
Fam. **Turdidae**	**702**	82.	*Turdus musicus* L. Thrush England (N) 14. vi. 01	•		•	•	•	
	703	194.	*Turdus musicus* (Garrood) 28. viii. 01	•	•	•	•		•
	704	195.	*Turdus musicus* (R) 16. viii. 01		•	•		•	/
	705	318.	*Turdus musicus* (R) 2. xii. 1901	•	•	•	•	•	•
	706	319.	*Turdus viscivorous.* Mistletoe Thrush England (R) 30. xi. 01	•		•	•	•	•
	707	465.	*Turdus pilaris.* Fieldfare England (Bird) xii. 01	•	•	•	•	•	•

688 Foam-test doubtful, all reactions negative.   **690** • with anti-hyaena and hedgehog.

Mammalia						Aves				Reptilia		Amphibia	Crustacea
Mexican-deer	Antelope	Sheep	Ox	Horse	Wallaby	Fowl	Ostrich	Fowl-egg	Emu-egg	Turtle	Alligator	Frog	Lobster
·	·	·	·	·	·	*210 d	*240 d	·	*240			·	
·	·	·	·	·	·	*	·	·		·			
		·	·	·	·	+ b	*tr?	·	*tr?	·	·		·
			·	·		*180 tr		·					
·	·	·	·	·	·	*d	*d		*240	·		·	
·	·	·	·	·	·	+60 d	·	·		·		·	·
·	·	·	·	·	·	*30 d	*d	·		·	·	·	·
·	·	·	·	·	·	*30 d	*d	·		·	·	·	·
·	·	·	·	·	·	*b	*d	·	·	·		·	
·	·	·	·	·	·	*210 d	*d	·	*? 240			·	
·					tr?	/	*30	·					
·					·	/	*30	·					
·			·	·		/		·					
	·	·	·	·	·		*30	·	·	·	·	·	·
·	·	·	·	·	·	*d	*	*	·	·	·	·	·
·	/	·	·	·	·	+d	/	*	/	·	·	·	·
·	·	·	·	·	·	+b	*tr?	·	·	·	·	·	·
·	·	·	·	·	·	+b	*tr?	×	·	·	·	·	·
·	·	·	·	·	*?	*d	*d	·	·	·	·	·	·

692, 699 Foam-test doubtful, no reactions.   **694** · with anti-seal, hyaena, hedgehog, llama.

Class AVES
II. *Division* CARINATAE
11. *Order* PASSERES
ANTISERA FOR.........

Fam.			Species	Man	Monkey	Cat	Dog	Pig	Hog-deer
Turdidae	**708**	191.	*Turdus merula* L. Blackbird England (Garrood) 22. viii. 01	•	•	•	•	•	/
	709	326.	*Turdus merula* L. England (R) 1. xii. 01	•	•	•	•	•	
	710	869.	*Cisticola cisticola brunneiceps.* Fan-Tail Okenawa, Japan (R) 14. iii. 02						
	711	868.	*Cettia cantans* Okenawa, Japan (R) 6. iii. 02						
	712	662.	*Luscinia luscinia* L. Nightingale England (B) 12. iv. 02	•		•	•	•	
	713	876.	*Geocichla varia.* Missel Thrush Okenawa, Japan (R) 10. iii. 02						
	714	880.	*Merula chrysolaus.* Thrush Okenawa, Japan (R) 15. iii. 02						
	715	864.	*Merula pallida.* Thrush Okenawa, Japan (R) 14. iii. 02						
	716	872.	*Merula pallida* 13. iii. 02						
	717	768.	*Pratincola rubetra.* Whinchat England (B) 23. iv. 02	•		•	•	tr?	•
	718	668.	*Saxicola oenanthe.* Wheatear England (B) 8. iv. 02	•		•	•	•	•
	719	670.	*Phylloscopus rufus.* Chiffchaff England (B) 11. iv. 02	•		•	•	•	•
	720	268.	*Accentor modularis.* Hedge Sparrow England (R) 1. xi. 01	•	•	•	•	•	•
	721	335.	*Accentor modularis* (Bird) 25. xi. 01	•		•	•	•	•
	722	774.	*Sylvia hortensis* Lath. Garden Warbler England (B) 6. v. 02	•		•	•		•
	723	169.	*Sylvia curruca* L. Germany (N) 1. ix. 01	•		•	•		
	724	166.	*Erithacus rubecula* L. Robin Germany (N) 25. viii. 01	•		•	•		•
	725	314.	*Erithacus rubecula* England (Bird) 9. xi. 01	•		•	•		•
	726	661.	*Ruticilla phoenicurus.* Redstart England (B) 14. iv. 02	•		•	•	•	
Hirundinidae	**727**	65.	*Chelidon urbica.* Swallow Ireland (N) 4. iv. 01	•		•	•	•	•
	728	773.	*Cotile riparia* L. Bank Swallow England (B) 23. iv. 02	•		•	•	•	•
Ampelidae	**729**	456.	*Ampelis garrulus.* Waxwing England (Clarke) 20. xii. 01	•		•	•	•	
Laniidae	**730**	777.	*Lanius (collurio ?).* Shrike England (Mitchell) 26. v. 02	•		•	•	•	
	731	752.	*Pityriasis gymnocephala* Borneo (Hose) p. 10. iv. 02						
	732	816.	*Hemipus picatus.* Tomtit Ceylon (R) 31. i. 02	•	•	•	•	•	

710, 711, 714, 715, 716 Foam-test doubtful, no reactions. **713** Foam-test negative, no reactions.
723 Very dilute solution.

	Mammalia						Aves				Reptilia		Amphibia	Crustacea
	Mexican-deer	Antelope	Sheep	Ox	Horse	Wallaby	Fowl	Ostrich	Fowl-egg	Emu-egg	Turtle	Alligator	Frog	Lobster
	•	/	•	•	•	•	*d	*? tr	*	/	•	•	•	•
	•	•	•	•	•	•	+b	*120	•	•	•	•		•
	•	•	•	•	•	•	•		•	•	•	•	•	•
	•	•	•	•	•	•	*45 tr		•		•		•	
	•	•	•	•	•	•	+60 d	tr	•				•	•
	•	•	•	•	•	•	tr	tr	•	•	•		•	
	•	•	•	•	•	•	*tr	*tr?	•	•	•	•	•	•
	•	•	•	•	•	•	*?	tr	•	•	•	•	•	•
	•	•	•	•	•	•	*90 tr		•		•		•	
	•	•	•	•	•	•	•	tr	•	•	•		•	•
	•	•	•	•	•	•	d	*tr	•	•	•	•	•	•
	•	•	•	•	•	•	*d	*tr	•	•	•	•	•	•
	•	•	•	•	•	•	+b	•	•	•	•	•	•	•
	•	•	•	•	•	•	*90 tr	•			•			•
	•	•	•	•	•	•	*d	*d	•	•	•	•	•	•
	•	•	•	•	•	•	tr	tr?			•			
	•					•	/	*30	•					

729 • with antisera for seal, hyaena, hedgehog, llama.   730 Foam-test doubtful.
731 Foam-test doubtful, no reactions.

293

				Mammalia			
		Man	Monkey	Cat	Dog	Pig	Hog-deer
Fam. Laniidae — **733** 107. *Gymnorhina leuconota.* White-backed Piping Crow. S. Australia (Z) 13. viii. 01		•	•	•	•	•	•
Fam. Sittidae — **734** 320. *Sitta caesia.* Nuthatch. England (R) 30. xi. 01		•					
Fam. Paridae — **735** 860. *Parus atriceps.* Tomtit. Okenawa, Japan (R) 9. iii. 02							
736 870. *Parus atriceps* 15. iii. 02							
737 269. *Parus ater.* Coletit. England (R) 1. xi. 01		•		•	•	•	
738 321. *Parus caeruleus.* Blue Tit. England (R) 30. xi. 01		•		•	•	•	
739 306. *Parus major.* Great Tit. England (R) 9. xi. 01		•		•	•		
Fam. Panuridae — **740** 839. *Parnurus biarmicus.* Bearded Tit. England (R) 11. iii. 02							
Fam. Corvidae — **741** 536. *Corvus macrorhynchus.* Indian Corby or Jungle Crow. Bombay, India (Phipson) i. 02		•		•	•		•
742 713. *Corvus macrorhynchus* Bhadarwa, Kashmir (Donald) 12. iii. 02		•		•	•		•
743 353. *Corvus cornix.* Hooded Crow. England (R) 14. xi. 01		•		•	•	•	•
744 824. *Corvus corone* L. Carrion Crow. Wales (Clarke) 11. iv. 02		•		•	•	•	
745 680. *Corvus corone* Mecklenburg (Kuse) 17. ii. 02		•		•	•		•
746 765. *Corvus corone* England (B) 8. v. 02		•		•	•	•	
747 90. *Corvus monedula.* Jackdaw. England (N) 24. vi. 01		•		•	•	•	•
748 711. *Corvus monedula* Bhadarwa, Kashmir (Donald) 25. iii. 02							
749 66. *Corvus frugilegus* L. Rook. England (N) 20. v. 01		•		•	•	•	
750 186. *Corvus frugilegus* Germany (Schlieffen) ix. 01		•		•	•	•	
751 679. *Corvus frugilegus* (Kuse) 28. x. 01		•		•		•	
752 303. *Corvus splendens.* Indian Crow. India (Z) 20. xi. 01		•		•	•	•	
753 558. *Corvus splendens* Bombay, India (Mason) 4. ii. 02		•		•	•		
754 764. *Corvus corax* L. Raven. Scotland (B) 23. iv. 02		•		•	•	•	

735, 736, 740, 748 Foam-test doubtful, no reactions.

Mammalia						Aves				Reptilia		Amphibia		Crustacea
Mexican-deer	Antelope	Sheep	Ox	Horse	Wallaby	Fowl	Ostrich	Fowl-egg	Emu-egg	Turtle	Alligator	Frog		Lobster
•	•	•	•	•	•	*	* 30 tr?	•	•	•	•	•		•
•	•	•	•	•	•	+	*	•	•	•	•	•		•
•	•	•	•	•	•	* tr	•	•	•	•	•	•		•
•	•	•	•	•	•	+	•	•	•	•	•	•		•
•	•	•	•	•	•	* d	tr	?	•	•	•	•		•
•	•	•	•	•	•	* D	* D	•	tr?			•		
•	•	•	•	•	•	* 60		•		•				
•	•	•	•	•	•	+ D	× tr	•	tr?		•			•
•	•	•	•	•	•	* ? 180 tr		•				•		
•	•	•	•	•	•	+ 60 d	* d		•	•		•		
•	•	•	•	•	•	× 45 tr		•		•				•
•	•	•	•	•	•	•	+ d	•	•	•	•	•		•
•	•	•	•	•	•	+	+ 30 d	•	•	•	•	•		•
•	•	•	•	•	•	• d	× tr	•	•	•	•	•		•
•	•	•	•	•	•	+ 60 d	• d	•	•	•		•		
•	•	•	•	•	•	* d	* tr	*	•	•	•	•		•
* ? 120	•	•	•	•	•	+ d	* d	•	•	•		•		•
•	•	•	•	•	•	* 45 tr	•	•		•				•

738 Foam-test doubtful when tested with anti-ostrich. **745** • with anti-tortoise.

Class AVES
II. *Division* CARINATAE
11. *Order* PASSERES
ANTISERA FOR.........

				Mammalia					
Fam.				Man	Monkey	Cat	Dog	Pig	Hog-deer
Corvidae	**755**	222.	*Cissa venatoria.* Hunting Crow India (Z) 24. viii. 01	•	•	•	•	•	•
	756	165.	*Garrulus glandarius* L. Jay Germany (N) 2. ix. 01	•	•	•	•	•	•
	757	333.	*Garrulus glandarius* England (Leighton) 26. xi. 01	•	•	•	•	•	•
	758	57.	*Nucifraga caryocatactes.* Nutcracker England (Lane) 28. iv. 01	•					
	759	352.	*Pica rustica.* Magpie England (R) 18. xi. 01	•		•	•	•	
	760	642.	*Pica rustica* (B) 14. xii. 01	•		•	•	•	
	761	822.	*Pica rustica* (Clarke) 23. i. 02	*?30	•	•	•	•	
	762	823.	*Pica rustica* (R) 5. iii. 02	•	•	•	•	•	
	763	681.	*Pica rustica* Mecklenburg (Kuse) 19. xii. 01	•		•	•	•	
Fam.	**764**	189.	*Sturnus vulgaris* L. Starling England (Garrood) 8. viii. 01	•		•	•	•	•
Sturnidae	**765**	725.	*Acridotheres tristis.* Common Minah Bhadarwa, Kashmir (Donald) 28. ii. 02						
Fam.	**766**	106.	*Icterus jamaica.* Brazilian Hangnest Brazil (Z) 13. viii. 01	•	•			•	•
Icteridae									
Fam.	**767**	664.	*Fringilla coelebs* L. Chaffinch England (B) 8. iv. 02	•		•	•	•	•
Fringillidae	**768**	56.	*Fringilla coelebs* (N) 28. iv. 01	•			•		
	769	323.	*Fringilla coelebs* (R) 30. xi. 01	•	•	•	•	•	
	770	844.	*Chrysomitris spinus.* Siskin England (R) 20. ii. 02	•	•	•	•	•	•
	771	663.	*Ligurinus chloris.* Greenfinch England (B) 8. iv. 02	•		•	•	•	•
	772	324.	*Ligurinus chloris* (R) 30. xi. 01	•	•	•	•	•	
	773	351.	*Fringilla chloris* × *Carduelis elegans.* Hybrid Greenfinch-Goldfinch England (B) 10. xii. 01	•		•	•	•	•
	774	350.	*Fringilla montifringilla.* Brambling England (B) 25. xi. 01	•	•	•	•	•	
	775	840.	*Carduelis elegans.* Goldfinch England (R) 7. ii. 02	*?30	•	•	*?30	•	
	776	144.	*Paroaria cucullata.* Red-crested Cardinal S. America (Z) 9. viii. 01	•	\	\	•	•	•
	777	455.	*Coccothraustes vulgaris* Pall. Hawfinch England (Clarke) 21. xii. 01	•		•	•	•	•
	778	348.	*Coccothraustes vulgaris* (B) 30. xi. 01	•	•	•	•	•	

765 Foam test doubtful, no reactions.

	Mammalia						Aves				Reptilia		Amphibia	Crustacea
Mexican-deer	Antelope	Sheep	Ox	Horse	Wallaby		Fowl	Ostrich	Fowl-egg	Emu-egg	Turtle	Alligator	Frog	Lobster
•	•	•	•	•	•		* d	* tr	•	•		•	tr	•	•
•	•	•	•	•	•		* d	* tr	•	•		•	•	•	•
•	•	•	•	•	•		+	× tr	•	•		•	•	•	•
		•	•	•	•									
•	•	•	•	•	•		+	× tr	•	•		•	•	•	•
•	•	•	•	•	•		* 	* 60	•	•				•	•
•					•		* 5 d	•	•						
					•		/	•	•					
•	•	•	•	•	* ? 240 •		* 60 d	d	•			•		•	
•	•	•	•	•	•		d	tr	•	•		•	•	•	•
•	•	•	•	•	•		•	* 30 tr	•	•		•	•	•	•
•	•	•	•	•	•		* 60	•	•	•		•	•	•	•
•	•	•	•	•	•		+ * 35	*	•	•		•	•	•	•
•	•	•	•	•	•		* 60 +	*	•	•		•	•	•	•
•	•	•	•	•	•		+	* tr	*	* tr?		•	•	•	•
•	•	•	•	•	•		+	•	•	* tr		•	•	•	•
		•		•	•		• tr	•	•						
\	•	•	•	•	\		* d	•	* ? tr	* ?		•	•	tr?	tr?
•	•	•	•	•	•		* d	d	•	•		•	•	•	•
•	•	•	•	•	•		+	* tr?	*	* tr?		•	•	•	•

297

Class AVES
II. *Division* CARINATAE
11. *Order* PASSERES
ANTISERA FOR.........

			Mammalia					
			Man	Monkey	Cat	Dog	Pig	Hog-deer
Fam. Fringillidae	**779**	643. *Loxia curvirostra* Gm. Crossbill England (B) 10. i. 02	•		•	•	•	•
	780	192. *Pyrrhula europaea.* Bullfinch England (Garrood) 28. viii. 01	•		•	•	•	
	781	772. *Pyrrhula europaea* (B) 26. iv. 02	•		•	•	•	
	782	665. *Linota cannabina.* Linnet England (B) 8. iv. 02	•		•	•	•	
	783	267. *Passer domesticus* L. House Sparrow England (R) 1. xi. 01	•		•	•	•	
	784	769. *Passer domesticus* (B) 3. v. 02	•		•	•	tr	•
	785	322. *Passer domesticus* (R) 30. xi. 01	•		•	•	•	•
	786	813. *Passer domesticus indicus.* Sparrow Kandy, Ceylon (R) 29. i. 02	•		•	•	•	
	787	861. *Passer montanus saturatus.* Sparrow Okenawa, Japan (R) 14. iii. 02						
	788	875. *Passer montanus saturatus* Okenawa, Japan (R) 6. iii. 02						
	789	838. *Plectrophanes nivalis* L. Snow-Bunting England (Clarke) 8. ii. 02	•	•				
	790	667. *Emberiza miliaria.* Corn-Bunting England (B) 9. iv. 02	•				•	
	791	771. *Emberiza schoeniclus.* Reed-Bunting England (B) 30. iv. 02	•			•	•	
	792	770. *Emberiza citrinella* L. Yellow Hammer England (B) 3. v. 02	•		•	•	•	
	793	325. *Emberiza citrinella* (R) 30. xi. 01	•	•	•	•	•	
	794	878. *Emberiza personata* Okenawa, Japan (R) 14. iii. 02	•		•	•	•	
	795	143. *Crithagra sulphuratus.* Sulphury Seed-Eater S. Africa (Z) 9. viii. 01	•		•	•	•	•

EGGS

			Man	Monkey	Cat	Dog	Pig	Hog-deer
	796	402. *Gallus domesticus.* Fowl England (N) 26. x. 01. *Egg-white*, dried on paper	•		•	•	•	•
	797	632. *Euplocomus (Phasianus) nycthemerus* L. Silver Pheasant China (Z) 2 eggs laid ca. 11. iv. 02. *Egg-white*, fluid	•	•		•	•	
	798	593. *Balearica regulorum* Bennett. Cape Crowned Crane S. Africa (Z) 13. iii. 02. *Egg-white*, fluid	•		•	•	•	
	799	439. *Dromaeus novae-hollandiae* Vieill. Emu Australia (Z) rec'd. 18. i. 02. *Egg-white*, fluid	•	•	•	•	•	

787, 788 Foam-test doubtful, all reactions negative.

Mammalia						Aves				Reptilia		Amphibia	Crustacea
Mexican-deer	Antelope	Sheep	Ox	Horse	Wallaby	Fowl	Ostrich	Fowl-egg	Emu-egg	Turtle	Alligator	Frog	Lobster
•	•	•	•	•	•	*	•	•	•	•		•	
•	•	•	•	•	•	d	* tr	•	•	•		•	•
•	•	•	•	•	•	* 90 tr + 60 d	•	•	•	•		•	•
•	•	•	•	•	•	* d	* tr	•	•	•		•	•
•	tr ?	•	•	•	•	* 45 tr	•	•	•	•		•	•
•	•	•	•	•	•	+	* ?	•	•	•		•	•
•	•	•	•	•	•	/	* 30	•	•			•	
•	•	•	•	•	•	* 180 tr	* 30	•	•	•		•	•
•	•	•	•	•	•	+ 60 d	* 60 d	•	•	•		•	•
•	tr ?	•	•	•	•	* 45 tr	•	•	•	•		•	•
•	tr ?	•	•	•	•	* ? tr	•	•	•	•		•	•
•	•	•	•	•	•	+	*	•	•	•	•	•	•
•	•	•	•	•	•	* d	•	* ? d	* ?	•	•	tr ?	tr ?
* ? 265	•	•	•	•	•	× d	*	**+** **D**	× tr	× d	× d	•	•
•	•	•	•	•	•	•	•	+ d	+ d	•		•	
•	•	•	•	•	•	•	•	+ D	+ D	•	•	•	•
•	•	•	•	•	•	•	* 30 cl	+ D	**+** **D**	•	•	•	•

794 Foam-test doubtful, no reactions.   797 • with anti-seal, hyaena, hedgehog, llama.

Class REPTILIA

1. Subclass CHELONIA
Order THECOPHORA
ANTISERA FOR.........

			Man	Monkey	Cat	Dog	Pig	Hog-deer
Fam. Testudinidae	800	805. *Chrysemys picta.* Painted terrapin N. America (Jordan) 22. v. 02	•	•	•	•	•	
	801	806. *Chrysemys elegans* N. America (Jordan) 11. v. 02	•	•	•	•	•	
	802	804. *Cistudo carolina* N. America (Jordan) 30. iv. 02	•	•	•	•	•	
	803	807. *Graptemys pseudogeographicus* N. America (Jordan) 17. v. 02	•	•	•	•		•
	804	425. *Testudo ibera.* Tortoise Greece (killed Cambridge, N) 14. i. 02, *fl.*	•	•	•	•	•	
	805	793. *Testudo elephantopus.* Giant Tortoise (N. York Zoo., Langmann) 14. vi. 02	•	•	•	•	•	
	806	904. *Testudo elephantopus* Galapágos (Z) 15. x. 02, *fl.*	•	•	•	•	•	
	807	78. *Testudo inepta.* Clumsy Tortoise Mauritius (Z) 20. v. 01	•	•	•	•	•	
	808	818. *Testudo* spec. Tortoise, egg-white (Leighton) egg laid 2. vii. 02, *fl.*	•	•	•	•	•	
Fam. Chelonidae	809	343. *Chelone midas.* Green Turtle Trop. Seas (killed Cambridge) 4. xii. 01, *fl.*	•	•	•	•	•	
	810	209. *Chelone midas* (Z) 9. x. 01	•	•	•	•	•	
	811	628. *Chelone viridis ?* Turtle (B) 13. xii. 01	•	•	•	•	•	
	812	615. "Water Turtle" Upper Shiré R., C. Africa (Dodds) p. 8. ii. 02	•	•	•	•	•	

2. Subclass CROCODILIA
Order EUSUCHIA

			Man	Monkey	Cat	Dog	Pig	Hog-deer
Fam. Crocodilidae	813	338. *Alligator mississippiensis* N. America (Z) 27. xi. 01, *fl.*	•	•	•	•	•	
	814	101. *Alligator sinensis* China (Z) 13. viii. 01	•	•	/	•	•	/
	815	616. Crocodile Upper Shiré R., C. Africa (Dodds) p. 8. ii. 02	•		•	•	•	

3. Subclass SAURIA
Order LACERTILIA

			Man	Monkey	Cat	Dog	Pig	Hog-deer
Suborder Geckones	816	153. *Gecko guttatus.* Gecko S. Sylhet, India (Dalgetty) 2. viii. 01	•	•	•	•	•	•
Suborder Lacertae	817	555. *Calotes versicolor.* Indian Tree-Lizard W. Deccan, India (Phipson) 12. i. 02	•	•	•	•	•	•
	818	911. *Uromastix spinipes* Egypt (G. Elliot Smith) 7. vi. 02	•	•			•	

806 + with anti-tortoise, last 3 tested by Dr Graham-Smith.

Mammalia						Aves				Reptilia		Amphibia	Crustacea
Mexican-deer	Antelope	Sheep	Ox	Horse	Wallaby	Fowl	Ostrich	Fowl-egg	Emu-egg	Turtle	Alligator	Frog	Lobster
·			·	·	·		·	·					
·			·	·	·		·	·					
·			·	·	·		·	·					
			·	·			·	·					
·	·	·	·	·	·	·	*tr	+D	·	+D	*d	·	·
	·		·	·	·		·	·					
·		·	·	·	·	·				+D	+d		
·	·	·	·	·	·	·	·	·	·	*	*d	·	*
·			·	·	·	·	*? 30	·			d		
·	·	·	·	·	·	·	*tr	*d	*d	+D	*d	·	·
·	·	·	·	·	·	·	tr?	*d	*d	+d	·	·	·
·	·	·	·	·	·	·	·	·	·	·	·	·	·
·	·	·	·	·		·	*120 cl	*d	·	*	+D	·	·
/	/	·	·	·	/	·	/	+D	/	·	+α	·	*
·	·	·	·	·	·	·	·	·		·	·		
·	·	·	·	·	·	·	·	·	·	·	·	·	·
·	·	·	·	·	·	·	·	·	·	·	·	·	·
			·	·		·							

812 Dried sample, note source and see text, p. 63

Class REPTILIA
3. *Subclass* SAURIA
1. *Order* LACERTILIA
ANTISERA FOR.........

Suborder / No.	Cat.	Description	Man	Monkey	Cat	Dog	Pig	Hog-deer
Lacertae 819	437.	*Iguana (tuberculata* Laur. ?). Iguana Trinidad (J. P. Tulloch) 1. i. 02	•	\	\	•	•	•
820	627.	*Anguis fragilis* L. Slow-Worm Herefordshire, England (Leighton) 31. iii. 02	•	•	•	•	•	•
821	635.	*Varanus salvator* Laur. Two-banded Monitor E. India (Z) 6. iv. 02	•	•	•	•	•	•
822	206.	*Varanus griseus.* Grey Monitor N. Africa (Z) 30. ix. 01	•	•	•	•	/	/
823	912.	*Lacerta agilis.* Sand Lizard England (G.-S.) 3. v. 02	•	•	•	•	•	
824	205.	*Scincidae* spec. ? Lizard S. Sylhet, India (Dalgetty) 1. ix. 01			•	•	•	/
825	570.	"Land Lizard," 2 ft. 6 in. long Orange River, S. Africa (Parkinson) 13. xii. 01	•		•	•	*? 60	•

2. *Order* OPHIDIA

No.	Cat.	Description	Man	Monkey	Cat	Dog	Pig	Hog-deer
826	434.	*Python molurus* Linn. Indian Python India (Z) 11. i. 02	•	•	•	•	•	•
827	280.	*Python seboe.* Python W. Africa (Z) 8. xi. 01	•	•	•	•	•	•
828	607.	*Corallus caninus* L. "Hundskopfschlange" Madagascar ?...(Lühe) 20. ii. 02	•	•	•	•	•	•
829	208.	*Boa constrictor.* Common Boa S. America (Z) 7. x. 01	•	•	•	•	•	•
830	802.	*Boa constrictor* Trinidad, W. Indies (Tulloch) 4. vi. 02	•	•	•	•	•	•
831	913.	*Eryx jaculus* Egypt (G. Elliot Smith) 19. vi. 02	•	•	•	•	•	•
832	760.	*Tropidonotus natrix* Gesn. Green Snake England (B) 28. iv. 02	•		•	•	•	•
833	68.	*Tropidonotus natrix* (N) 2. v. 01	•	•	•	•	•	•
834	909.	*Tropidonotus natrix* (Graham-Smith) 18. x. 02, *fl. ser.*	•		•	•	•	•
835	914.	*Zamenis diadema* Egypt (G. Elliot Smith) 11. vi. 02	•	•	•	•	•	•
836	915.	*Zamenis varageus* var. *nummifer* Egypt (G. Elliot Smith) 19. vi. 02	•	•	•	•	•	•
837	618.	*Dendrophis liocercus.* Whip Snake Trinidad, W. Indies (Tulloch) 2. iii. 02	•	•	•	•	•	•
838	554.	*Cerberus rhynchops.* Indian Estuary Snake Bombay, India (Phipson) 17. i. 02	•	•	•	•	•	•
839	574.	*Cerberus rhynchops* Calcutta, India (Rogers) 8. x. 01	•	•	•	•	•	•
840	883.	*Naja tripudians* Merr. Cobra Rajputana, India (MacWatt) 21. vi. 02	•	•	•	•	•	•
841	884.	*Naja tripudians* 23. vi. 02	•	•	•	•	•	•
842	553.	*Naja tripudians* Bombay, India (Phipson) 4. ii. 02	•	•	•	•	•	•

822 Another sample (920) proved as insoluble as the following.

Mammalia							Aves					Reptilia			Amphibia		Crustacea
Mexican-deer	Antelope	Sheep	Ox	Horse	Wallaby		Fowl	Ostrich	Fowl-egg	Emu-egg		Turtle	Alligator		Frog		Lobster
\	·	·	·	·	\		·	·	·	·		·	·		·		·
·	·	·	·	·	·		·	·	·	·		·	·		·		
·	·	·	·	·	·		·	·	·	·		·	·		·		
·	/	·	·	·	·		/	/	/	/		/	/		/		/
·	/	·	·	·	·		·	/	·	/		·	·		·		·
·	·	·	·	·	·		·	·	·	·		·	·		·		
·	·	·	·	·	·		·	·	·	·		·	·		·		·
·	·	·	·	·	·		·	·	·	·		·	·		·		·
·	·	·	·	·	·		·	·	·	·		·	·		·		·
·	·	·	·	·	·		·	tr ?	·	·		d ?	·		·		·
·	·	·	·	·	·		·	·	·	·		·	·		·		·
·	·	·	·	·	·		*?45 ·	·	·	·		·	·		·		·
·	·	·	·	·	·		·	·	·	·		·	·		·		·
·	·	·	·	·	·		·	·	·	·		·	·		·		·
·	·	·	·	·	·		·	·	·	·		·	·		·		
·	·	·	·	·	·		·	·	·	·		·	·		·		
·	·	·	·	·	·		·	·	·	·		·	·		·		
·	·	·	·	·	·		·	*180 d	·	·		·	·		·		

823 Insoluble in saline after 30 hours at 37° C.

303

Class REPTILIA
3. Subclass SAURIA
2. Order OPHIDIA
ANTISERA FOR.........

			Man	Monkey	Cat	Dog	Pig	Hog-deer
843	157.	*Naja tripudians* Merr. Cobra India (Z) 6. viii. 01	•	•	•	•	•	•
844	916.	*Naja naje.* African Hooded Cobra (Aspis) Egypt (G. Elliot Smith) 19. vi. 02	•	•			•	
845	685.	*Pseudechis porphyriacus.* Black Snake Oberon, N.S.W. (Cashman) 21. xii. 01	•	•	•	•	•	•
846	142.	*Bites arietans.* Puffadder S. Africa (Z) 6. viii. 01	•	•	•	•	•	•
847	787.	Puffadder S. Nyassa, Africa (Dodds) 27. iii. 02	/	/	/	/	/	•
848	686.	*Vipera berus.* Adder N. Devon, England (Leighton) iv. 02	•	•	•	•	•	•
849	689.	*Vipera berus* Dumfriesshire, Scotland (Leighton) iv. 02	•			•	•	•
850	688.	*Vipera berus* Devon, England (Leighton) 18. iv. 02	•	•		•	•	•
851	898.	*Vipera (russelli ?).* Himalayan Viper Bhadarwa, Kashmir State (Donald) 4. iv. 02			•		•	
852	378.	*Crotalus* spec. Rattlesnake N. America (Salmon) 6. xi. 01	•	•	•	•	•	•
853	613.	"A poisonous serpent" nr. Lake Nyassa, C. Africa (Dodds) p. 8. ii. 02	/	/	/			

844 Solution foamed well.

304

| Mammalia | | | | | | | Aves | | | | | Reptilia | | | Amphibia | | Crustacea |
Mexican-deer	Antelope	Sheep	Ox	Horse	Wallaby		Fowl	Ostrich	Fowl-egg	Emu-egg		Turtle	Alligator		Frog		Lobster
•	•	•	•	•	•		*? d	*?	*	•		•	•		•		•
		•	•	•			•		•			•			•		
•	•	•	•	•	•		•		•			•			•		
•	•	•	•	•	•		•	•	*? tr?	•		•			•		•
/		•		•	/			/	/			•					
•	•	•	•	•	•		•			•		•			•		
		•	•	•	•		•			•		•					
•	•	•	•	•	•		•			•		•			•		
		•		•	•					•							
•	•	•		•			•	•	•	•		•			•		•
												•					

847 • with 4 antisera for seal, hyaena, hedgehog, llama.

Class AMPHIBIA
1. Order URODELA

ANTISERA FOR........

			Man	Monkey	Cat	Dog	Pig	Hog-deer
Fam. Amphiumidae	**854**	599. *Amphiuma means* Garden. Amphiuma N. America (Z) 19. iii. 02, *fl. ser.*	•	•	•	•	•	•
Fam. Salamandridae	**855**	365. *Amblystoma tigrinum* (larva). Axolotl Mexico (raised in England by E. Bles) 26. xii. 01	•					
	856	55. *Triton cristatus.* Crested Newt England (N) 26. iv. 01	•	/	•	•	•	/

2. Order ANURA

			Man	Monkey	Cat	Dog	Pig	Hog-deer
1. *Suborder* AGLOSSA	**857**	598. *Xenopus loevis* Daud. Smooth-clawed Frog E. Africa (Z) 19. iii. 02	•	•	•	•	•	•
2. *Suborder* PHANEROGLOSSA *Fam.* Bufonidae	**858**	81. *Bufo vulgaris* Laur. Common Toad England (N) 4. vii. 01	•		•	•	•	
	859	430. *Bufo mauritanica* Schlegel. Moorish Toad N. W. Africa (Z) 10. i. 02	•					
Fam. Engystomatidae	**860**	207. *Magalobatrachus maxima.* Gigantic Salamander Japan (Z) 23. ix. 01	•	•	•	•	•	/
Fam. Ranidae	**861**	54. *Rana temporaria.* Frog England (N) 28. iv. 01	•		•	•	/	•
	862	582. *Rana temporaria* 3. iii. 02, *fl.*	•		•	•		•
	863	406. *Rana temporaria* 11. xi. 01	•		•	•		•
	864	478. *Rana tigrina* Daud. Tigrine Frog India (Z) 25. i. 02	•	\	•	•		•
	865 **866**	479. *Rana tigrina* 3. ii. 02	•		•	•		•
	867	597. *Rana tigrina* 19. iii. 02	•	•	•	•	•	•
	868	732. *Rana spec.* Common Trinidad Frog " Crapaud " Trinidad, W. Indies (Tulloch) 27. iii. 02	/	/	•	/	•	/

865 · with 5 antisera for monkey, seal, hyaena, hedgehog, llama.

	Mammalia							Aves				Reptilia		Amphibia	Crustacea
Mexican-deer	Antelope	Sheep	Ox	Horse	Wallaby		Fowl	Ostrich	Fowl-egg	Emu-egg		Turtle	Alligator	Frog	Lobster
•	•	•	•	•	•		•	•		•		•	•	•	•
	•	•	•	•			•	•	•	•		•	•	•	*
•	/	•	•	•	•		•	/	•	/		•	•	•	•
•	•	•	•	•	•		•	•				•		•	•
•	•	•	•	•	•		*	•	•	•		•	•	•	*
•	•	•	•	•	•		•	•	•	•		•	•	•	•
•	/	•	•	•	•		•	/	•	/		•	•	•	•
•	/	•	•	•	•		•	/	•	/		•	•	+	•
•	•	•	•	•	•		•	•	•	•		•	•	α+	•
														α+	
														α	
\	•	•	•	\	\		•	•	•	•		•	•	* 60 d	•
•	•	•	•	•	•		•	•	•	•		•	•	* 60 d	•
•	•	•	•	•	•		•	•	•	•		•	•	* tr	•
•	/	/	/	•	•		/	•				•			

Class PISCES

The tests made with some of the antisera used in the preceding tables gave entirely negative results. The antisera used were the following:

1.	Anti-Man	10.	Anti-Lobster	19.	Anti-Cat
2.	,, Dog	11.	,, Alligator	20.	,, Wallaby
3.	,, Horse	12.	,, Turtle	21.	,, Monkey
4.	,, Ox	13.	,, Ostrich	22.	,, Seal
5.	,, Sheep	14.	,, Emu's egg	23.	,, Hedgehog
6.	,, Pig	15.	,, Antelope	24.	,, Hyaena
7.	,, Fowl	16.	,, Hog-deer	25.	,, Llama
8.	,, Fowl's egg	17.	,, Reindeer		
9.	,, Frog	18.	,, Mexican-deer		

For convenience the antisera are ordered in this case in the order in which they were produced, not in their zoological order. The fish bloods which proved insoluble after some hours in salt solution are not included in the list. With the exception of the blood of the dogfish (872) which was fluid, all of these bloods were received dried on filter-paper. The bloods tested were the following:

			Tested with Antisera
I. *Subclass* **PALAEICHTHYIES** *Order* **CHRONDROPTERYGII (Elasmobranchii)**			
Suborder **Plagiostomata** A. Selachoidei *Fam.* **Carchariidae**	**869**	490. *Zygaena malleus* ? Hammerhead Shark Trinidad, W. I. (Tulloch) 15. i. 02	1—8, 11—16, 18—25
	870	199. *Mustelus canis (vulgaris* ?). Dogfish New Jersey, U.S.A. (Silvester) viii. 01	6—12, 18—21
	871	197. *Squalus acanthias.* Dogfish New Jersey, U.S.A. (Silvester) 30. viii. 01	6—12, 18—21
Fam. **Scylliidae**	**872**	578. *Scyllium canicula.* Dogfish England (Plymouth Lab.) 27. ii. 02, *fl.*	1—8, 12—16, 18—20
B. Batoidei *Fam.* **Rajidae**	**873**	198. *Raja ocellata.* Skate New Jersey, U.S.A. (Silvester) viii. 01	1—12, 18—21
Fam. **Trigonidae**	**874**	200. *Dasyatus* spec. Sting-ray New Jersey, U.S.A. (Silvester) viii. 01	1—16, 18—21
Fam. **Myliobatidae**	**875**	203. *Rhinoptera bonasus.* Sting-ray New Jersey, U.S.A. (Silvester) 30. viii. 01	6—12, 18—21
II. *Subclass* **TELEOSTEI** *Order* **ACANTHOPTERYGII**			
Cottoscomberiformes *Fam.* **Carangidae**	**876**	803. *Naucrates ductor.* Pilot-fish Trinidad, W. I. (Tulloch) 29. v. 02	1—4, 6, 8, 13, 18—25
Fam. **Scomberidae**	**877**	619. *Scomber auxis.* Spanish Mackerel Trinidad, W. I. (Tulloch) 13. ii. 02	1—9, 12—20
	878	617. *Scomber carangus.* Mackerel Trinidad, W. I. (Tulloch) 28. ii. 02	1—9, 12—20
Muciliformes *Fam.* **Sphyraenidae**	**879**	620. *Sphyraena barracuda.* Barracouta Trinidad (Tulloch) 7. iii. 02	1—9, 12—20
Gobiesociformes *Fam.* **Gobiesocidae**	**880**	919. *Lepidogaster* spec. ? (Muscle extract) Guernsey (G.-S.) vi. 02	1, 3, 4, 6—8, 21, 23—25

Order PHYSOSTOMI

				Tested with Antisera
Fam. **Cyprinidae**	**881**	427.	*Carassius auratus.* Goldfish England (N) 15. i. 02	1—16
Fam. **Esocidae**	**882**	358.	*Esox lucius.* Pike England (B) 26. xi. 01	1—16, 18, 19, 21
	883	94.	*Esox lucius* Ireland (Dillon) 6. v. 01	1—16, 18, 19, 21
Fam. **Salmonidae**	**884**	210.	*Salmo fario.* Trout England (Leighton) 17. x. 01	1—12, 18—21
	885	92.	*Salmo salar.* Salmon Ireland (Dillon) 9. v. 01	1—16, 18—21
Fam. **Hyodontidae**	**886**	734.	*Hyodon tergisus ?* Moonfish Trinidad (Tulloch) 1. iv. 02	1—8, 12, 14—16, 18—20

Order PLECTOGNATHII

Fam. **Sclerodermi**	**887**	204.	*Aluton schoepfii.* File-fish New Jersey, U.S.A. (Silvester) viii. 01	18—21
Fam. **Gymnodontes**	**888**	202.	*Tetrodon turgidus ?* Toadfish New Jersey, U.S.A. (Silvester) viii. 01	1—16, 18—21

UNCLASSED FISHES

	889	201.	*Spheroides maculatus* New Jersey, U.S.A. (Silvester) 20. viii. 01	6—12, 18—20, 22
	890	735.	Grunt or "Oro-Oro" Trinidad, W. I. (Tulloch) 27. iii. 02	1—8, 12, 14—16, 18—20
	891	736.	"Green Chule" Trinidad, W. I. (Tulloch) 29. iii. 02	3, 6, 9, 12, 18—20
	892	468.	"Paona-fish" Trinidad, W. I. (Tulloch) 6. i. 02	1—16, 18—20
	893	467.	"Grouper" Trinidad, W. I. (Tulloch) 1. i. 02	1—16, 18—20

None of these bloods gave even a trace of reaction with any of the antisera used.

Class **CRUSTACEA**

Order **DECAPODA**

ANTISERA FOR.........

		No.		Man	Monkey	Cat	Dog	Pig	Hog-deer
Suborder **MACRURA**	**894**	426.	*Astacus fluviatilis.* Crayfish England (N) 14. i. 02	•	•	•	/	•	/
Fam. **Astacidae**	**895**	407.	*Homarus vulgaris* Bel. Lobster England (N) 14. xi. 01	•		•	•	•	•
Fam. **Loricata**	**896**	591.	*Palinurus vulgaris.* Crawfish or Rock-Lobster England (Plymouth Lab.) 26. ii. 02	•		•	•	•	•
Suborder **BRACHYURA**	**897**	592.	*Maia squinado.* Spider Crab England (Plymouth Lab.) 27. ii. 02	•		•	•	•	•
Fam. **Oxyrhyncha**	**898**	470.	*Carcinus maenas.* Common Shore Crab England (Plymouth Lab.) 25. i. 02	•		•	•	•	•
Fam. **Cyclometopa**	**899**	471.	*Portunus puber.* Velvet Fiddler Crab England (Plymouth Lab.) 25. i. 02	•	\	•	•	•	•
	900	469.	*Portunus depurator.* Common Swimming Crab England (Plymouth Lab.) 25. i. 02	•		•	•	•	•

897 Foam-test doubtful.

	Mammalia						Aves				Reptilia		Amphibia	Crustacea
	Mexican-deer	Antelope	Sheep	Ox	Horse	Wallaby	Fowl	Ostrich	Fowl-egg	Emu-egg	Turtle	Alligator	Frog	Lobster
	·	/	·	·	·	·	·	/	·	/	·	·	·	+D
	·	·	·	·	·	·	·	·	·	·	·	·	·	+D
	·	·	·	·	·	·	·	·	·	·	·	·	·	*tr
	·	·	·	·	·	·	·	·	·	·	·	·	·	·
	·	·	·	·	·	·	·	·	·	·	·	·	·	+D
	·	\	·	·	\	\	·	·	·	·	·	·	·	+D
	·	·	·	·	·	·	·	·	·	·	·	·	·	+D

SECTION VII.

ON THE RESULTS OF 500 QUANTITATIVE TESTS WITH PRECIPITATING ANTISERA UPON THE BLOODS OF PRIMATES, INSECTIVORA, CARNIVORA, UNGULATA, CETACEA, MARSUPIALIA, AND AVES.

By GEORGE H. F. NUTTALL, M.A., M.D., Ph.D.,

AND T. S. P. STRANGEWAYS, M.A., M.R.C.S.,

Demonstrator in Pathology, Cambridge.

IN view of the crudity of the methods employed and the many possibilities of error, it is not a little surprising that the figures obtained in the following quantitative tests are in such accord. The combined qualitative and quantitative tests demonstrate certain broad facts, namely, the persistence of blood affinities amongst groups of animals which have descended from a common stock. The difficulties in the way of determining *finer* differences appear to be considerable, owing to the variations in the reacting power of the rabbits treated with a given blood, possibly also to the nature of the animal yielding the blood injected (even if it be normal), but especially to much of the blood injected being derived from diseased animals. The bloods we have tested, also those we have used for the treatment of rabbits, with few exceptions, have been derived from animals dying at the Zoological Society's Gardens, London. These animals have died from various diseases, and in the majority of cases it has been impossible to determine the exact cause of death. In some cases the animals died naturally, and the blood was obtained from the cadaver at variable periods after death, in other cases they were killed and the blood collected immediately. In some cases it appears reasonable to suppose that fluids derived from extravasations into the body cavities have become mixed with the blood, although precautions were taken against such accidents. We know that the concentration of the blood may vary in disease and even in health, and it is possible that blood remaining in the blood vessels after death may undergo changes in concentration. Apart from this there are chemical changes taking place in the blood of which we have but slight knowledge at present, which may materially affect the quantitative results of the precipitin test. The tests were conducted upon bloods of very different ages, some preserved with chloroform, others filtered through porcelain and preserved pure. In most cases serum alone was used, in some instances clots were present in the serum. All the bloods were kept in the ice-chest until used. In a few cases putrefaction had already taken place when we received the bloods. The bloods were kept in tightly stoppered bottles, or in sealed tubes. In warm weather it is reasonable to suppose that a certain amount of evaporation may have taken place from the bloods which were allowed to clot in the covered vessels in which they were placed for obtaining serum. Where glass stoppers were used a small amount of vaseline was smeared upon the ground-glass surfaces. In the bloods first collected a slight loss of water by evaporation may well have taken place through the use of corks which loosened or were defective. All bloods of later date were kept in sealed tubes to prevent evaporation.

The quantitative method devised by Nuttall is also open to criticism. It appears *à priori* improbable that the physical character of the precipitum will be the same in different bloods to dilutions of which a given antiserum is added. We may assume the precipitum should "pack" to a greater extent in tubes containing the most and to various degrees in tubes of inconstant calibre. Finally, there is considerable chance of error in the measurement of such small quantities as have at times to be dealt with. A considerable number of bloods were tested to determine the range of experimental error, the result being that we found the error to be within 10%. Thus Graham-Smith and Sanger (1903, p. 264) obtained the following results:

"*Results of measurements of four samples of a* 1 *in* 21 *dilution of human blood and* 8 *samples of a similar dilution of ox serum.*

Human serum		Ox blood		
·0281 c.c.		·0215 c.c.	·0233 c.c.	
·0281 ,,	Mean ·0293 c.c.	·0225 ,,	·0233 ,,	Mean ·0233 c.c.
·0300 ,,		·0233 ,,	·0233 ,,	
·0309 ,,		·0233 ,,	·0262 ,,	

" The fluctuations above and below the means in the human series are ·0012 c.c. or 4 %, and in the ox series ·0029 c.c. and ·0018 c.c. respectively, or 12 % and 7 %.

" In order to arrive at the most trustworthy figures possible, in most experiments two observations were made under identical conditions in the hope that by this means the experimental error might be reduced to a minimum.

" In measurements of this kind with every precaution probably a margin of 10 % must be allowed for experimental error.

" Throughout these experiments the aim has been to indicate by measurements the effects of varying conditions on the formation of the precipitum as compared with controls. It must be stated, however, that with every precaution the measurements of the same set of materials on different days are not identical, although the proportions which the various members of the set bear to each other remain fairly constant. Hence though improvements in the technique of measurement may result in more accurate and constant figures, yet it is improbable that the general results will be materially altered."

Measurements could be made with reasonable accuracy down to ·001 c.c. but of course it would be a mistake to attach importance to minute reactions of 10 % and under, unless results of repeated tests are concordant.

The results in spite of all these possible sources of fallacy certainly speak for themselves, and show beyond question that much may be accomplished by the use of the precipitin tests in the study of animal relationships. We believe that we have but entered on the threshold of a method of investigations which is bound with time to yield useful data. This investigation must necessarily be regarded as preliminary in character. A careful selection of material to work upon, especially material which is readily obtained, together with care directed to excluding the sources of error above mentioned, are indicated in the future.

In view of the extent of the work, Dr Graham-Smith, whose paper follows, was entrusted with similar investigations upon the bloods of lower animals than those here considered (see p. 336).

The tests here recorded were made upon the bloods of Primates, Insectivora, Carnivora, Ungulata, Cetacea, Marsupialia, and Aves by means of the following antisera. The method of obtaining antisera has already been described elsewhere by Nuttall (p. 51).

1. Antisera for *Primates*:   Man (2 series)
                                    Ourang
2.   ,,   *Insectivora*:   Hedgehog
3.   ,,   *Carnivora*:   Cat
                                      Dog
4.   ,,   *Ungulata*:   Sheep (3 series)
                                      Ox
                                      Antelope (Cobus unctuosus)
                                      Cariacus mexicanus
                                      Cervus porcinus
                                      Rangifer tarandus
                                      Pig (3 series)
                                      Horse
                                      Donkey   } several series
                                      Zebra (E. Grevyi)
5.   ,,   *Marsupialia*: Onychogale unguifera
6.   ,,   *Aves*:   Struthio molybdophanes
                                      Gallus domesticus

Before proceeding to describe the different series of tests it appears desirable to give briefly the method of testing quantitatively, quoting Nuttall's description which appeared in the *British Medical Journal* for April 5th, 1902, since his original method has not been modified.

Method for the Measurement of the Degree of Reaction.

" Dilutions of the serums to be tested quantitatively are made with 0·6 per cent. normal salt solution to the degree desired—namely, 1 : 100, 1 : 200, etc. The solutions must be perfectly clear and free from suspended matter. Accurately-measured quantities—usually 0·5 cm. of the dilution—are transferred into a series of clean, dry test-tubes, having a capacity of about 1 cm. and a lumen of about 5 mm. The dilution is allowed to run into the tubes, along a fine, freshly-drawn, sealed capillary tube in order to avoid the formation of bubbles or of overflow in filling the tubes. Each tube is supplied with a sealed capillary, which is left in place after the introduction of the dilution, subsequently serving a similar purpose when adding the antiserum. An accurately-measured quantity of antiserum is now added to the dilution, in the proportion of 20 : 1 to 200 : 1 or more of the blood in dilution. After the antiserum has been added to the series of tubes the capillaries are removed and the mixture of dilution and antiserum is effected. This can be done very satisfactorily by applying the tip of the clean finger to the mouth of the tube and shaking it up and down.

By using the fingers of both hands in rotation eight tubes can be shaken in succession, after which the hands must be cleansed before proceeding to the next set of tubes. This is only mentioned because it saves time. With a little practice it will be found easy to bring about complete mixture without the formation of troublesome air-bubbles, and if the fluid is shaken down to the bottom of the tube before removing the finger the loss of fluid entailed by contact with the finger is practically negligible, the finger-tip remaining apparently dry after some practice. The mixture is now left standing at a desired temperature, the rate of precipitum formation, etc. being noted.

After 24 hours have elapsed the precipitum has become deposited upon the bottom of the tube. Should any adhere to the walls of the vessel it may be dislodged by gently tapping or rotating the tube upon its vertical axis. The particles thus displaced soon sink to the bottom of the tube, provided it does not contain too much serum, which through its viscidity impedes their gravitation. The clear supernatant fluid is now pipetted off by means of a conveniently bent bulb capillary pipette, the tip of which is allowed to slide down along the inner wall of the test-tube, the precipitum not being disturbed, the clear fluid being aspirated into the bulb as the tip of the pipette slides down a little beneath the receding surface of the column of fluid in the tube. The bulk of pre-cipitum and clear fluid left is so small that it can now be readily drawn up into the capillaries.

The capillaries used for measuring the precipitum volumetrically are made of fairly thick glass, are about 12 cm. long, and have a lumen of about 1 mm. The ends should be cut off square by means of a diamond, this being essential, as it is thus possible to practically remove all the fluid and precipitum in the bottom of the small test-tube. The end of the capillary is placed against the bottom of the tube, and the deposit and fluid are well mixed by being drawn backwards and forwards into the capillary. It is imperative that bubbles should be avoided. The fluid is now drawn up a little way into the capillary, and the dry end of this is drawn out and sealed in a very small flame. The sealed end should be blunt to avoid possibility of breakage; the other end of the tube, which can now be cut down to a convenient length, is left open. After a number of capillaries have been thus prepared they are placed verti-cally in a rack in a cool place. After 24 hours the height of the column of precipitum is marked upon the capillary with a glass pencil. After 48 to 72 hours it will be seen that the column of precipitum remains constant. The amount of precipitum may now be measured.

The accompanying figure shows a little apparatus which I have made in the laboratory, and have found very convenient for making these measurements of the amount of precipitum. A fairly large thermometer tube (A) is used, which is graduated in tenths of a c.c., the calibre being chosen so that a fine steel scale of 10 cm. (E) subdivided into 0·5 mm. corresponds to the length of column of fluid measuring 0·1 cm. in the tube. The tube is filled with pure water up to the graduation mark (not shown in the figure), the mouth being wiped dry with filter-paper. The tube is fixed in a spring clip which is attached to the wooden rod (D) so that it pivots in a vertical plane, thus permitting the tube to rest in a horizontal position or at any desired angle. The rod (D) slides horizontally between the rods C and C', which are nailed to a solid block of wood beneath. Two small brass plates attached to rod C prevent rod D from being displaced. The steel cm. measure is attached to a metal support, which is screwed to the block beneath. By a lucky chance I happened to have an old "peep-sight" from a sporting rifle which served admirably for carrying the metal scale, as it can be inclined at any angle and be screwed up or down. By pivotting the scale on a screw passing through the eyehole of the sight, the scale can be placed and retained at any angle on its support, thus making it possible to closely apply the fine scale to the thermometer tube in any position desired.

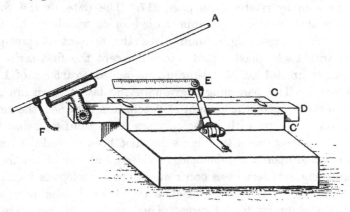

Fig. 5.

The height of the volume of precipitum having been marked upon the capillary tube, the latter is cut square by means of a diamond at the lower meniscus, a small quantity of air having been left in the lower end of the tube when it was sealed. The contents of the tube are blown out, the tube is drained as far as possible by blowing out the moisture it contains on to filter-paper, or it can be washed out and dried. The mouth of the capillary is now brought in contact with the end of the thermometer tube, which has been slightly raised at one end (A), and fluid is allowed to pass into the capillary up to the pencil mark previously noted. The amount of fluid which has passed into the capillary is now read off on the scale. The thermometer tube is moved by means of the handle (F) so as to prevent expansion of the fluid through the heat of the hand. Where it is desired to know the actual quantity of precipitum, it is necessary to note the temperature and calculate accordingly. This may be disregarded when a simple comparison is being made with different quantities of precipitum in a given series, the measurements being made in rapid succession under similar outward conditions. If it is desired to make simply comparative measurements a diamond scratch on any suitable tube is sufficient, only the measurements on the steel scale being recorded."

In the tables which follow the bloods are arranged in groups according to the orders to which the animals yielding them belong. In most instances the scientific names of the animals are given, this is however omitted with the commoner domesticated or laboratory animals. The numbers preceding the names of the animals refer to the order in which they were collected by Nuttall and tested qualitatively, as will be seen by reference to page 217. The date in the column following is that on which the animal died, or on which its blood was collected. The succeeding column gives the amount of precipitum obtained with each blood tested. In all except the first series with antihuman serum 0·1 c.c. of antiserum was added to 0·5 c.c. of 1 : 100 blood dilution. The percentages given in the last column are calculated upon the basis of the amount of precipitum obtained, the amount given by an antiserum with its homologous blood dilution being taken as 100 %. In most cases this figure has not been exceeded when the antiserum acted upon a non-homologous blood, but notably in the case of the anti-ungulate sera cases occur where non-homologous bloods give higher figures. In some cases comments alongside the percentages sufficiently account for high figures. Thus, the serum may have been concentrated, or the precipitum loose in character, and consequently with difficulty measureable. Brief comments in the text also serve to explain and summarize the results.

1. Quantitative Tests with Anti-Primate Sera.

Tests with Antihuman Serum (11. III. '02). *Series* I.

The antiserum here used was obtained in the usual way by treating a rabbit with human blood serum. The 45 bloods tested had been preserved for various lengths of time in the refrigerator with the addition of a small amount of chloroform. The only difference in this series is that 1 : 40 blood-dilutions were used instead of the usual 1 : 100 dilutions, the usual amount of antiserum 0·1 c.c. being added to ·5 c.c. of the dilution.

Blood of	Date	Precipitum amount	Percentage
Primates			
Man	1. iii. 02	·031	100
298 Chimpanzee	xi. 01	·04	130 (loose precipitum)
299 Gorilla	xi. 01	·021	64
254 Ourang	30. x. 01	·013	42
364 Cynocephalus mormon	10. xii. 01	·013	42
501 ,, sphinx	14. ii. 02	·009	29
431 Ateles geoffroyi	28. xii. 01	·009	29
Insectivora			
433 Centetes ecaudatus	3. i. 02	·0	0
Carnivora			
359 Canis aureus	3. xii. 01	·003	10 (loose precipitum)
,, familiaris	4. iii. 02	·001	3
579 Lutra vulgaris	4. iii. 02	·003	10 (concentrated serum)
367 Ursus tibetanus	27. xii. 01	·0025	8
362 Genetta tigrina	9. xii. 01	·001	3
Felis domesticus	3. xi. 01	·001	3
300 ,, caracal	15. xi. 01	·0008	3
103 ,, tigris	13. viii. 01	·0005	2
Ungulata			
Ox	3. ii. 02	·003	10
Sheep	—	·003	10
486 Cobus unctuosus	8. ii. 02	·002	7
502 Cervus porcinus	15. ii. 02	·002	7
483 Rangifer tarandus	4. ii. 02	·002	7
243 Capra negaceros	23. x. 01	·0005	2
Equus caballus	10. vi. 01	·0005	2
Sus scrofa	26. ii. 02	·0	0
Rodentia			
436 Dasyprocta cristata	6. i. 02	·002	7 (concentrated serum clots)
Guinea-pig	—	·0	0
Rabbit	2. xii. 01	·0	0
Marsupialia			
Seven bloods (7 species, see list on p. 331 under anti-pig series No. 3)		·0	0

Among the Primate bloods that of the Chimpanzee gave too high a figure, owing to the precipitum being flocculent and not settling well for some reason which could not be determined. The figure given by the Ourang is somewhat too low, and the difference between Cynocephalus sphinx and Ateles is not as marked as might have been expected in view of the qualitative tests and the series following. The possibilities of error must be taken into account in judging of these figures, repeated tests should be made to obtain something like a constant. Other bloods than those of Primates give small reactions or no reactions at all. The high figures (10 %) obtained with two Carnivore bloods can be explained by the fact that one gave a loose precipitum, and the other was a somewhat concentrated serum.

Tests with Antihuman Serum (2. xi. '02). *Series* ii. *and* iii.

Antiserum from two rabbits treated with human serum.

Blood of	Date	Percentages of precipitum	
		Antiserum I.	Antiserum II.
Man	8. viii. 02	100 (·018)	100 (·010)[1]
906 Ourang	14. x. 02	47	80
364 Cynocephalus mormon	10. xii. 01	30	50
742 Cercopithecus petaurista	10. vi. 02	30	50
697 Ateles vellerosus	11. iv. 02	22	25
784 Cynocephalus porcarius	4. vii. 02	61	70 } note text
908 Macacus rhesus	14. x. 02	72	90 }

[1] The succeeding amounts of precipitum are purposely omitted from the table.

The results with the two antisera appear divergent. *The first antiserum was nearly twice as powerful as the second,* and appears to cause less precipitation proportionately in the non-homologous bloods. The bloods Nos. 784 and 908 were from animals which had died of intussusception and dysentery respectively, these diseases, as we know, tending to produce concentration of the serum during life; and this accounts fully for the very high percentage of precipitum obtained. On the other hand, if we take the percentages obtained with the blood of the Ourang as 100, the figures agree fairly well between the two series, if we except that of Ateles, thus:

With Antiserum I.		Antiserum II.	The sera of the diseased animals considered separately gave with		
				Antiserum I.	Antiserum II.
Ourang	100 %	100 %		No. 784 100 %	100 %
364 Cynocephalus	63 ,,	64 ,,		No. 908 118 ,,	128 ,,
742 Cercopithecus	63 ,,	64 ,,			
697 Ateles	31 ,,	48 ,,			

Of other mammalian sera of approximately the same age several reacted as follows, Human serum being taken to give a precipitum of 100 (Reactions with antiserum I.):

Carnivora : Cat and Dog 11 %, Hyaena striata 8 %, Seal but a trace.
Ungulata : Zebra 11 %. *Insectivora :* Hedgehog 3 %, Mole bnt a trace.
Marsupialia : Petrogale 0 %.

Tests with Anti-Ourang Serum (7. I. '03).

The rabbit yielding the antiserum was treated with the serum of *Simia satyrus* (906).

Blood of	Date	Precipitum amount	Percentage
906 Simia satyrus	14. x. 02	·008	100 (2 samples gave same result)
Man	fresh	·006	75
907 Macacus rhesus	26. ix. 02	·005	62 (4 samples from 2 monkeys gave same result)

Owing to the weakness of this antiserum the measurements of the smaller amounts of precipitum given by the serum of one other mammal cannot be considered accurate. Thus the blood of the Ox (1. I. '02) gave the percentage 34 %. The sera of all other Mammalia gave no measurable precipitum, the bloods tested being as follows: *Insectivora :* Hedgehog and Mole. *Carnivora :* Cat and Dog.—*Ungulata :* Sheep, Deer, Llama, Horse, Donkey, Pig. *Cetacea :* Balaenoptera (two examples) and Phocaena communis.—*Rodentia :* Rabbit.

It will be seen that the Primate bloods stand in the same relation to each other as in the series with anti-human serum, and what is more that the percentages approximate closely to those obtained with anti-human serum No. II.

2. Quantitative Tests with Anti-Insectivore Serum.

Tests with Anti-Hedgehog Serum. (18. x. '02).

In view of the limited number of Insectivore bloods and the negative results obtained by qualitative tests with this antiserum upon other bloods than those belonging to this order, only a small series was tested. Although very powerful, this antiserum produced practically no effect on any other blood:

Blood of	Date	Precipitum amount	Percentage
783 Erinaceus europeus	23. vi. 02	·022	100
782 Talpa europea	23. vi. 02	·0015	7
Bos bovis	1. i. 02	·001	4

N.

21

No trace of reaction was obtained with corresponding dilutions of the bloods of Man, Cat, Hyaena, Dog, Seal, and Wallaby. Further investigations will be required with antisera for Insectivora.

3. Quantitative Tests with Anti-Carnivore Sera.

Tests with Anti-Cat Serum (1. vi. '02).

Antiserum from rabbit treated with the serum of *Felis domesticus*.

Blood of	Date	Precipitum amount	Percentage	
739 Felis domesticus	8. v. 02	·005	100	another cat's serum gave an identical result
579 Lutra vulgaris	4. iii. 02	·002	40*⎫	see text
367 Ursus tibetanus	27. xii. 01	·001	20*⎭	
738 Felis tigris	21. v. 02	·001	20	another tiger (103) gave 16 % reaction
Canis familiaris	—	·0008	16	

The antiserum used in this case was weak. The only bloods which gave measurable quantities of precipitum were those from Carnivora. The blood of Felis caracal (300) gave no precipitum, this being probably due to an excess of chloroform having been added to it, causing some precipitation; it however reacted with a distinct cloudiness upon the addition of antiserum. The bloods of the Jackal, Raccoon and Genetta gave no reaction. The high figures given by the bloods of Lutra* and Ursus* are attributable to these bloods being somewhat concentrated through keeping, as noted elsewhere. Owing to the very small amount of precipitum with the bloods of Ursus, Canis, and Felis tigris, these measurements cannot be considered accurate. Of 11 bloods of Carnivora examined with this weak antiserum, seven gave a precipitum, one a clouding only, and three a negative result. Twenty-eight other mammalian bloods were tested at the same time with uniformly negative results, the bloods examined being from Primates (7), Ungulates (12), Cetacea (1), Marsupialia (8).

Tests with Anti-Dog Serum (25. III. '02).

Antiserum from rabbit treated with the serum of *Canis familiaris*.

Blood of	Date	Precipitum	Percentage	
397 Canis familiaris	2. xii. 01	·015	100	
359 Canis aureus	3. xii. 01	·0035	24	
367 Ursus tibetanus	27. xii. 01	·0022	15	
739 Felis domesticus	8. v. 02	·0021	14	
362 Genetta tigrina	9. xii. 01	·002	13	
579 Lutra vulgaris	4. iii. 02	·0018	12	
103 Felis tigris	13. viii. 01	·0016	10·6	
738 „ „	21. v. 02	·0012	8	average 9·3 °/₀
366 Procyon lotor	27. xii. 01	·001	7	

No other bloods than those of Carnivora were tested. A canine blood, that of the Jackal (359), gave 24 % reaction; the four succeeding non-canine bloods gave fairly uniform results, a somewhat lower figure being obtained with the last three. Owing however to the small amount of precipitum obtained in these cases too much reliance cannot be placed upon the figures.

4. Quantitative Tests with Anti-Ungulate Sera.

Tests with Anti-Sheep Serum (18. II. '02). *Series I.*

Antiserum from rabbit (F) treated with the serum of the domesticated sheep (*Ovis aries*).

In this series bloods of approximately the same age were taken for testing. The sera of five sheep gave the following amounts of precipitum: ·0325, ·0285, ·0330, ·0285, ·0280, the average amount ·030 being obtained with 1 : 100 sheep serum dilution. The figures obtained with 1 : 200 dilutions were somewhat lower, as might be expected, but also remarkably uniform: ·0215, ·025, ·0265, ·026, ·026, ·0235, the average here being ·024. The results obtained were as follows. With 1 : 100 blood dilutions taking the average amount obtained with the blood of the five sheep (·03) as 100:

Blood of	Date	Precipitum	Percentage
5 sheep	—	·030	100
502 Cervus porcinus	15. ii. 02	·0135	45
Horse	—	·0045	15
Cat	3. x. 01	·0035	12
Dog	10. xii. 01	·002	7
488 Wallaby	3. ii. 02	·0015	5

Here an immediate and marked reaction only took place with sheep blood, a marked clouding in that of Cervus, a faint clouding in that of the dog. The figure for the cat appears too high.

Tests with Anti-Sheep Serum (15. II. '02). Series II.

Antiserum from another rabbit (B) treated as before.

In this series the bloods of two sheep (4. I. and 10. II. '02) gave a precipitum of ·0185 and ·022 respectively in 1 : 100 dilution (average ·02), the corresponding figures in 1 : 200 dilution being ·0135 and ·011. Tests conducted as before, ·02 being taken as 100 gave the following results :

Blood of	Date	Percentage
Ovis aries	—	100
Bos bovis	1. i. 02	75
Cobus unctuosus	8. ii. 02	67
Rangifer tarandus	4. ii. 02	35

			in 1·100 dilution	in 1·200 dilution	
Sheep serum dried in scales	19. iii. 01 gave		·017	·015	
,, ,, fluid, with chloroform	,,	,,	·025	·0125	c.c. precipitum
,, ,, ,, sealed in tubes	,,	,,	·026	·0155	

Tests with Anti-Sheep Serum (11. III. '02). Series III.

Antiserum from a third rabbit treated as before.

	Blood of	Date	Precipitum amount	Percentage
	Ovis aries	—	·051	100
243	Capra megaceros	23. x. 01	·02	39
	Bos bovis	3. ii. 02	·019	37
486	Cobus unctuosus	8. ii. 02	·017	33
483	Rangifer tarandus	4. ii. 02	·015	29
502	Cervus porcinus	15. ii. 02	·011	22
	Sus scrofa	26. xii. 01	·002	4
	Equus caballus	10. vi. 01	·0015	3

The antiserum in this case was exceptionally powerful, and it appears due to this fact that such a contrast is noted in related Ungulate bloods. The results with the bloods of other Bovidae and the Cervidae appear fairly uniform, whereas very little reaction was obtained with the bloods of the pig and horse.

The results with 28 other mammalian bloods were as follows:

Primates. 8 bloods examined from 7 species: Man (2 samples gave 0 and 3 % reaction), Ourang and Chimpanzee (0 %), Mandrill and Baboon (4 %), Gorilla and Ateles (6 and 10 % respectively, but both these sera were thick and certainly somewhat concentrated through keeping).

Carnivora. Bloods from 9 species : Tiger, Caracal, Genetta, Dog (0 %), Raccoon (3 %), Cat and Bear (4 %), Otter and Jackal (7 and 5 % respectively). It was noted at the time that the last three bloods were viscid or thick, doubtless concentrated.

Rodentia. Three bloods examined, from the Agouti, Rabbit, and Guinea-pig (0 %).

Insectivora. One blood examined, that of the Tenrec (0 %).

Marsupialia. Seven bloods examined (0 %).

If we exclude the bloods which were regarded as concentrated we have 23 non-ungulate bloods of which 3 gave a 4 % reaction, and 2 a 3 % reaction, all the others yielding no precipitum.

Tests with Anti-Ox Serum (15. II. '02).

Antiserum from rabbit treated with the serum of *Bos bovis*, the domesticated Ox.

Blood of	Date	Precipitum amount	Percentage
Bos bovis	1. i. 02	·011	100
Cobus unctuosus	8. ii. 02	·005	45
Ovis aries		·004	36

In the above tests three sheep sera gave the same figure, these having been obtained on different dates (19, III. '01, sealed in tubes, 4, I., and 10, I. '02). Two other sheep sera preserved since 19, III. '01, the one dried in scales and brought into solution for the experiment, the other preserved with chloroform, gave a precipitum of ·003 and ·0035 respectively. The ox and sheep sera were from animals slaughtered in Cambridge and presumably normal. Whereas granules formed in the solution of ox serum soon after the addition of antiserum, a cloudiness persisted for over three hours in the other bloods. Compare these percentages with those of the corresponding series with anti-antelope serum, which follows.

Tests with Anti-Antelope Serum (6. VI. '02).

Antiserum from rabbit treated with the serum of *Cobus unctuosus* (486).

	Blood of	Date	Precipitum amount	Percentage
486	Cobus unctuosus	8. ii. 02	·055	100
243	Capra megaceros	23. x. 01	·031	56
631	Cariacus mexicanus	9. iv. 02	·028	51
	Bos bovis	—	·025	45
596	Capra hircus	13. iii. 02	·021	38
483	Rangifer tarandus	4. ii. 02	·020	36
	Ovis aries	—	·019	35
595	Ovis burrhel	10. iii. 02	·012	22
502	Cervus porcinus	15. ii. 02	·008	15
698	Balaenoptera rostrata	31. iv. 02	·0045	8
	Horse (2 samples average)[1]	20. iv. 02	·0025	5
	Sus scrofa („ „ „)[1]	29. iv. 02	·0015	3

[1] Date of but one sample of each pair given.

From the above it will be seen that the larger reactions are all confined to Bovidae and Cervidae, although from some cause the last two of these (Nos. 595 and 502) gave low figures. These measurements were unfortunately not made in duplicate consequently there is no control. Nevertheless the result agrees fairly with those obtained with other antisera of this class. It will be seen that the whale (698) gives a higher figure than either the horse or pig, although the difference is very slight.

Tests with Anti-Mexican-Deer Serum (1. VI. '02).

Antiserum from rabbit treated with the serum of *Cariacus mexicanus* (631).

Blood of	Date	Precipitum amount	Percentage	
Ungulata				
631 Cariacus mexicanus	9. iv. 02	·015	100	Died from phthisis, pleurisy.
483 Rangifer tarandus	4. ii. 02	·022	146	? serum with clots, probably concentrated.
502 Cervus porcinus	15. ii. 02	·009	60	
486 Cobus unctuosus	8. ii. 02	·018	120	
Ovis (domestic)	—	·018	120	
Bovis (domestic)	—	·016	107	
596 Capra hircus	13. iii. 02	·012	77	
595 Ovis burrhel	10. iii. 02	·01	67	
243 Capra megaceros	23. x. 01	·0095	63	
696 Sus scrofa	29. iv. 02	·006	40	
,, ,,	—	·0055	37	
694 Equus caballus	20. iv. 02	·006	40	
,, ,,	—	·005	33	
Cetacea				
698 Balaenoptera rostrata	30. iv. 02	·006	40	

Of other mammalian bloods, those of *Carnivora* (11) gave 15 % reaction, those of *Primates* (7) gave 18 % reaction, the percentage given being the average for the groups. On the other hand the bloods of *Marsupialia* (8) all gave no trace of reaction.

From the above we see that the high percentages are confined to Cervidae and Bovidae amongst the Ungulata. The high percentages obtained with non-homologous bloods in these groups is difficult to interpret, the blood-relationship appears very close in any case. The bloods of Sus and Equus give results remarkably in accord, and both are practically the same as with the Rorqual (Balaenoptera). A very moderate reaction is observed with the bloods of Carnivora and Primates, both being fairly equal, whereas the distance separating the Marsupialia is clear from the total absence of reaction with their bloods.

Tests with Anti-Hog-Deer Serum (6. VI. '02).

Antiserum from rabbit treated with the serum of *Cervus porcinus* (502).

Blood of	Date	Precipitum amount	Percentage
502 Cervus porcinus	15. ii. 02	·0063	100
631 Cariacus mexicanus	9. iv. 02	·01	159
Bos bovis	—	·0095	151
Ovis (domestic)	2. ii. 02	·008	127
596 Capra hircus	13. iii. 02	·0078	124
595 Ovis burrhel	10. iii. 02	·006	95
486 Cobus unctuosus	8. ii. 02	·004	64
243 Capra megaceros	23. x. 01	·004	64
698 Balaenoptera rostrata	30. iv. 02	·0038	60
696 Sus scrofa	29. iv. 02	—	35 ⎫ average of 2 observations
694 Equus caballus	20. iv. 02	—	24 ⎭

Tests with Anti-Reindeer Serum (6. VI. '02).

The antiserum used in this series was obtained by treating a rabbit with the serum of *Rangifer tarandus* (483). Only the bloods of Ungulata and that of the whale were tested. Antiserum weak.

Blood of	Date	Precipitum amount	Percentage
483 Rangifer tarandus	4. ii. 02	·004	100
596 Capra hircus	13. iii. 02	·0015	38
631 Cariacus mexicanus	9. iv. 02	·0014	35
502 Cervus porcinus	15. ii. 02	·0013	33
595 Ovis burrhel	10. iii. 02	·0013	33
243 Capra megaceros	23. x. 01	·0007	18
486 Cobus unctuosus	8. ii. 02	·0006	15
Ovis aries	2. ii. 02	·0006	15
Equus caballus	10. vi. 01	·0004	10
Sus scrofa	26. ii. 02	·0003	8
698 Balaenoptera rostrata	30. iv. 02	·0011	28

The above figures are fairly comparable with those obtained with other antisera for this group. The figures for Capra, Cobus, and Ovis (towards the end of the series) are too low, judging from what has been obtained in other series. The whale on the other hand again gives a remarkably high figure, this being in accord with the qualitative tests on three samples of cetacean blood.

Tests with Anti-Pig Serum (18. ii. '02). *Series I.*

Antiserum from rabbit treated with the serum of *Sus scrofa* (domesticated Pig).

The figures given here were published in part by Nuttall (5, iv. 1902) in the *British Medical Journal*. Of five normal sheep sera four gave a precipitum of ·006, one of ·0055 with this antiserum, the sheep having been slaughtered 20 hours previously. The reactions obtained were as follows:

Blood of	Date	Precipitum amount	Percentage
Pig (fresh)	—	·045	100
Horse	—	·0075	16
Cervus porcinus	15. ii. 02	·0065	14
Sheep	—	·006	13
Cat	3. x. 01	·0065	14
Dog	10. xii. 01	·006	13
Wallaby	3. ii. 02	·0025	5 [1]

[1] It may be noted that such a reaction was only obtained with a marsupial blood with two very powerful anti-pig sera (Series I and III) whereas in Series II less precipitum was obtained although an apparently even more powerful antiserum was used. Series II was tested 4 months later than Series I. The most powerful anti-sheep serum (Series I) also gave a 5 % reaction with a marsupial blood.

It will be noted that the antiserum in this case was very powerful (precipitum of ·045). It produced an immediate reaction with pig serum dilution, a deposit being formed, and the supernatant fluid clear within 30 minutes. In the other bloods the clouding took place more slowly as also did deposition. The Wallaby blood dilution only began to show faint clouding after 30 minutes. This antiserum demonstrates the "mammalian reaction," and does not show any distinct difference between the behaviour of the other non-homologous bloods if we except that of the Wallaby.

Tests with Anti-Pig Serum (11. vi. '02). *Series II.*

In this series another antiserum to the preceding was tried, being obtained in the same way from a rabbit. In this case a larger series of bloods was tested, with the following results:

Blood of	Date	Precipitum amount	Percentage
Ungulata			
696 Pig	29. iv. 02	·055	100 *
694 Horse	20. iv. 02	·006	12
596 Goat	13. iii. 02	·005	9
483 Reindeer	4. ii. 02	·0045	8
595 Burrhel's sheep	10. iii. 02	·004	7
Ox	—	·004	7
631 Cariacus mexicanus	9. iv. 02	·0035	6
486 Cobus unctuosus	8. ii. 02	·003	6
502 Cervus porcinus	15. ii. 02	·0022	4
243 Capra megaceros	23. x. 01	·001	2
Cetacea			
698 Balaenoptera rostrata	30. iv. 02	·003	6
Primates			
364 Cynocephalus mormon	10. xii. 02	·004	7
501 ,, sphinx	14. ii. 02	·003	6
703 ,, babuin	10. v. 02	·002	4
Homo sapiens	—	·003	6
254 Simia satyrus	30. x. 01	·001	2
697 Ateles vellerosus	11. iv. 02	·002	4
431 ,, geoffroyi	28. xii. 01	·0011	2
Carnivora			
739 Felis domesticus	8. v. 02	·005	9
,, ,,	—	·004	7
Canis familiaris	—	·0035	6
738 Felis tigris	21. v. 02	·003	6
103 ,, ,,	13. viii. 01	·0022	4
579 Lutra vulgaris	4. iii. 02	·0025	5
359 Canis aureus	3. xii. 01	·0021	4
366 Procyon lotor	27. xii. 01	·0021	4
367 Ursus tibetanus	27. xii. 01	·002	4
300 Felis caracal	15. xi. 01	·0015	3
362 Genetta tigrina	9. xii. 01	·0014	3
Rodentia			
436 Dasyprocta cristata	6. i. 02	·0005	1
Rabbit	—	·0	0
Guinea-pig	—	·0	0
Marsupialia			
611 Onychogale unguifera	22. iii. 02	·0012	2
488 ,, ,,	3. ii. 02	·0001	0·2
503 ,, ,,	13. ii. 02	·0005	1
459 ,, frenata	23. i. 02	·0001	0·2
487 Petrogale xanthopus	10. ii. 02	·0006	1·1
484 ,, penicillata	4. ii. 02	·0	0
512 Macropus bennetti	22. ii. 02	·0001	0·2
482 Thylacinus cynocephalus	1. ii. 02	·0001	0·2

* The serum of another pig of older date (26. ii. 02) gave a precipitum of ·07 = 127 %. Pig serum No. 696 had been used for treating the rabbit yielding the antiserum.

The results recorded here will be considered at the end of the succeeding series.

Tests with Anti-Pig Serum (11. iii. '02). Series III.

The following series is recorded for the reason that it shows how *an antiserum may at times give apparently aberrant results*. The antiserum was obtained exactly like the preceding ones by injecting a rabbit intra-peritoneally with pig serum. Why the antiserum should have behaved in this manner we cannot explain unless it be that it belonged to the class we have described as "milky," although we should not have used it if it had shown any marked opacity. The antiserum reacted most promptly with its homologous blood and would be classed as very powerful. These results should serve as a warning to always use controls in medico-legal work with the precipitin method.

Blood of	Date	Precipitum amount	Percentage
Ungulata			
Pig	26. ii. 02	·044	100
Horse	10. vi. 01	·012	27
Ox	3. ii. 02	·009	20
486 Cobus unctuosus	8. ii. 02	·009	20
Sheep	—	·006	14
483 Rangifer tarandus	4. ii. 02	·006	14
502 Cervus porcinus	15. ii. 02	·0035	8
243 Capra megaceros	23. x. 01	·0015	3
Primates			
298 Chimpanzee	xi. 01	·012	27
299 Gorilla	xi. 01	·012	27
Man	1. iii. 02	·011	25
„	5. iii. 02	·005	11
254 Ourang	30. x. 01	·006	14
364 Cynocephalus mormon	10. xii. 01	·013	30
501 „ sphinx	14. ii. 02	·01	23
431 Ateles geoffroyi	28. xii. 01	·0125	28
Carnivora			
362 Genetta tigrina	9. xii. 01	·011	25
579 Lutra vulgaris	4. iii. 02	·009	20 (concentr. serum)
Felis domesticus	3. xi. 01	·0075	17
367 Ursus tibetanus	27. xii. 01	·006	14
359 Canis aureus	3. xii. 01	·004	9
Canis familiaris	4. iii. 02	·0035	8
103 Felis tigris	13. viii. 01	·003	7
300 Felis caracal	15. xi. 01	·003	7
366 Procyon lotor	27. xii. 01	·0	0
Insectivora			
433 Centetes ecaudatus	3. i. 02	·011	25

Blood of	Date	Precipitum amount	Percentage
Rodentia			
Guinea-pig	—	·002	5
436 Dasyprocta cristata	6. i. 02	·012	27 (concentr. clots)
Rabbit	2. xii. 02	·0	0
Marsupialia			
487 Petrogale xanthopus	10. ii. 02	·003	7
484 ,, penicillata	4. ii. 02	·001	2
459 Onychogale frenata	23. i. 02	·0007	2
488 ,, unguifera	3. ii. 02	·0005	1
503 ,, ,,	13. ii. 02	·0005	1
512 Macropus bennetti	22. ii. 02	·0	0
482 Thylacinus cynocephalus	1. ii. 02	·0	0

Tests with Anti-Horse Serum (1. vi. '02).

The following tests were conducted with anti-horse serum obtained from a rabbit killed 48 hours previously.

Blood of	Date	Precipitum amount	Percentage
Equus caballus	10. vi. 01	·012	150
*694 ,, ,,	20. iv. 02	·008	100
502 Cervus porcinus	15. ii. 02	·003	38
483 Rangifer tarandus	4. ii. 02	·0015	19
696 Sus scrofa	29. iv. 02	·001	13
,, ,,	26. ii. 02		0

Also negative with the following Ungulate Bloods: Ox, Sheep, Antelope, Markhoor, Burrhel Sheep, Goat, Mexican Deer.

Cetacea :   Result negative with the blood of Balaenoptera rostrata (30. iv. 02).

Primates :   Result negative with the bloods of Man, Baboon, Ourang, Mandrill and two species of Ateles. The blood of a Cynocephalus (703) gave a curiously high figure, viz. 25 %.

Rodentia :   Result negative with the bloods of Agouti, Rabbit, and Guinea-pig.

Carnivora :   Result negative with the bloods of Otter, Genetta, Raccoon, Jackal, Caracal, and one Tiger. On the other hand a curious result was obtained with some bloods which gave the following percentages : two Cats (31 and 25 %), Dog (31 %), Bear (14 %), Tiger (13 %).

Marsupialia :   Result negative with 8 bloods recorded on p. 329.

* The blood used for treating the rabbit which yielded the antiserum.

The antiserum used in this case was of moderate power only, and the precipitum amounts obtained were so small in most cases as to permit of no small error in stating the percentages. A considerable distance separates the blood of the horse from that of other animals, a relatively high figure being only obtained with the blood of Cervus porcinus amongst the Ungulata, others giving less or no reaction. Of the Primate bloods only that of a Cynocephalus gave any reaction, and then one of 25 %, which we cannot explain, having been unable to repeat the test. Possibly the result was due to error. It will be noted however that two cats and a dog, amongst the Carnivora, also gave relatively high figures.

Tests with Anti-Donkey Serum (8. i. '03).

Antiserum from rabbit treated with serum of donkey (929).

Blood of	Date	Precipitum amount 2 tests	Percentages
929 Equus asinus	3. xi. 02	·010—·009	100—90
930 ,, ,,	5. xi. 02	·008—·008	80—80
Equus caballus	17. xii. 02	·009—·009	90—90
,, ,,	10. vi. 01	·007—·007	70—70
903 Equus grevyi	13. x. 02	·004—·004	40—40

The immediate reaction observed on the addition of anti-donkey serum to donkey blood dilutions was very marked. A faint clouding appeared in the dilution of the more recent horse-blood. After 90 minutes granules and deposit had formed in these solutions, whilst the horse serum of older date and that of the zebra still showed but a marked clouding. In other words, there was a well-marked sequence in the reaction.

Parallel Tests with Antisera for Horse, Donkey, and Zebra.

The anti-horse serum used was taken from two different rabbits (A and B). The anti-donkey serum was a different one from that used in the preceding series. The anti-zebra serum as in the other cases was obtained by treating rabbits with the corresponding serum. The order in which the immediate reactions took place corresponded exactly to the order of precipitum amounts, that is, the dilutions giving the largest amounts of precipitum reacted first, those giving the smallest amounts reacted last. Corresponding appearances were noted after 90 minutes, the precipitum had accumulated entirely at the bottom of the tubes giving the highest precipitum amounts, partially at the bottom and on the walls of the tubes giving lesser reactions, whilst the solutions in the tubes giving the least reactions sometimes only showed a clouding or granulation when deposition was complete in the first of the series. Consequently the *rate* at which reaction takes place is of value in judging the result. Only two horse-bloods (*a*, 10. vi. '01 and *b*, 10. vii. '02), two donkey-bloods (*a*, 3, and *b*, 5, XI. '02) and one zebra blood (13, X. '02, No. 903, see above) were tested.

Anti-horse serum A and B			A %	B %	Anti-donkey		%	Anti-zebra		%
Blood of	Precip.	Precip.			Blood of	Precip.		Blood of	Precip.	
Horse *a*	·013	·017	100	100	Donkey *a*	·017	100	Zebra	·012	100
Donkey *a*	·011	·010	84	59	Donkey *b*	·012	70	Donkey *b*	·010	83
Donkey *b*	·012	·011	85	66	Horse *a*	·010	59	Donkey *a*	·008	66
Zebra	·008	·005	58	30	Zebra	·006	35	Horse	·008	66

5. Quantitative Tests with Anti-Marsupial Serum.

Tests with Anti-Wallaby Serum (1. VI. '02).

The antiserum was obtained by treating a rabbit with the serum of *Onychogale unguifera* (No. 611). The antiserum was weak. Nine marsupial bloods were tested, of which 8 reacted as follows :

	Blood of		Date	Precipitum amount	Percentage
611	Onychogale	unguifera	22. iii. 02	·007	100
488	,,	,,	3. ii. 02	·0025	36
503	,,	,,	13. ii. 02	·002	29
459	,,	frenata	23. i. 02	·002	29
512	Macropus	bennetti	22. ii. 02	·005	71
487	Petrogale	xanthopus	10. ii. 02	·004	57
484	,,	penicillata	4. ii. 02	·0025	36

482 Thylacinus cynocephalus 1. ii. 02 yielded no precipitum, as did also the blood of 35 non-marsupial mammalia, including *Primates* (7), *Carnivora* (11), *Ungulata* (13), *Cetacea* (1), *Rodentia* (3).

The serum which gave the greatest precipitum with antiserum was the one (611) used for treating the rabbit. In the absence of records regarding the diseases from which the animals suffered it is impossible to offer an explanation of the low figures observed in the precipitum in bloods 611, 488, 503. Nevertheless we see the antiserum excluding all other bloods than those of the Marsupialia, excepting that of the carnivorous Thylacine, upon which it failed to act.

6. Quantitative Tests with Anti-Avian Sera.

Tests with Anti-Ostrich Serum (11. III. '02).

Antiserum from rabbit treated with the serum of *Struthio molybdophanes*.

	Blood of		Date	Precipitum amount	Percentage
740	Struthio molybdophanes.	Ostrich	20. xii. 01	·042	100
277	Casuarius bicurunculatus.	Cassowary	14. xi. 01	·026	62
342	Rhea americana.	Rhea	25. xi. 01	·017	41
511	Pelecanus onocrotalus.	Pelican	18. ii. 02	·017	41
369	Sarcorhamphus gryphus.	Condor	27. xii. 01	·015	36
	Gallus domesticus.	Fowl	25. ii. 02	·007	17
590	Anas boscas.	Duck.	7. iii. 02	·007	17
281	Nyctea scandiaca.	Owl.	14. xi. 01	·007	17

From the above it will be seen that a considerable reaction was obtained with all the avian bloods tested, the percentage falling in the order given. It must be noted that the blood of Rhea (342) was putrid when received, and that it contained a considerable amount of deposit due possibly to the chloroform, this may account for the low percentage obtained in this case. Although all avian bloods react apparently to anti-avian sera (see qualitative tests) it is evident that there are quantitative differences which will require further study.

Tests with Anti-Fowl Serum (8. VI. '02).

Antiserum from rabbit treated with the serum of *Gallus domesticus*.

	Blood of	Date	Precipitum amount	Percentage
	Gallus domesticus	—	·035	100
590	Anas boscas	7. iii. 02	·012	34 (average of 2 tests)
277	Casuarius	14. xi. 01	·0095	27
	Columba	vi. 02	·008	23
281	Nyctea	14. xi. 01	·0072	21
369	Sarcorhamphus	27. xii. 01	·0065	19
342	Rhea	25. xi. 01	·0065	19
511	Pelecanus	18. ii. 02	·004	11

Note the comments in connection with the preceding tests.

Conclusions relating to Quantitative Tests.

The 500 quantitative tests recorded above give results in substantial accord with those obtained by the qualitative method, they however exclude the personal element which necessarily must enter into records of tests done qualitatively. The summary of conclusions following the qualitative tests (p. 214) therefore applies equally to the quantitative tests. The reader will be able by glancing down the column of percentages to rapidly gather an impression of the results obtained.

SECTION VIII.

BLOOD-RELATIONSHIP AMONGST THE LOWER VERTE-BRATA AND ARTHROPODA, ETC., AS INDICATED BY 2,500 TESTS WITH PRECIPITATING ANTISERA.

BY G. S. GRAHAM-SMITH, M.A., M.B., D.P.H. (CAMB.).

IN the foregoing pages Nuttall has described in detail the various methods of collecting and preserving sera, and of preparing antisera, together with the precautions which have to be adopted in testing, both by the qualitative and quantitative methods.

These experiments were conducted by the methods he has described, and it is therefore unnecessary again to describe them.

On reference to Nuttall's tables it will be seen that his work was principally carried out on Mammalian and Avian sera, though Reptilian, Amphibian, Fish, and Crustacean sera as well as a few Egg-albumins were tested. In order to render these experiments more complete it was thought desirable to further study the relationships indicated by this test amongst the latter groups of animals, and Egg-albumins. For this purpose Dr Nuttall kindly placed at my disposal all his specimens of blood of the Reptilia, Amphibia, Pisces, and Crustacea, and these together with a few others have been tested with antisera derived from animals of their own classes.

Methods.

Production of Antisera.

In these experiments some of the antisera were prepared by the intraperitoneal method, and some by the intravenous, but the injection of much smaller quantities than those usually employed was found to suffice (see table, p. 337). Continental workers have used quantities

TABLE OF ANTISERA §.

Material injected	Method of Injection	Animal treated	Weight in grammes at commencement and end of treatment	Quantity injected on each occasion. Intervals in days between injections*.	Total quantity injected in c.c.	Duration of treatment in days.	Number of injections	When bled after last injections, days	Power of Antiserum obtained †
Duck's Egg-Albumin	Peritoneal	Rabbit	2800—2730	5, 2, 10, 7, 8, 10, 14, 18 / 9, 2, 5, 5, 3, 4, 7	74	31	8	5	powerful
Fowl's Egg-Albumin I.	,,	,,	1965—2480	5, 5, 15, 5, 20, 20, 20, 15 / 4, 2, 2, 4, 2, 3, 3	90	19	8	6	,,
,, ,, II.	,,	Guinea-pig	213—252	4, 6, 8, 5, 5, 5, 5, 5, / 2, 2, 10, 3, 4, 3, 4	43	27	8	6	moderately powerful
Crane's Egg-Albumin	,,	Rabbit	2120—2140	3·5, 2, 4, 12, 12, 8 / 2, 2, 2, 2, 5	41·5	14	6	7	,,
Tortoise Serum (T. vicina)	Intravenous	,,	2030—2050	1, ·5, 1·5, 2, 1·5, 3·5, 2·5, 1·5, 3 / 2, 4, 2, 3, 3, 2, 3, 2	17	31	9	6	powerful
Tortoise Serum (T. ibera)	Peritoneal	,,	1650—1730	5, 5, 3, 8, 9, 3, 10 / 6, 2, 4, 3, 2, 3	43	19	7	5	,,
Turtle Serum	,,	,,	1850—2080	4, 8, 4·5, 4, 4·5, 14, 10, 18 / 4, 7, 3, 2, 2, 3, 5	67	26	8	5	,,
Lizard Serum (Uromastix spinipes)	Intravenous	,,	2000—2130	1·5, 2, 2, 1, 3, 2·5, 3, 3 / 6, 2, 3, 2, 3, 2, 2	18	20	8	9	very feeble
Lizard Serum (Varanus griseus)	Peritoneal	Guinea-pig	414—480	6, 2·5, 3, 4·5, 5, 5, 4·5 / 4, 4, 3, 11, 8, 5, 7	33·5	44	8	4	powerful
Snake Serum (Tropidonotus natrix)	,,	Rabbit	2720—2560	5, 2, 8, 5, 5, 5, 13, 13, 10 / 5, 2, 2, 2, 3, 4, 1, 2	66	22	9	4	,,
Snake's Egg-Albumin (Tropidonotus natrix)	,,	,,	2250—2400	5, 5, 4, 5, 10, 10, 10, 16, 20 / 3, 2, 2, 3, 2, 4, 1, 2	85	20	9	4	,,
Python Serum	,,	Guinea-pig	402—410	3, 1·5, 3, 2·5, 1·5 / 4, 3, 3, 10	8·5	20	5	10	feeble
Ammocoetes Serum	,,	Rabbit	1690—1820	4, 4, 4, 9, 10, 8, 8, 8, 5 / 2, 3, 2, 2, 3, 3, 5, 3	60	21	9	5	moderately powerful
Ascidian Extract	,,	Guinea-pig	448—420	5, 5, 5, 5, 8, 10 / 3, 9, 6, 7, 2	28	30	6	6	very feeble
Crab Serum	Intravenous	Rabbit	1950—1890	2, ·5, 1·5, 2 / 4, 2, 2	6	8	4	3	powerful
Limulus Serum I.	Peritoneal	,,	2880—2950	20, 5, 8, 10, 10 / 4, 3, 4, 9	53	20	5	10	extremely powerful
,, ,, I.	Intravenous	,,	2405—2600	2, 2·5, 3, 3, 1, 4, 2, 4, 4 / 2, 4, 3, 3, 2, 2, 3, 3	25·5	22	9	8	,,

* The upper figures in column 5 give the amount of each injection in c.c., the lower figures the intervals in days between each injection.
† Measurements of precipita obtained are given on p. 339. § Compare with Nuttall's table, p. 54.

N.

22

extending to hundreds of c.c., and have frequently found their animals stood the operation badly, whereas most of the animals treated in these experiments continued to gain weight, and appeared to be healthy.

In the preceding table the methods of production, animals treated, their weight before and after treatment, the number of injections, and the quantities used in each case, together with the power of the resulting antisera are given.

Methods of Testing.

In the *Qualitative* Experiments the method which has already been fully described (p. 62 *et seq.*) was followed.

Throughout the subsequent tables the following signs have been used to denote the degree of clouding, and quantity of precipitum after 24 hours.

+ = very marked cloud—full reaction.	**D** = very large deposit after 24 hours.
+ = marked cloud.	D = large ,, ,, ,,
× = medium cloud.	d = small ,, ,, ,,
* = small cloud.	tr = trace of a deposit.
*? = very slight, or doubtful, clouding.	• = no reaction.

The very marked cloud, followed by a very large deposit, was seldom obtained except on the addition of an antiserum to its homologous blood dilution. In a few instances closely allied sera also produced this condition. Marked cloudings followed by large deposits as indicated by the second signs occurred on the addition of antisera to nearly related bloods. In some cases the antisera only gave this amount of reaction with their homologous sera. Medium clouds occurred under the same conditions. Powerful antisera, however, occasionally produced reactions of this degree with distantly related sera. Small cloudings indicated in the tables by the sign * were usually obtained when antisera reacted with distantly related bloods. Such cloudings were occasionally followed by small deposits (d), or traces of deposits (tr), but not infrequently after the lapse of 24 hours no precipitum was found.

The sign *? indicates that there was a haze at the line of junction of the serum and antiserum. In most cases this result was probably due to some cause other than a reaction between the serum and antiserum, but where an antiserum produces several such reactions on the sera of a group of animals a remote relationship is probably indicated. For this reason in the following tables summarizing the results of the tests with the various antisera, these very small reactions have been inserted, though they are probably only of value where the antiserum reacts in this way with several members of a group. In every table the number of reactions of each kind has been given, and the results of tests with each antiserum either including, or omitting, these reactions of doubtful value can be seen.

Several instances occurred in which though no clouding appeared in two hours yet a precipitum was present after 24 hours. The majority of these are probably true reactions, and are inserted in the tables at the end, but are not included in the summarised results of the individual antisera.

In *Quantitative* experiments the method devised by Nuttall (5. VI. '02) and described elsewhere (p. 315) was employed.

In all cases dilutions of the strength of 1 : 21 in ·6 % salt solution of Egg-albumins or sera to be tested were used.

In similar observations on Human and other sera it had been found that with every precaution probably a margin of 10 % must be allowed for experimental error (p. 313).

In the following table the antisera are placed in order according to their strength, as shown by the amount of precipitum produced by each with its homologous serum when tested by the quantitative method.

Anti-Limulus (I)............... ·0980 c.c.   Anti-Fowl's-egg (I)........... ·0159 c.c.
     ,,   (II)............... ·0560 ,,   Anti-Snake serum .... ·0159 ,,
Anti-Tortoise (T. ibera)...... ·0394 ,,   Anti-Crab serum............... ·0140 ,,
Anti-Duck-egg .... ·0384 ,,   Anti-Fowl's egg (II)............ ·0122 ,,
Anti-Turtle .... ·0300 ,,   Anti-Crane's egg .... ·0073 ,,
Anti-Emu-egg .... ·0281 ,,   Anti-Uromastix.................trace
Anti-Tortoise (T. vicina) ... ·0234 ,,   Anti-Python .... ,,

The anti-ammocoetes, and anti-snake's-egg sera produced well-marked precipita, but the strengths of the dilutions tested were unknown. These, however, were certainly not as strong as 1 : 21.

Anti-lizard (*Varanus griseus*) was a powerful antiserum, but no quantitative tests were made, as the quantity of it was very limited.

Materials used in these experiments.

All the egg-albumins were kept in a fluid condition in 1 : 21 dilution with ·6 % salt solution. Fluid sera, reptilian, amphibian, and crustacean, were similarly preserved. In many cases a few drops of chloroform were added to prevent putrefaction. Before testing, small quantities of the dilutions were allowed to stand in watch-glasses in order to allow the chloroform to evaporate off. In the case of the egg-albumins accurate dilutions of 1 in 21 are difficult to make, and consequently the results of quantitative experiments are not so trustworthy as those done on sera. The sera were of various ages, but the majority were derived from animals which were killed for the purpose. Some, however, came from animals which had died from disease. The great majority of the specimens tested were bloods dried on filter-paper, and preserved in this way for various periods of time. Some of these were apparently insoluble in salt solution and failed to react with any antisera, but the rest gave tinted solutions, which foamed well on shaking. Nearly all these bloods had been tested by Dr Nuttall with various mammalian and avian antisera. The results of these experiments have been tabulated by him in the preceding pages (pp. 300 *et seq.*), and in the following tables the number attached to each specimen by him is quoted, so that the results of tests with mammalian, avian, egg-albumin, reptilian, amphibian, and crustacean antisera can be followed.

In a few cases, in which it was found impossible to obtain blood serum, extracts of muscle in salt solution were made. After standing the supernatant fluid was usually clear and slightly tinted, and produced foam on shaking. The few specimens obtained in this way are marked 'extracts' in the tables.

It may here be stated that the specimens were not tested in their zoological order, but in the order in which they were obtained. Consequently it was only after the observations had been reduced to zoological order that any conclusions could be arrived at as to the general results of the experiments.

The results of the tests with each antiserum upon all the specimens are given in the tables at the end of this section. The following short tables summarize the actions of the antisera upon zoological groups of egg-albumins and bloods.

I. Antisera for Aves.

A. *Qualitative Tests with Antisera to Birds' Egg-Albumins.*

(1) *Tests with Anti-Emu Egg Serum.*

This antiserum was made by the repeated intraperitoneal injection into a Rabbit of Emu egg-albumin (*Dromaeus novae-hollandiae*).

Material tested	•	*?	*	×	+	✚	Percentage of positive reactions
22 *Egg-Albumins*							
(a) 15 Birds' egg-albumins	1 (6 %)	•	1 (6 %)	•	4 (26 %)	9 (60 %)	93
(b) 1 Reptile ,,	1	•	•	•	•	•	0
(c) 1 Amphibian ,,	1	•	•	•	•	•	0
(d) 5 Fish ,,	5	•	•	•	•	•	0
69 *Blood Sera*							
35 Reptilia { 10 Chelonia	9	•	1	•	•	•	10
3 Crocodilia	3	•	•	•	•	•	0
8 Lacertilia	8	•	•	•	•	•	0
14 Ophidia	14	•	•	•	•	•	0
10 Amphibia	10	•	•	•	•	•	0
14 Pisces	14	•	•	•	•	•	0
10 Arthropoda	10	•	•	•	•	•	0

This antiserum was made by Dr Nuttall, who has recorded the results of tests therewith upon Mammalian and Avian sera. When the above tests were made the antiserum had lost some of its power; but it, nevertheless, gave well-marked reactions with most of the egg-albumins. One egg-albumin, that of the chaffinch, for some unknown reason failed to react with any of the anti-egg sera. A moderate reaction was obtained also with the blood serum of one Chelonian (tortoise).

(2) *Tests with Anti-Duck's-Egg Serum.*

This antiserum was made by repeated intraperitoneal injections of the egg-albumin of the domestic Duck into a Rabbit.

Material tested	•	*?	*	×	+	✚	Percentage of positive reactions
27 *Egg-Albumins*							
(a) 15 Birds' egg-albumins	1 (6 %)	•	3 (20 %)	•	3 (20 %)	8 (53 %)	93
(b) 2 Reptile ,,	1	•	1	•	•	•	50
(c) 3 Amphibian ,,	3	•	•	•	•	•	0
(d) 7 Fish ,,	7	•	•	•	•	•	0
129 *Blood Sera*							
60 Reptilia { 15 Chelonia	12	3 (13 %)	1 (7 %)	•	•	•	20
4 Crocodilia	4	•	•	•	•	•	0
13 Lacertilia	13	•	•	•	•	•	0
28 Ophidia	28	•	•	•	•	•	0
18 Amphibia	18	•	•	•	•	•	0
39 Pisces	39	•	•	•	•	•	0
12 Arthropoda	12	•	•	•	•	•	0

This antiserum was decidedly more powerful than the last. The maximum reactions were confined to the birds' egg-albumins, but small reactions were obtained with the egg-albumin of a tortoise, and with a Chelonian blood (tortoise). Three very slight cloudings were also produced with the sera of three tortoises. No sign of a reaction was seen with any other specimen of egg-albumin or serum tested.

(3) *Tests with Anti-Fowl's-Egg Serum* (I).

This antiserum was prepared by intraperitoneal injections of Fowl's egg-albumin into a Rabbit.

Material tested	.	*?	*	×	+	✚	Percentage of positive results
24 *Egg-Albumins*							
(a) 13 Birds' egg-albumins	1 (7%)	.	3 (23%)	3 (23%)	3 (23%)	3 (23%)	92
(b) 2 Reptile ,,	1	.	1	.	.	.	50
(c) 3 Amphibian ,,	3	.	.	.	.	.	0
(d) 6 Fish ,,	6	.	.	.	.	.	0
128 *Blood Sera*							
60 Reptilia ⎰ 15 Chelonia	9	3 (20%)	2 (13%)	.	1 (7%)	.	40
4 Crocodilia	2	1	.	.	1	.	50
13 Lacertilia	12	1	.	.	.	.	7
28 Ophidia	26	1	1	.	.	.	6
18 Amphibia	17	1	.	.	.	.	5
38 Pisces	33	.	.	.	.	.	0
12 Arthropoda	12	.	.	.	.	.	0

Reptilia: 18%

This was one of the most powerful of the antisera to egg-albumins. Its maximum effects were produced on the Avian egg-albumins, but well-marked reactions occurred with the egg-albumin of the grass snake, and with the blood sera of three Chelonians (Tortoise (2), Turtle), one crocodile, and one snake. Very feeble reactions occurred with three Chelonians, one lizard, one snake, and one Amphibian (Toad).

(4) *Tests with Anti-Fowl's-Egg Serum* (II).

This antiserum was prepared by intraperitoneal injections of Fowl's egg-albumin into a Guinea-pig.

Material tested	.	*?	*	×	+	✚	Percentage of positive reactions
8 Birds' egg-albumins	3 (37%)	.	3 (37%)	.	2 (25%)	.	62

This was a weak serum, and only a few birds' egg-albumins were tested with it. A few quantitative tests were also made.

(5) *Tests with Anti-Crane's-Egg Serum.*

This antiserum was prepared by intraperitoneal injections of the egg-albumin of the Cape Crowned Crane (*Balearica regulorum*) into a Rabbit.

Material tested		•	*?	*	×	+	✚	Percentage of positive reactions
27 *Egg-Albumins*								
(a) 15 Birds' egg-albumins		4	•	2 (13%)	•	5 (33%)	4 (26%)	73
(b) 2 Reptile ,,		2	•	•	•	•	•	0
(c) 3 Amphibian ,,		3	•		•	•	•	0
(d) 7 Fish		7	•	•	•	•	•	0
129 *Blood Sera*								
60 Reptilia	15 Chelonia	13	2 (13%)	•	•	•	•	13
	4 Crocodilia	4	•	•	•	•	•	0
	13 Lacertilia	13	•	•	•	•	•	0
	28 Ophidia	28	•	•	•	•	•	0
18 Amphibia		18	•	•	•	•	•	0
39 Pisces		39	•	•	•	•	•	0
12 Arthropoda		12	•	•	•	•	•	0

This was a rather weak antiserum and failed entirely to react with four birds' egg-albumins, and gave only two very feeble reactions with two tortoise bloods.

B. *Qualitative tests with Antiserum for Fowl's Blood.*

This antiserum was prepared by Dr Nuttall, and with it only the following tests on birds' egg-albumins were made to render the experiments already carried out more complete.

Material tested	•	*?	*	×	+	✚	Percentage of positive reactions
13 Birds' egg-albumins	3 (23%)	1 (8%)	5 (38%)	2 (14%)	2 (14%)	•	77

Summary of results of Qualitative Tests with Antisera to Avian Egg-Albumins.

With the five antisera for Avian egg-albumins 66 tests were made with dilutions of birds' egg-albumins and 85 % of distinct positive reactions were obtained, of which 62 % were very well marked. One egg-albumin dilution, namely that of the chaffinch, failed to react with any of these antisera, but reacted with antiserum to fowl's blood.

Seven tests were carried out with dilutions of Reptile egg-albumins of which 28 % were positive.

No positive reactions were obtained with 10, and 25 tests, respectively with Amphibian, and Fish egg-albumins.

The 215 tests with Reptilian sera yielded noteworthy results. All the antisera tested reacted to some extent with Chelonian bloods, but only anti-fowl's egg serum (I) gave any positive results with the other Reptilia.

Reptilian Sera	Tests	Positive reactions	Good reactions
Chelonia	55	23·5 %	9 %
Crocodilia	15	13 ,,	7 ,,
Lacertilia	47	2 ,,	0 ,,
Ophidia	98	2 ,,	1 ,,

Out of 64 experiments with Amphibian sera only one showed any trace of reaction, namely, a very feeble clouding when anti-fowl's-egg serum was added to the diluted blood of the Toad.

130 tests with Fish, and 46 with Crustacean sera gave no indications of positive reactions.

These experiments, therefore, indicate a relationship between the Aves and Chelonia, and a less marked one between the Aves and Crocodilia. As indicated by this test the relationship between the Aves and Lacertilia and Ophidia is very remote (see pp. 200—207, 216, 300—305).

C. *Quantitative Measurements with Antisera to Birds' Egg-albumin.*

These experiments were conducted by the method which has already been described (p. 315), and the results of the measurements of precipita in each case are stated in c.c. in the second column. The percentages given in the last column are calculated on the basis of the amount of precipitum obtained, the amount given by an antiserum with its homologous blood being taken as 100 %. In most cases this figure has not been exceeded when the antiserum acted on a non-homologous blood.

(1) *Tests with Anti-Emu-Egg Serum.*

Material tested		Amount of precipitum	Percentage
Emu's	egg-albumin	·0281	100
Duck's	,,	·0234	80
Blackbird's	,,	·0131	46
Crane's	,,	·0122	43
Fowl's	,,	·0122	43
Greenfinch's	,,	·0122	43
Silver Pheasant's	,,	·0112	40
Pheasant's	,,	·0073	26
Moorhen's	,,	·0054	20
Thrush's	,,	·0054	20
Hedge-Sparrow's	,,	·0018	6
Chaffinch's	,,	trace	?
Turtle	serum	·0018	6

The following sera were also tested but gave no trace of precipitum: Tortoise, Alligator, Frog, and Dogfish.

The egg-albumins shown by qualitative tests to produce full reactions occupy the first positions on this table. Those which produced marked reactions with the exception of the Greenfinch egg-albumin occupy the lower places.

(2) *Tests with Anti-Duck's-Egg Serum.*

Material tested		Amount of precipitum	Percentage
Duck's	egg-albumin	·0384	100
Pheasant's	,,	·0328	85
Fowl's	,,	·0234	61
Silver Pheasant's	,,	·0140	36
Blackbird's	,,	·0065	15
Crane's	,,	·0051	14
Moorhen's	,,	·0046	12
Thrush's	,,	·0046	12
Emu's	,,	·0018	5
Hedge-Sparrow's	,,	trace	?
Chaffinch's	,,	·	0
Tortoise serum		trace	?
Turtle	,,	,,	?
Alligator	,,	·	0

Frog, Amphiuma, and Dogfish sera, as well as Tortoise and Dogfish egg-albumins, were also tested, with negative results.

In this as in the last table the results of quantitative and qualitative experiments closely correspond. Egg-albumins from the Duck's to the Moorhen's in the above list were shown by qualitative tests to produce full reactions. By the latter method a medium clouding occurred with Thrush egg-albumin, and small cloudings with the albumins of the Emu and Hedge-sparrow.

(3) *Tests with Anti-Fowl's-Egg Serum.*

Material tested			Amount of precipitum	Percentage
Fowl's	egg-albumin (old)		·0159	100
,,	,,	(fresh)	·0140	88
Silver Pheasant's	,,		·0075	47
Pheasant's	,,		·0075	47
Crane's	,,		·0046	29
Blackbird's	,,		·0046	29
Duck's	,,		·0037	23
Moorhen's	,,		·0028	18

Thrush, Emu, Greenfinch, and Hedge-sparrow egg-albumins were tested and gave traces of precipita, as also did Tortoise and Turtle sera. The egg-albumins of the Tortoise, Frog, Skate, and two species of Dogfish did not react. Alligator, Frog, Amphiuma and Dogfish sera also yielded no results.

As before the quantitative and qualitative methods closely agree. By the former method full reactions were recorded with Fowl and Silver Pheasant egg-albumins, marked reactions with Pheasant, Crane and Blackbird, medium with Duck, Thrush and Greenfinch, and small with Moorhen, Emu and Hedge-sparrow.

(4) *Tests with Anti-Crane's-Egg Serum.*

Material tested		Amount of precipitum	Percentage
Crane's	egg-albumin	·0073	100
Greenfinch's	,,	·0054	75
Silver Pheasant's	,,	·0018	24
Moorhen's	,,	·0018	24
Emu's	,,	·0018	24
Hedge-sparrow's	,,	·0009	12
Turtle serum		·0018	24

Tortoise, Alligator, Frog, Dogfish and Salmon sera were also tested but failed to give any traces of precipita.

By the qualitative method full reactions were recorded with Crane and Silver Pheasant egg-albumins, marked reactions with Greenfinch, Moorhen, and Hedge-sparrow, and small with Emu.

The quantitative method fails to bring out the delicate reactions which can be determined by the qualitative method, but in the case of the major reactions differences can be recognised which are not apparent by the latter method. By its means closely allied sera and egg-albumins can be differentiated, which cannot be done with certainty by the qualitative method unless low dilutions are made use of. The results of the few quantitative experiments which have been made are exactly in accord with those obtained by the qualitative method.

II. Antisera for Reptilia.

Qualitative Tests with Anti-Reptilian Sera.

A. *Anti-Chelonian Sera.*

(1) *Tests with Anti-Tortoise Serum.*

This antiserum was prepared by intraperitoneal injections of the serum of the Tortoise (*Testudo ibera*) into a Rabbit. The serum used was obtained from healthy tortoises.

Material tested	•	*?	*	×	+	**+**	Percentage of positive reactions
26 *Egg-Albumins*							
(a) 15 Birds' egg-albumins	8	2 (13 %)	5 (33 %)	•	•	•	46
(b) 2 Reptile ,,	•	1	1	•	•	•	100
(c) 3 Amphibian ,,	2	1	•	•	•	•	33
(d) 6 Fish ,,	6	•	•	•	•	•	0
129 *Blood Sera*							
60 Reptilia ⎰ 15 Chelonia	2	•	5 (33 %)	•	2 (13 %)	6 (40 %)	87
4 Crocodilia	3	•	1	•	•	•	25
13 Lacertilia	13	•	•	•	•	•	0
28 Ophidia	26	1	1	•	•	•	6
18 Amphibia	17	1	•	•	•	•	5
39 Pisces	39	•	•	•	•	•	0
12 Arthropoda	12	•	•	•	•	•	0

This was the most powerful anti-chelonian serum made. Large reactions occurred with 87 % of the Chelonian bloods tested, but slight reactions only with a few other specimens of Reptilian blood. 46 % of the Birds' egg-albumins tested gave some indications of clouding.

(2) *Tests with Anti-Tortoise Serum.*

This antiserum was prepared by intravenous injections of the serum of a giant Tortoise (*Testudo vicina*) into a Rabbit. The serum used was collected after the death of the tortoise from natural causes.

Material tested	•	*?	*	×	+	**+**	Percentage of positive reactions
24 *Egg-Albumins*							
(a) 13 Birds' egg-albumins	10	•	2 (14 %)	1 (7 %)	•	•	23
(b) 2 Reptile ,,	2	•	•	•	•	•	0
(c) 3 Amphibian ,,	3	•	•	•	•	•	0
(d) 6 Fish ,,	6	•	•	•	•	•	0
128 *Blood Sera*							
59 Reptilia ⎧15 Chelonia	5	•	5 (33 %)	2 (13 %)	•	3 (20 %)	66
⎪ 4 Crocodilia	2	1	1	•	•	•	50
⎨12 Lacertilia	11	1	•	•	•	•	8
⎩28 Ophidia	24	2	2	•	•	•	12
18 Amphibia	18	•	•	•	•	•	0
39 Pisces	39	•	•	•	•	•	0
12 Arthropoda	12	•	•	•	•	•	0

Only a small quantity of the tortoise serum was available for preparing this antiserum, which was considerably weaker than the preceding one. The results of the tests are, however, very much alike. In this case large reactions with Chelonian sera were obtained in 66 % as against 87 % in the former, and in 23 % as against 46 % of Birds' egg-albumins. Again, the few reactions obtained amongst the rest of the Reptilia were feeble, and the results of tests with Reptile egg-albumins were negative.

(3) *Tests with Anti-Turtle Serum.*

This antiserum was prepared by the intraperitoneal injection of the blood of a Turtle (*Chelone midas*) into a Rabbit. The turtle was healthy.

Although this was a very powerful antiserum, giving well-marked reactions with 66 % of the Chelonia, yet only 20 % of the Birds' egg-albumins gave any indication of clouding, and in each case the reaction was very feeble. Fewer positive results were obtained with the other specimens of Reptilian bloods than with either of the anti-tortoise sera.

Omitting very feeble cloudings the anti-chelonian sera give well-marked reactions with the Chelonia and moderate reactions with the Crocodilia. With Birds' egg-albumins moderate cloudings are obtained, indicating a relationship with the Chelonia. The few positive results obtained with Lacertilia and Ophidia were feebly marked.

Material tested	•	*?	*	×	+	**+**	Percentage of positive reactions
26 *Egg-Albumins*							
(a) 15 Birds' egg-albumins	12	2	1	•	•	•	20
(b) 2 Reptile ,,	1	•	•	1	•	•	50
(c) 3 Amphibian ,,	3	•	•	•	•	•	0
(d) 6 Fish ,,	6	•	•	•	•	•	0
129 *Blood Sera*							
15 Chelonia	5	•	1 (7%)	1 (7%)	3 (20%)	5 (33%)	66
4 Crocodilia	3	•	•	1	•	•	25
60 Reptilia 13 Lacertilia	13	•	•	•	•	•	0
28 Ophidia	28	•	•	•	•	•	0
18 Amphibia	18	•	•	•	•	•	0
39 Pisces	39	•	•	•	•	•	0
12 Arthropoda	12	•	•	•	•	•	0

B.   *Anti-Crocodilian Sera.*

Tests with Anti-Alligator Serum.

This antiserum was made by Dr Nuttall with the serum of the Chinese Alligator (*Alligator sinensis*).

Material tested	•	*?	*	×	+	**+**	Percentage of positive reactions
8 *Egg-Albumins*							
(a) 4 Birds' egg-albumins	3	•	1	•	•	•	25
(b) 1 Reptile ,,	•	•	1	•	•	•	100
(c) 3 Amphibian ,,	3	•	•	•	•	•	0
59 *Blood Sera*							
11 Chelonia	5	1	2 (18%)	•	1 (9%)	2 (18%)	54
4 Crocodilia	•	•	1	•	1	2	100
26 Reptilia 5 Lacertilia	5	•	•	•	•	•	0
6 Ophidia	6	•	•	•	•	•	0
12 Amphibia	12	•	•	•	•	•	0
12 Pisces	12	•	•	•	•	•	0
9 Arthropoda	9	•	•	•	•	•	0

This was a moderately powerful antiserum. It showed maximum reactions with the Crocodilia and Chelonia. One out of the four Birds' egg-albumins tested gave a slight reaction. The tests on the Lacertilia, Ophidia, Amphibia, Pisces, and Arthropoda were carried out on the best material obtainable, but all with negative results.

C. *Anti-Lacertilian Sera.*

(1) *Tests with Anti-Lizard Serum.*

This antiserum was obtained by the intravenous injection into a Rabbit of the filtered serum of a Lizard (*Uromastix spinipes*). This serum had undergone slight putrefaction, and concentration, and had to be filtered through porcelain before treatment was commenced.

Material tested	•	*?	*	×	+	✚	Percentage of positive reactions
25 *Egg-Albumins*							
(a) 14 Birds' egg-albumins	14	•	•	•	•	•	0
(b) 2 Reptile ,,	•	2	•	•	•	•	100
(c) 3 Amphibian ,,	3	•	•	•	•	•	0
(d) 6 Fish ,,	6	•	•	•	•	•	0
128 *Blood Sera*							
15 Chelonia	14	•	1	•	•	•	7
4 Crocodilia	4	•	•	•	•	•	0
60 Reptilia 13 Lacertilia	11	•	•	1	1	•	15
28 Ophidia	28	•	•	•	•	•	0
17 Amphibia	17	•	•	•	•	•	0
39 Pisces	39	•	•	•	•	•	0
12 Arthropoda	12	•	•	•	•	•	0

This antiserum was very feeble, giving good reactions only with two specimens of the blood of Uromastix, moderate with the serum of *Testudo vicina,* and very feeble with the two reptile egg-albumins. All the other tests were negative.

(2) *Tests with Anti-Lizard Serum.*

This antiserum was obtained by the intraperitoneal injection into a Guinea-pig of a solution of the dried serum of a Lizard (*Varanus griseus*).

Material tested	•	*?	*	×	+	✚	Percentage of positive reactions
23 *Egg-Albumins*							
(a) 15 Birds' egg-albumins	15	•	•	•	•	•	0
(b) 1 Reptile ,,	•	1	•	•	•	•	100
(c) 1 Amphibian ,,	1	•	•	•	•	•	0
(d) 6 Fish ,,	6	•	•	•	•	•	0
84 *Blood Sera*							
15 Chelonia	8	3 (20%)	2 (13%)	1 (7%)	1 (7%)	•	47
4 Crocodilia	•	2	1	•	1	•	100
60 Reptilia 13 Lacertilia	•	1	3 (23%)	6 (46%)	•	3 (23%)	100
28 Ophidia	16	4 (14%)	6 (21%)	2 (7%)	•	•	42
18 Amphibia	18	•	•	•	•	•	0
6 Pisces	6	•	•	•	•	•	0

This antiserum was a powerful one, but was very slightly opalescent. As shown in the above table, however, it was entirely confined to the Reptilia in its reactions. The maximum reactions occurred amongst the Lacertilia, though fairly well-marked clouding and precipita were obtained with the Ophidia as well as the Chelonia and Crocodilia. It is unfortunate that both the Lacertilian antisera were somewhat unsatisfactory, the first being very weak, and the second slightly opalescent, rendering the determination of slight reactions difficult. In the latter case all very slight cloudings have been disregarded, and a clouding usually described by the sign ∗ has here been marked as ∗?. The larger reactions were however well marked and unmistakeable.

D. *Anti-Ophidian Sera.*

a. Qualitative Tests.

(1) *Tests with Anti-Snake Serum.*

This antiserum was obtained by the intraperitoneal injection into a Rabbit of Snake serum (*Tropidonotus natrix*). The serum had been previously heated to 55° C. in order to destroy its toxic properties.

Material tested	.	∗?	∗	×	+	**+**	Percentage of positive reactions
22 *Egg-Albumins*							
(a) 10 Birds' egg-albumins	9	.	1	.	.	.	10
(b) 2 Reptile ,,	.	.	1	.	1	.	100
(c) 3 Amphibian ,,	3	.	.	.	.	.	0
(d) 7 Fish ,,	7	.	.	.	.	.	0
127 *Blood Sera*							
59 Reptilia ⎰ 15 Chelonia	13	1	1	.	.	.	13
4 Crocodilia	2	1	1	.	.	.	50
12 Lacertilia	7	.	5	.	.	.	40
28 Ophidia	15	.	7	4	1	1	46
17 Amphibia	17	.	.	.	.	.	0
39 Pisces	39	.	.	.	.	.	0
12 Arthropoda	12	.	.	.	.	.	0

This was a moderately powerful antiserum, producing its maximum effects amongst the Ophidia. The few reactions obtained with Chelonian, Crocodilian and Lacertilian bloods were all feeble. One bird's egg-albumin (Moorhen) gave a small reaction, but a well-marked clouding followed by a deposit was obtained with a dilution of snake's egg-albumin, and a smaller clouding with tortoise egg-albumin.

(2)   *Tests with Antiserum to Snake's Egg-albumin.*

This antiserum was obtained by the intraperitoneal injection into a Rabbit of diluted Snake's egg-albumin (*Tropidonotus natrix*).

Material tested		.	*?	*	×	+	**+**	Percentage of positive reactions
23 *Egg-Albumins*								
(a) 11 Birds' egg-albumins		7	.	4	.	.	.	36
(b) 2 Reptile ,,		.	.	1	.	.	1	100
(c) 3 Amphibian ,,		3	.	.	.	.	.	0
(d) 7 Fish ,,		7	.	.	.	.	.	0
127 *Blood Sera*								
59 Reptilia	15 Chelonia	9	5 (33 %)	1 (7 %)	.	.	.	40
	4 Crocodilia	3	.	1	.	.	.	25
	12 Lacertilia	4	3 (15 %)	5 (41 %)	.	.	.	66
	28 Ophidia	14	1	11 (39 %)	2 (7 %)	.	.	50
17 Amphibia		17	.	.	.	.	.	0
39 Pisces		38	1	.	.	.	.	2
12 Arthropoda		12	.	.	.	.	.	0

This was a very powerful antiserum giving an immediate dense clouding, followed by a very large precipitum, with a dilution of snake serum. Moderately well-marked reactions were obtained with specimens from all the sub-classes of the Reptilia, especially the Ophidia, as well as with the diluted egg-albumin of the tortoise. 36 % of the Birds' egg-albumins also reacted with this antiserum.

(3)   *Tests with Anti-Python Serum.*

This antiserum was obtained by the intraperitoneal injection into a Guinea-pig of Python's serum (*Python molurus*). The serum was obtained from an animal which had died, and was very slightly contaminated with bile.

Material tested		.	*?	*	×	+	**+**	Percentage of positive reactions
18 *Egg-Albumins*		18	.	.	.	.	.	0
(11 Bird, 1 Reptile, 6 Fish)								
127 *Blood Sera*								
58 Reptilia	15 Chelonia	14	1	.	.	.	.	7
	4 Crocodilia	4	.	.	.	.	.	0
	11 Lacertilia	11	.	.	.	.	.	0
	28 Ophidia	24	.	4	.	.	.	14
18 Amphibia		18	.	.	.	.	.	0
39 Pisces		39	.	.	.	.	.	0
12 Arthropoda		12	.	.	.	.	.	0

This antiserum was very feeble and failed to react except with the blood of four Snakes and one Chelonian (*Testudo ibera*). Even with its homologous serum the reaction was feeble.

Summary of results of Qualitative Tests with Anti-Reptilian Sera.

The anti-reptilian sera may be divided into two groups according to their reactions. The first group, comprising the antisera for Chelonia and Crocodilia, shows its maximum reactions with the sera of these two sub-classes, and few and feeble reactions with the bloods of the Lacertilia and Ophidia. This group also gave moderate results with the Avian egg-albumins. These experiments confirm those done with the antisera to Avian egg-albumins.

The second group, comprising the antisera for the Lacertilia and Ophidia, exerts its maximum effects on the Lacertilia and Ophidia. Moderate reactions were obtained with Chelonian and Crocodilian sera. With one exception no positive results were obtained with Avian egg-albumins. The antiserum for Snake's egg-albumin, though behaving otherwise like anti-snake serum, gave 36 % of moderately well-marked reactions with Avian egg-albumins.

All anti-reptilian sera reacted well with dilutions of Reptilian egg-albumins, but gave negative results with Amphibian and Fish egg-albumins, and Amphibian, Fish and Crustacean sera.

The results of these experiments indicate a close relationship between the Chelonia and Crocodilia and between the Lacertilia and Ophidia. A distant relationship is shown between the Aves and the former group, and a more distant one between them and the latter group. This last is best seen in the tests with the antiserum to Snake's egg-albumin. Anti-Chelonian and Crocodilian sera show very little affinity between these animals and the Lacertilia and Ophidia, but antisera to the latter indicate a relationship between them and the former.

The following table summarises the results of tests with anti-Reptilian sera. The percentages of the better marked reactions only are given, and the experiments with anti-Lizard I (*Uromastix spinipes*) and anti-Python are not included.

Material tested	Marked positive results with Antisera to				
	Chelonia	Crocodilia	Lacertilia	Ophidia	Ophidian egg-alb.
Egg-Albumins					
Birds' egg-albumins	21 %	25 %	0 %	10 %	36 %
Reptile ,,	33 ,,	100 ,,	100 ,,	100 ,,	100 ,,
Amphibian ,,	0 ,,	0 ,,	0 ,,	0 ,,	0 ,,
Fish ,,	0 ,,	0 ,,	0 ,,	0 ,,	0 ,,
Blood Sera					
Reptilia Chelonia	73 ,,	45 ,,	27 ,,	7 ,,	7 ,,
Crocodilia	25 ,,	100 ,,	50 ,,	25 ,,	25 ,,
Lacertilia	0 ,,	0 ,,	92 ,,	40 ,,	40 ,,
Ophidia	3 ,,	0 ,,	30 ,,	46 ,,	46 ,,
Amphibia	0 ,,	0 ,,	0 ,,	0 ,,	0 ,,
Pisces	0 ,,	0 ,,	0 ,,	0 ,,	0 ,,
Arthropoda	0 ,,	0 ,,	0 ,,	0 ,,	0 ,,

Quantitative measurements with Anti-Reptilian Sera.

The methods used in tabulating these results are the same as those made use of in recording similar observations with the antisera to egg-albumins.

(1) *Tests with Anti-Tortoise Serum (T. ibera)*.

Material tested	Amount of precipitum	Percentage
Turtle serum	·0463	117
Tortoise ,,	·0394	100
Alligator ,,	·0018	7
Greenfinch egg-albumin	·0046	12
Silver Pheasant's ,,	·0041	10
Moorhen's ,,	·0037	9
Hedge-Sparrow ,,	·0018	5

Frog, Amphiuma, and Dogfish sera failed to react. The few quantitative tests made with this antiserum are in accord with the qualitative tests. By the latter method Tortoise and Turtle sera gave full reactions. This is the only instance in which a non-homologous serum gave a higher reading than the homologous serum, and the result is probably due to some error in measurement, or dilution. By the qualitative test Alligator serum, Greenfinch, Silver Pheasant, and Moorhen's egg-albumins gave slight reactions, and Hedge-sparrow egg a doubtful reaction.

(2) *Tests with Anti-Tortoise Serum (T. vicina).*

Material tested	Amount of precipitum	Percentage
Tortoise serum (T. vicina)	·0234	100
,, ,, (T. ibera)	·0054	24
Turtle ,,	·0028	12
Alligator ,,	·	0
Tortoise egg-albumin	·	0
Duck's ,,	·0056	24

Traces of precipita were given by Emu's, Fowl's, Silver Pheasant's, Pheasant's, Greenfinch's and Thrush's egg-albumins, but Crane's, Blackbird's, Moorhen's, Hedge-sparrow's, Chaffinch's, Skate's and Dogfish egg-albumins, as well as Frog, Amphiuma, and Dogfish sera, failed to react.

The results of these tests again agree fairly closely with those done by the qualitative method. In these a full reaction was indicated with Tortoise serum (*T. vicina*) and small reactions with Tortoise (*T. ibera*), and Turtle. Alligator serum showed a doubtful reaction, whereas Duck's egg-albumin gave a medium precipitum. No signs of reaction were obtained qualitatively with Fowl's, Pheasant's and Thrush's egg-albumins, but quantitatively they yielded traces of precipita.

(3) *Tests with Anti-Turtle Serum.*

Material tested	Amount of precipitum	Percentage
Turtle serum	·0300	100
Tortoise ,,	·0018	6
Alligator ,,	·0009	3
Emu's egg-albumin	·0037	12
Pheasant's ,,	·0028	9
Duck's ,,	trace	?
Fowl's ,,	,,	?
Moorhen's ,,	,,	?

The results in this case differ slightly from the qualitative experiments. In the latter, Tortoise serum showed a full reaction and the quantity of precipitum yielded ought accordingly to be more than 6 % of that given by Turtle serum. Fluid Alligator serum on the other hand failed to react. Emu's and Pheasant's egg-albumins showed small reactions, but Duck's, Fowl's and Moorhen's egg-albumins failed to react, although by the quantitative method they showed traces of precipita.

These quantitative Tests, as in the case of those made with antisera to Avian egg-albumins, are almost entirely in accord with the qualitative experiments.

III. Anti-Amphibian Serum.

Tests with Anti-Frog Serum.

This antiserum was prepared by Dr Nuttall by the intraperitoneal injection of Frog's blood (*Rana temporaria*) into Rabbits.

Materials tested	•	*?	*	×	+	**+**	Percentage of positive reactions
18 *Egg-Albumins* (1 Amphibian)	18	•	•	•	•	•	0
70 *Blood Sera*							
31 Reptilia	31	•	•	•	•	•	0
15 Amphibia { 4 Urodela	4	•	•	•	•	•	0
{11 Anura	5	•	4 (36 %)	•	•	2 (18 %)	54
15 Pisces	15	•	•	•	•	•	0
9 Crustacea	9	•	•	•	•	•	0

This was a moderately powerful antiserum. Good reactions occurred only with two specimens of the blood of the Frog (*Rana temporaria*) and moderate with four examples of Frog's blood (1 *Rana temporaria* and 3 *R. tigrina*). All other tests were negative (see p. 210).

IV. Anti-Fish Serum.

Tests with Anti-Ammocoetes Serum.

This antiserum was prepared by the intraperitoneal injection into a Rabbit of Ammocoetes serum, which had been dried on filter-paper, and dissolved out by means of salt solution.

Material tested	•	*?	*	×	+	**+**	Percentage of positive reactions
23 *Egg-Albumins*	23	•	•	•	•	•	0
131 *Blood Sera*							
60 Reptilia	60	•	•	•	•	•	0
17 Amphibia	17	•	•	•	•	•	0
{ 12 Chondropterygii	11	1	•	•	•	•	8
39 Pisces { 26 Teleostei	24	2	•	•	•	•	8
{ 1 Cyclostomata	•	•	•	1	•	•	100
15 Arthropoda	15	•	•	•	•	•	0

This antiserum was only of moderate power, producing with its homologous serum a medium clouding, followed by a well-marked precipitum. Only three doubtful cloudings were, however, obtained with other fish sera, and other tests made with it were entirely negative. Several tests were made with Limulus, and Arthropod sera, but with negative results.

V. Anti-Ascidian Serum.

Tests with Anti-Ascidian Serum.

This antiserum was obtained by the intraperitoneal injection into a Guinea-pig of Ascidian extract. The extract was made by cutting large fixed Ascidians into small pieces, and extracting these with salt solution. The supernatant fluid which was used in these experiments was clear and of a light yellow tint.

Material tested	.	*?	*	×	+	**+**	Percentage of positive reactions
14 *Egg-Albumins*	14	.	.	.	.	.	0
22 *Blood Sera*							
6 Amphibia	6	.	.	.	.	.	0
8 Pisces	8	.	.	.	.	.	0
13 Arthropoda	13	.	.	.	.	.	0
2 Ascidian extracts	.	.	1	.	1	.	100

This antiserum produced a well-marked precipitum with its homologous extract, and a slight reaction with an extract of a species of small fixed Ascidian, but failed to give any trace of reaction with any of the blood sera or egg-albumins.

VI. Anti-Crustacean Sera.

(1) Tests with Anti-Lobster Serum.

This antiserum was made by Dr Nuttall by the injection of Lobster (*Homarus vulgaris*) serum into Rabbits.

Material tested	.	*?	*	×	+	**+**	Percentage of positive reactions
14 *Egg-Albumins*							
(a) 13 Birds' egg-albumins	13	.	.	.	.	.	0
(b) 1 Reptile ,,	1	.	.	.	.	.	0
53 *Blood Sera*							
20 Reptilia	18	2	.	.	.	.	11
12 Amphibia	10	2	.	.	.	.	20
12 Pisces	12	.	.	.	.	.	0
9 Crustacea	2	.	1	.	5 (55%)	1	77

This was a powerful antiserum producing well-marked results with the Crustacean sera. Slight indications of a reaction were obtained with two specimens of Reptilian, and two of Amphibian sera, probably due to some fault in the dilutions (see pp. 211, 217, 310).

(2) *Tests with Anti-Crab Serum.*

This antiserum was obtained by the intravenous injection into a Rabbit of the blood of a Crab (see pp. 211, 217, 310).

Material tested	•	*?	*	×	+	**+**	Percentage of positive reactions
21 *Egg-Albumins*							
(*a*) 13 Birds' egg-albumins	13	•	•	•	•	•	0
(*b*) 1 Reptile ,,	1	•	•	•	•	•	0
(*c*) 1 Amphibian ,,	1	•	•	•	•	•	0
(*d*) 6 Fish ,,	6	•	•	•	•	•	0
128 *Blood Sera*							
60 Reptilia	60	•	•	•	•	•	0
18 Amphibia	18	•	•	•	•	•	0
38 Pisces	38	•	•	•	•	•	0
12 Arthropoda { 1 Xiphosura	1	•	•	•	•	•	0
11 Decapoda	•	•	6	3	1	1	100

This was a very powerful antiserum giving well-marked reactions with all specimens of the Crustacea. The serum of the King-crab (*Limulus polyphemus*) failed to react. Tests with all other blood sera and egg-albumins were negative.

VII. Anti-Xiphosura Serum.

Tests with Anti-Limulus Serum.

These tests were conducted with two different antisera for the serum of the King-crab (*Limulus polyphemus*). The first was made by intra-peritoneal, and the second by intravenous injections into Rabbits. In the following table the tests made with both antisera are recorded.

Both these antisera were extremely powerful, the first giving more than three times as much precipitum as any other antiserum. The second was not quite so powerful. The first antiserum produced a small reaction with crab's serum and with spider serum (very dilute), and three doubtful cloudings with fish, and two with Crustacean sera. The second antiserum gave two medium reactions with spider serum. All other tests were negative.

Material tested		•	*?	*	×	+	✚	Percentage of positive reactions
46 *Egg-Albumins*								
(a) 26 Birds' egg-albumins	26	•	•	•	•		•	0
(b) 4 Reptile ,,	4	•	•	•	•		•	0
(c) 3 Amphibian ,,	3	•	•	•	•		•	0
(d) 12 Fish ,,	12	•	•	•	•		•	0
222 *Blood Sera*								
120 Reptilia	120	•	•	•	•	•		0
30 Amphibia	30	•	•	•	•		•	0
46 Pisces	43	3	•	•	•		•	6
26 Arthropoda 〈 1 Xiphosura	•		•		•		1	100
(Serum No. I) 〈 11 Decapoda	8	2	1	•	•		•	27
3 Arachnida	•		•	1	2		•	100
(Serum No. II) 〈 1 Xiphosura	•		•		•		1	100
10 Decapoda	10	•	•	•	•		•	0

Numerous experiments were made with these antisera on dilutions of Ammocoetes blood to determine whether this test would bring forward any evidence in support of Gaskell's[1] hypothesis on the origin of the Vertebrates. On one occasion a small clouding resulted on the addition of the first antiserum to a very strong solution of Ammocoetes blood. All subsequent experiments were negative. Anti-Ammocoetes serum also failed on all occasions to react with dilutions of Limulus serum. By this means therefore no evidence can as yet be produced in support of this hypothesis.

Unlike Mammalian and Avian sera Limulus serum when passed through a porcelain filter loses its (blue) colour, and no longer produces any precipitum when tested with anti-Limulus serum.

Summary of Results with Anti-Arthropod Sera.

Well-marked reactions were obtained with the anti-lobster serum amongst the Crustacea. Limulus serum was not tested. No positive results of any importance were produced by tests on other sera.

Anti-crab serum though producing very marked reactions with specimens of the Decapoda, failed entirely to react with the blood of Limulus, or any of the other sera, or egg albumins.

Of the extremely powerful anti-Limulus sera only one showed any

[1] Gaskell, W. H. "On the Origin of the Vertebrates," *Journal of Anatomy and Physiology*, 1898, *et seq*.

positive results with the Decapoda and these were very slight. Good reactions were however obtained with dilutions of spider sera.

The results of these tests indicate a fairly close relationship between the King-crabs and the Spiders, and a very distant one between the King-crabs and the Decapoda. These sera were much more powerful than the anti-egg, and anti-Chelonian sera which interacted, and it might have been expected that even with distant relationship the sera of the Decapoda would have produced at least small reactions.

During its collection the anti-crab serum behaved in a rather curious manner. The blood clotted rapidly, and a very milky serum separated from the clot. The latter in a short time also coagulated, and from this coagulum a clear serum eventually came off.

Anti-Limulus I behaved in the same way, but only a very partial coagulation of the serum first separated took place.

In the previous tables the reactions produced with all specimens, whether preserved in a fluid or dried condition, have been given. Some of the latter, owing to overheating during the process of drying in the sun, or some other cause, failed to go into solution and consequently the tables of percentages of positive results are probably somewhat under-estimated. The results of the actions of the antisera to the Avian egg-albumins and Reptilian sera with the egg-albumins, and *fluid* Reptilian sera are, therefore, given below. In the other groups very few fluid sera were used, and their reactions with the various antisera may be seen in the tables at the end.

Fluid materials tested	Avian egg albumins (5)	Chelonia (3)	Crocodilia (1)	Lacertilia (1)	Ophidia (2)
15 Birds' egg-albumins	85 %	21 %	25 %	0 %	10 %
2 Reptile ,,	28 ,,	33 ,,	100 ,,	100 ,,	100 ,,
4 Chelonian sera	16 ,,	100 ,,	75 ,,	25 ,,	0 ,,
1 Crocodilian ,,	0 ,,	50 ,,	100 ,,	100 ,,	100 ,,
2 Lacertilia ,,	0 ,,	0 ,,	0 ,,	100 ,,	100 ,,
3 Ophidian ,,	0 ,,	0 ,,	0 ,,	0 ,,	100 ,,

Percentage of well-marked positive reactions with Antisera to

The present experiments can only be regarded as of a preliminary nature, and indicate the interesting results which might be obtained by systematically following up this line of research. For more accurate and trustworthy results both quantitative and qualitative tests would have to be undertaken with fluid sera obtained from healthy animals and repeated two or three times. Most of the experiments, which are the subject of this paper, have only been done once. Had it been possible to repeat them, however, the results would probably not have been materially affected.

In the following tables the name of the collector of the serum is given with the majority of specimens. Dr Nuttall acknowledges the source of all specimens examined by him at the end of this book. After all these samples the numbers given in Nuttall's tables are quoted, so that the results obtained by him with a large number of anti-Mammalian and anti-Avian sera may also be followed.

The Limulus and Ammocoetes sera were procured for me by Dr W. H. Gaskell, F.R.S., to whom I am much indebted for these very interesting bloods.

The previous work on these lines has already been fully considered in the foregoing monograph, and for this reason no attempt has been made here to summarize the results of other observers.

The following works have been used in the construction of the annexed tables :

Evans, A. H. "Birds," *Cambridge Natural History*, Vol. ix. 1900.
Gadow, H. "Amphibia and Reptiles," *Cambridge Natural History*, Vol. viii. 1901.
Günther, A. C. L. G. *The Study of Fishes.* 1880.

General Conclusions.

(1) Powerful antisera to *Avian egg-albumins* produce very large reactions with dilutions of Birds' egg-albumins. They also give distinct positive reactions with Chelonian and Crocodilian sera, as well as with dilutions of Reptile egg-albumins. No reactions of any importance were obtained with Amphibian or Fish egg-albumins or with Lacertilian, Ophidian, Amphibian, Fish, or Crustacean sera. These tests, therefore, show a distinct relationship between the Aves and Chelonia and Crocodilia.

(2) Powerful *anti-Chelonian sera* gave maximum reactions with the Chelonia and Crocodilia. Smaller reactions were obtained with Reptile egg-albumins and Avian egg-albumins, but with Lacertilian and Ophidian sera the results were almost negative. No positive reactions occurred with other egg-albumins or blood sera.

These results, showing a well-marked relationship between the Chelonia and Crocodilia, and distinct one with the Aves, confirm the above conclusion.

(3) A moderately powerful *anti-Crocodilian serum* behaved like the anti-Chelonian sera.

(4)　A powerful *anti-Lacertilian serum* produced its maximum reactions with the Lacertilia, and also reacted well with the Ophidia. Smaller reactions were obtained with the Crocodilia and Chelonia, but none with the Avian egg-albumins or other albumins or blood sera.

(5)　A powerful *anti-Ophidian serum* gave well-marked results with the Ophidia, and also reacted well with the Lacertilia. Reactions with the Crocodilia and Chelonia were much smaller. A moderate reaction was obtained with one Avian egg-albumin. All other tests were negative.

(6)　A very powerful antiserum to *Ophidian egg-albumin* showed its maximum reactions with the Ophidia, good reactions with the Lacertilia, and smaller ones with the Chelonia and Crocodilia. 36 % of distinct positive reactions were obtained with Avian egg-albumins. Other tests were negative.

The results of experiments with anti-Lacertilian and anti-Ophidian sera show a well-marked relationship between these two groups, a more distant relationship between them and the Chelonia and Crocodilia, and a still more distant one between them and the Aves. The latter is most markedly shown by the anti-snake-egg serum.

(7)　A moderately powerful *anti-Frog serum* reacted only with the Anura. No reactions were obtained with the Urodela or any other serum or egg-albumin tested.

(8)　A moderately powerful *anti-Ammocoetes serum* failed to show any affinity between this animal and any other.

(9)　A weak *anti-Ascidian serum* reacted only with its homologous extract.

(10)　Two powerful *anti-Crustacean* (*Decapoda*) *sera* reacted well with the sera of Decapoda, but failed to react with the serum of the King-crab (*Limulus polyphemus*). Tests with egg-albumins and other sera were all negative.

(11)　Two extremely powerful *anti-Limulus sera* reacted well with dilutions of Limulus serum, and produced moderate reactions with spider sera. One showed a small reaction with a crab serum, but the other did not. Other tests with Crustacean sera were negative. Experiments with Ammocoetes and other Vertebrate sera and egg-albumins were negative. These tests, therefore, show a distinct relationship between Limulus and the Arachnida, and a doubtful one between Limulus and the Decapoda.

BLOOD-RELATIONSHIP AMONGST THE LOWER VERTE-BRATA AND ARTHROPODA, ETC., AS INDICATED BY 2,500 TESTS WITH PRECIPITATING ANTISERA.

By G. S. GRAHAM-SMITH, M.A., M.B., D.P.H. (Camb.).

TABLES.

EGG ALBUMINS
Class AVES
Subclass NEORNITHES

	No.	Species	Anti-Emu-egg	Anti-Duck's-egg	Anti-Fowl's-egg I	Anti-Fowl's-egg II	Anti-Crane's-egg	Anti-Fowl's-serum
Division I. **Ratitae**	1	*Dromaeus novae-hollandiae.* Emu — Australia (Z) 18. i. 02 (No. 799)	+D	* d	* tr		* d	* d
Division II. **Carinatae** *Order* **Anseriformes**	2	*Anas boscas.* Common Duck — Cambs. 24. v. 02	+D	+D	× d	*	+d	·
Order **Galliformes**	3	*Gallus domesticus.* Domestic Fowl — Cambs. 24. v. 02	+D	+D	+D	+D	·	×d
	4	*Gallus domesticus* — Cambs. 30. xi. 01	+D	+D	+D	+D	·	+D
	5	*Phasianus colchicus.* Pheasant — Sussex 26. v. 02	+D	+D	+d	*	+D	*
	6	*Euplocamus nycthemerus.* Silver Pheasant — China (Z), laid 11. iv. 02 (No. 797)	+D	+D	+D		+d	*D
Order **Gruiformes** *Fam.* **Rallidae**	7	*Gallinula chloropus.* Moorhen — Sussex 26. v. 02	+D	+D	*d		+d	*d
Fam. **Gruidae**	8	*Balearica regulorum.* Cape Crowned Crane — S. Africa (Z) 13. iii. 02 (No. 798)	+D	+d	+tr	·	+D	+tr
Order **Passeriformes.** Oscines *Fam.* **Turdidae**	9	*Turdus merula.* Blackbird — Cambs. 13. v. 02	+D	+D	+d	·	+d	·
	10	*Turdus musicus.* Song Thrush — Cambs. 13. v. 02	+d	+d	×tr	*tr	+d	·
	11	*Accentor modularis.* Hedge Sparrow — Cambs. 12. v. 02	+d	*d	*tr		+d	* ?
	12	*Sylvia atricapilla.* Blackcap — Sussex 26. v. 02	*	+D			·	
Fam. **Fringillidae**	13	*Ligurinus chloris.* Greenfinch — Sussex v. 02	+d	+D	×tr		+d	*
	14	*Fringilla coelebs.* Chaffinch — Cambs. 12. v. 02	·	·	·	·	·	×d
	15	*Linota cannabina.* Linnet — Cambs. 13. v. 02	+D	*			*	

| | Anti-Reptilian Sera | | | | | | | | | | | | Anti-Arthropod Sera | | | | |
| Chelonia | | | | Lacertilia | | Ophidia | | | | | | | | | | | |
Anti-Tortoise (T. ibera)	Anti-Tortoise (T. vicina)	Anti-Turtle	Anti-Alligator	Anti-Lizard (Uromastix)	Anti-Lizard (Varanus)	Anti-Snake serum	Anti-Snake-egg	Anti-Python	Anti-Frog	Anti-Ammocoetes	Anti-Ascidian	Anti-Lobster	Anti-Crab	Anti-Limulus	Anti-Ox	Anti-Hedgehog
*	* tr	*	·	·	·	·	·					·	·	·	·	·
*	× D	·		·	·	·	*		·	·	·	·	·	·	·	·
*?	·	·	* d	·	·	·	*	·	·	·	·	·	·	·	·	·
·	·	·		·	·	·	*	·	·	·	·	·	·	·	·	·
*	tr	·	·	·	·	·	·	·	·	·	·	·	·	·	·	·
*	·	·	·	·	·	*	·	·	·	·	·	·	·	·	·	·
·	·	·	·	·	·	·	·	·	·	·	·	·	·	·	·	·
·	* tr	·	·	·	·	·	·	·	·	·	·	·	·	·	·	·
·	·	*?	·	·	·	·	*	·	·	·	·	·	·	·	·	·
*?	·	·	·	·	·	·	·	·	·			·	·	·	·	·
·	·	·	·	·	·	·	·	·	·			·	·	·	·	·
*	tr	·	·	·	·	·	·	·	·			·	·	·	·	·
·	tr ·	·	·	·	·	·	·	·	·			·	·	·	·	·
·	·	·	·	·	·	·	·	·	·							

EGG ALBUMINS
Class REPTILIA

	No.	Species	Anti-Emu-egg	Anti-Duck's-egg	Anti-Fowl's-egg I	Anti-Fowl's-egg II	Anti-Crane's-egg	Anti-Fowl's serum
Subclass **CHELONIA** *Fam.* **Testudinidae**	16	*Testudo ibera.* Tortoise (Leighton) laid 2. viii. 02 (No. 808)	•	*	•		•	
Subclass **SAURIA** *Fam.* **Colubridae**	17	*Tropidonotus natrix.* Grass Snake Cambs. vii. 02		•	*		•	
Class AMPHIBIA								
Fam. **Salamandridae**	18	*Triton cristatus.* Crested Newt Cambs. 8. v. 02	•	•	•		•	•
	19	*Triton vulgaris.* Common Newt Cambs. 8. v. 02		•	•		•	
Fam. **Ranidae**	20	*Rana temporaria.* Common Frog Cambs. v. 02		•	•			
Class PISCES								
Subclass **PALAEICHTHYES** *Fam.* **Scylliidae**	21	*Scyllium canicula.* Dog-fish Plymouth 27. vi. 02	•	•	•		•	
	22	*Scyllium catulus.* Dog-fish Plymouth 27. vi. 02	•	•	•		•	
Fam. **Rajidae**	23	*Raja batis ?* Skate Plymouth vi. 02		•	•		•	
Subclass **TELEOSTEI** *Fam.* **Scomberidae**	24	*Scomber scomber.* Mackerel	•	•	•		•	
Fam. **Cottidae**	25	*Cottus scorpius.* Bull-head Guernsey vi. 02		•			•	
Fam. **Gobiidae**	26	*Gobius ruthenspari* Plymouth v. 02	•	•	•		•	
Fam. **Salmonidae**	27	*Salmo salar.* Salmon 02		•			•	

| Anti-Reptilian Sera | | | | | | | | | | | | Anti-Arthropod Sera | | | | |
| Chelonia | | | | Lacertilia | | Ophidia | | | | | | | | | | |
Anti-Tortoise (T. ibera)	Anti-Tortoise (T. vicina)	Anti-Turtle	Anti-Alligator	Anti-Lizard (Uromastix)	Anti-Lizard (Varanus)	Anti-Snake serum	Anti-Snake-egg	Anti-Python	Anti-Frog	Anti-Ammocoetes	Anti-Ascidian	Anti-Lobster	Anti-Crab	Anti-Limulus	Anti-Ox	Anti-Hedgehog
*d	•	•	*	*?	*?	*	*	•	•	•	•	•	•	•		
*?	•	×		*?		+d	+D		•					•		
•	•	•	•	•	•	•	•	•	•	•	•		•	•		
•	•	•	•	•	•		•		•					•		
*?	•	•	•	•	•	•	•	•		•	•			•		
•	•	•	•	•	•	•	•	•	•	•	•		•	•		
•	•	•	•	•	•	•	•	•	•	•	•		•	•		
•	•	•	•	•	•	•	•	•	•	•	•		•	•		
•	•	•	•	•	•	•	•	•	•	•	•		•	•		
						•	•	•	•					•		
•	•	•	•	•	•	•	•	•	•	•	•		•	•		
•	•	•	•	•	•	•	•	•	•		•		•	•		

SERA
Class REPTILIA

			Anti-Emu-egg	Anti-Duck's-egg	Anti-Fowl's-egg I	Anti-Crane's-egg
Subclass **CHELONIA** *Fam.* **Testudinidae**	28	*Chrysemys picta.* Painted Terrapin N. America (Jordan) 22. v. 02 (No. 800)	•	•	•	•
	29	*Chrysemys elegans* N. America (Jordan) 11. v. 02 (No. 801)		•	•	•
	30	*Cistudo carolina* N. America (Jordan) 30. iv. 02 (No. 802)		•	•	•
	31	*Graptemys pseudogeographica* N. America (Jordan) 17. v. 02 (No. 803)		•	•	•
	32	*Testudo ibera.* Tortoise (fluid) Greece, killed Cambs. 14. i. 02 (No. 804)	•	*?	*? tr	*?
	33	*Testudo ibera* (fluid) 13. v. 02	•	*	*	*?
	34	*Testudo ibera*	* d	•	+ d	•
	35	*Testudo vicina.* Giant Tortoise (N. York Zoo., Langmann) 14. vi. 02 (No. 805)		•	*?	•
	36	*Testudo vicina* (fluid) Galapagos (Z) 20. x. 02 (No. 806)	•	*?	•	•
	37	*Testudo inepta.* Clumsy Tortoise Mauritius (Z) 20. v. 02 (No. 807)	•	•	•	•
Fam. **Chelonidae**	38	*Chelone midas.* Green Turtle (fluid) King's Coll., Camb. 14. xii. 02 (No 809)	•	•	*? tr	•
	39	*Chelone midas* Tropical seas (Z) 9. x. 01 (No. 810)	•	•	•	•
	40	*Chelone midas* 4. xii. 02	• tr	•	•	•
	41	*Chelone viridis* ? Turtle Brighton 13. xii. 01 (No. 811)	•	•	*	•
	42	*Chelone* ? Water Turtle L. Nyassa (Dodds) 8. ii. 02 (No. 812)	•	•	•	•
Subclass **CROCODILIA**	43	*Alligator mississippiensis.* Alligator (fluid) N. America (Z) 27. xi. 01 (No. 813)	•	•	•	•
	44	*Alligator mississippiensis.* Alligator N. America (Z) 27. viii. 01 (dried)	•	•	*?	•
	45	*Alligator sinensis.* Chinese Alligator China (Z) 13. viii. 01 (No. 814)			+	
	46	Crocodile L. Nyassa (Dodds) 8. ii. 02 (No. 815)	•	•	•	•
Subclass **SAURIA** *Order* **Lacertilia** *Suborder* **Geckones** *Fam.* **Geckonidae**	47	*Gecko guttatus.* Gecko India (Dalgetty) 24. viii. 02 (No. 816)	•	•	*?	•

| Anti-Reptilian Sera | | | | | | | | | | | Anti-Arthropod Sera | | |
| Chelonia | | | | Lacertilia | | Ophidia | | | | | | | |
Anti-Tortoise (T. ibera)	Anti-Tortoise (T. vicina)	Anti-Turtle	Anti-Alligator	Anti-Lizard (Uromastix)	Anti-Lizard (Varanus)	Anti-Snake serum	Anti-Snake-egg	Anti-Python	Anti-Frog	Anti-Ammocoetes	Anti-Lobster	Anti-Crab	Anti-Limulus
·	·	·		·	·	·	·	·	·			·	·
·	tr	·		·	*?	·	·	·	·			·	·
*	×	·		·	*	·	*?	·				·	·
*	*tr	·		·	·	·	·	·				·	·
+D	*d	+D	* tr	·	·	*?	*?	·	·			·	·
+D	*d	+D	+D	d	+d	·	d	+?	·			·	·
+D	+D	+D	+D	·	*?	·	*?	·	·			·	·
+	+D	×d	·	*	×	·	·	·	·			·	·
d													
+D	+D	+D	+	·	·	·	tr	·	·			·	·
+D	×D	*	*	·	·	·	·	·	·		*?	·	·
+D	*D	+D	tr	tr	·	·	*?	·	·			·	·
*	·	+D	·	·	*	·	·	·	·			·	·
+d	*	+D	·	·	*?	·	·	·	·			·	·
*tr	·	+d	·	·	·	·	·	·	·			·	·
*	·	·	*?	·	·	·	*?	·	·			·	·
*tr	*?	·	+D	·	+D	*	*	·	·			·	·
·	·	×	+D	·	*?	·	·	·	·			·	·
·	*	·	+D	·	*	·	·	·	·		*?	·	·
·	·	·	*	·	*?	*?	·	·	·			·	·
·	·	·	·	·	×	·	*?	·	·			·	·

SERA
Class REPTILIA

			Anti-Avian Sera			
			Anti-Emu-egg	Anti-Duck's-egg	Anti-Fowl's-egg I.	Anti-Crane's-egg
Subclass **SAURIA**	**48**	*Calotes versicolor.* Indian Tree-lizard India (Phipson) 12. i. 02 (No. 817)	•	•	•	•
Order **Lacertilia**	**49**	*Uromastix spinipes* Cairo (Elliot-Smith) 7. vi. 02 (No. 818)		•	•	•
Suborder **Lacertae** *Fam.* **Agamidae**	**50**	*Uromastix spinipes* (fluid)	•			•
Fam. **Iguanidae**	**51**	*Iguana tuberculata.* Iguana Trinidad (Tulloch) 1. i. 02 (No. 819)	•	•		•
Fam. **Anguidae**	**52**	*Anguis fragilis.* Slow-Worm Hereford (Leighton) 31. iii. 02 (No. 820)	•	•		•
Fam. **Varanidae**	**53**	*Varanus salvator.* Two-banded Monitor India (Z) 6. vi. 02 (No. 821) *Varanus salvator* (fluid)	• •	• •	• •	• •
	54 **55**	*Varanus griseus.* Grey Monitor. N. Africa (Z) 30. ix. 01 (No. 822) *Varanus griseus* (fluid)		• •	• •	•
Fam. **Lacertidae**	**56**	*Lacerta agilis.* Sand Lizard Cambs. 3. v. 02 (No. 823)		•	•	•
Fam. **Scincidae**	**57** **58**	Spec. ? Indian Lizard India (Dalgetty) 1. ix. 02 (No. 824) Land Lizard Natal (Parkinson) 13. xii. 02 (No. 825)	• 	• •	• •	• •
Order **Ophidia** *Fam.* **Boidae**	**59** **60**	*Python molurus.* Indian Python India (Z) 11. i. 02 (No. 826) *Python molurus* (fluid)	• 	• •	• •	• •
	61	*Python sebae.* West African Python W. Africa (Z) 8. xi. 01 (No. 827)	•	•	•	•
	62	*Boa constrictor.* Common Boa S. America (Z) 7. x. 01 (No. 829)	•	•	•	•
	63	*Boa constrictor* (Marajuel) Trinidad (Tulloch) 4. vi. 02 (No. 830)		•	•	•
	64	*Eryx jaculus* Cairo (Elliot-Smith) 19. vi. 02 (No. 831)		•	•	•
	65	*Corallus caninus* L. "Hundskopfschlange" Madagascar (Lühe) 20. ii. 02 (No. 828)	•	•	•	•
Fam. **Colubridae** *Series* A **Aglypha**	**66**	*Tropidonotus natrix.* Grass Snake Cambs. 28. iv. 02 (No. 832)	•	•	•	•
	67	*Tropidonotus natrix* (fluid) 2. v. 02 (No. 833)	•	•	* ?	•

| Anti-Reptilian Sera | | | | | | | | | | | Anti-Arthropod Sera | | |
| Chelonia | | | | Lacertilia | | Ophidia | | | | | | | |
Anti-Tortoise (T. ibera)	Anti-Tortoise (T. vicina)	Anti-Turtle	Anti-Alligator	Anti-Lizard (Uromastix)	Anti-Lizard (Varanus)	Anti-Snake serum	Anti-Snake-egg	Anti-Python	Anti-Frog	Anti-Ammocoetes	Anti-Lobster	Anti-Crab	Anti-Limulus
•	•	•		•	×	•	•	•		•		•	•
•	•	•		×	×	•	*?	•		•		•	•
•	•	•	•	+/d	*/tr	•	•	•		•		•	•
•	•	•	•		*	•	*?	•	•	•		•	•
•					*?				•	•	•	•	•
•	•	•	•	•	×/d	•	*/d	•	•	•	•	•	•
•	•	•	•	•	+/**D**	*/d	*/d	•	•	•	•	•	•
•	•	•	•	• •	+/**D**	*/d	•	•	•	•	•	•	•
•	•	•	•	•	+/**D**	*/d	*/d	•	•	•	•	•	•
•	*?	•	•	•	*	•	•	•	•	•	•	•	•
•	•	•	•	•	×/tr	*	*	•	•	•	•	•	•
•	•	•	•	•	×/tr	*	*	•	•	•	•	•	•
•	•	•	•	•	•	*/d	*/tr	•	•	•	•	•	•
•	•	•	•	•	•	*/d	*/d	*/d	•	•	•	•	•
•	•	•	•	•	•	×/D	•/d	•	•	•	•	•	•
•	•	d	•	•	•	*/D	*/d	•	•	•	•	•	•
*	*	•	•	•	*	•	•	•	•	•	•	•	•
•	*?	•	•	•	*	•	•	•	•	•	•	•	•
•	•	•	•	•	*?	×/D	*/d	*	•	•	•	•	•
•	•	•	•	•	•	×/D	*/d	•	•	•	•	•	•
•	•	d	•	•	•	+/**D**	×/**D**	*	•	•	•	•	•

SERA
Class REPTILIA

	No.		Anti-Avian Sera			
			Anti-Emu-egg	Anti-Duck's-egg	Anti-Fowl's-egg I	Anti-Crane's-egg
Subclass **SAURIA**	68	*Tropidonotus natrix* (fluid) 18. x. 02 (No. 834)	·		·	·
Order **Ophidia**	69	*Zamenis diadema* Cairo (Elliot-Smith) 11. vi. 02 (No. 835)		·	·	·
	70	*Zamenis ravergieri*, var. *nummifere* Cairo (Elliot-Smith) 19. vi. 02 (No. 836)		·	·	·
Fam. **Colubridae**	71	*Dendrophis liocerus.* Whip Snake Trinidad (Tulloch) 2. iii. 02 (No. 837)	·	·	·	·
Series A Aglypha						
Series B Opisthoglypha	72	*Cerberus rhynchops.* Indian Estuary Snake Bombay (Phipson) 17. i. 02 (No. 838)	·	·	·	·
	73	*Cerberus rhynchops* Calcutta (Rogers) 8. x. 02 (No. 839)	·	·	·	·
Series C Proteroglypha	74	*Naja tripudians.* Cobra India (MacWatt) 23. vi. 02 (No. 841)	·	·	·	·
	75	*Naja tripudians* 21. vi. 02 (No. 840)	·	·	·	·
	76	*Naja tripudians* (Z) 6. viii. 01 (No. 843)		·	·	·
	77	*Naja tripudians* (Phipson) 4. ii. 02 (No. 842)		·	·	·
	78	*Naja haje.* Hooded Cobra of Africa (Aspis) Cairo (Elliott-Smith) 19. vi. 02 (No. 844)		·	·	·
	79	*Pseudechis porphyriaceus.* Black Snake N. S. Wales (Cashman) 31.xi. 01 (No. 845)	·	·	·	·
Fam. **Viperidae**	80	*Bitis arietans.* Puff Adder S. Africa (Z) 6. viii. 01 (No. 846)		·	* tr	·
	81	*Bitis arietans* L. Nyassa (Dodds) 27. iii. 02 (No. 847)	·	·	·	·
	82	*Viper berus.* Adder N. Devon (Leighton) iv. 02 (No. 848)	·	·	·	·
	83	*Viper berus* iv. 02 (No. 849)		·	·	·
	84	*Viper Russelli.* Himalayan Viper India (Donald) 4. vi. 02 (No. 851)		·	·	·
	85	*Crotalus horridus.* Rattlesnake United States (Salmon) 6. xi. 01 (No. 852)		·	·	·
	86	Poisonous Serpent L. Nyassa (Dodds) 8. ii. 02 (No. 853)		·	·	·

| Anti-Reptilian Sera | | | | | | | | | | | Anti-Arthropod Sera | | |
| Chelonia | | | | Lacertilia | | Ophidia | | | | | | | |
Anti-Tortoise (T. ibera)	Anti-Tortoise (T. vicina)	Anti-Turtle	Anti-Alligator	Anti-Lizard (Uromastix)	Anti-Lizard (Varanus)	Anti-Snake serum	Anti-Snake-egg	Anti-Python	Anti-Frog	Anti-Ammocoetes	Anti-Lobster	Anti-Crab	Anti-Limulus
·	·	·		·	·	+ / D	× / D	·	·	·		·	·
·	·	·		·	*	·	*?	·				·	·
·	·	·		·	×	·	*	·				·	·
·	·	·			·				·	·		·	·
·	·	·		·	*?	*d / ·	tr / ·	·				·	·
·	·	·		·		·	·					·	·
·	·	·		·	·	× / D / ·	d / ·	·	·	·	*	·	·
·	·	·		·	·	·	·	·				·	·
·	·	·		·	*	·	·	·				·	·
·	·	·		·	*?	·	*	·				·	·
·	·	·		·	*?	·	*	*				·	·
·	·	·		·	×	*	*	·				·	·
·	·	·		·	·	*	·	·				·	·
·	·	·		·	·	·	·	·				·	·
·	*?	·		·	tr	·	·	·				·	·
·	*	·		·	*·	·	*	·				·	·
·	·	·		·	*	*d	*d	·	·	·	*	·	·
*?	·	·		·	*	d / ·	d / ·	*	·			·	·

SERA
Class AMPHIBIA

			Anti-Avian Sera			
			Anti-Emu-egg	Anti-Duck's-egg	Anti-Fowl's-egg	Anti-Crane's-egg
Order **Urodela** *Fam.* **Amphiumidae**	**87**	*Amphiuma means.* Amphiuma (fluid) N. America (Z) 19. iii. 03 (No. 854)	·	·	·	·
Fam. **Salamandridae**	**88**	*Amblystoma tigrinum* (Axolotl) Bless. 26. xii. 01 (No. 855)	·	·	·	
	89	*Triton cristatus.* Crested Newt Cambs. 8. v. 02		·	·	
	90	*Triton cristatus* 6. v. 02			·	·
	91	*Triton cristatus* 29. iv. 02 (No. 856)			·	·
	92	*Triton vulgaris.* Common Newt Cambs. 8. v. 02			·	·
Order **Anura** *Suborder* **Aglossa**	**93**	*Xenopus laevis.* Smooth-clawed Frog E. Africa (Z) 19. iii. 02 (No. 857)	·	·	·	·
Suborder **Phaneroglossa** *Fam.* **Bufonidae**	**94**	*Bufo vulgaris.* Common Toad Cambs. 14. vii. 01 (No. 858)	·	·	* ?	·
	95	*Bufo vulgaris* (fluid) viii. 02		·	·	·
	96	*Bufo mauritanica.* Moorish Toad N. W. Africa (Z) 10. i. 02 (No. 859)	·	·	·	·
Fam. **Engystomatidae**	**97**	*Megalobatrachus maxima.* Gigantic Salamander Japan (Z) 23. ix. 01 (No. 860)	·	·	·	·
Fam. **Ranidae**	**98**	*Rana temporaria.* Common Frog Cambs. 29. iv. 02 (No. 861)	·	·	·	·
	99	*Rana temporaria* 11. xi. 01 (No. 863)		·	·	·
	100	*Rana temporaria* 11. iii. 02 (No. 862)		·	·	·
	101	*Rana tigrina.* Tigrine Frog India (Z) 19. iii. 02 (No. 867)	·	·	·	·
	102	*Rana tigrina* (Z) 3. iii. 02 (No. 865)	·	·	·	·
	103	*Rana tigrina* 25. i. 02 (No. 864)	·	·	·	·
	104	*Rana* (spec. ?). Trinidad Frog Trinidad (Tulloch) 27. iii. 02		·	·	·

| Anti-Reptilian Sera | | | | | | | | | | | | Anti-Arthropod Sera | | |
| Chelonia | | | | Lacertilia | | Ophidia | | | | | | | | |
Anti-Tortoise (T. ibera)	Anti-Tortoise (T. vicina)	Anti-Turtle	Anti-Alligator	Anti-Lizard (Uromastix)	Anti-Lizard (Varanus)	Anti-Snake serum	Anti-Snake-egg	Anti-Python	Anti-Frog	Anti-Ammocoetes	Anti-Ascidian	Anti-Lobster	Anti-Crab	Anti-Limulus
·	·	·	·	·	·				·	·		·	·	·
·		·	·		·				·			*?	·	·
·	·	·	·	·	·	·	·	·	·	·	·		·	·
·	·	·		·	·	·	·	·	·	·			·	·
*?	·	·		·	·	·	·	·	·	·			·	·
·	·	·	·	·	·	·	·	·	·	·	·	·	·	·
·	·	·	·	·	·	·	·	·	·	·	·	*?	·	·
·	·	·	·	·	·	·	·	·	·	·	·	·	·	·
·	·	·	·	·	·	·	·	·	·	·		·	·	·
·	·	·	·	·	·	·	·	·	+	·		·	·	·
·	·	·	·	·	·	·	·	·	+	·		·	·	·
·	·	·	·	·	·	·	·	·	*	·		·	·	·
·	·	·	·	·	·	·	·	·	tr	·	·	·	·	·
·	·	·	·	·	·	·	·	·	*D	·	·	·	·	·
·	·	·	·	·	·	·	·	·	*D	·		·	·	·

		Anti-Avian Sera		
SERA *Phylum* **ARTHROPODA.**		Anti-Emu-egg	Anti-Duck's-egg	Anti-Fowl's-egg	Anti-Crane's-egg
Xiphosura	105 *Limulus polyphemus.* King Crab Gaskell	•	•	•	•
Decapoda	106 *Crangon vulgaris.* Common Shrimp Guernsey vi. 02		•	•	•
	107 *Homarus vulgaris.* Lobster England 14. xi. 01 (No. 895)	•	•	•	•
	108 *Astacus fluviatilis.* Cray-fish Cambs. 14. i. 02 (No. 894)	•	•	•	•
	109 *Palinurus vulgaris.* Craw-fish Plymouth 26. ii. 02 (No. 896)	•	•	•	•
	110 *Pagurus Bernhardus.* Hermit Crab	•	•	•	•
	111 *Carcinus maenas.* Shore Crab Plymouth 25. i. 02 (No. 898)	•	•	•	•
	112 *Carcinus maenas* (fluid) Guernsey vi. 02		•	•	•
	113 *Cancer pagurus.* Crab	•	•	•	•
	114 *Portunus puber.* Velvet Fidler-Crab Plymouth 25. i. 02 (No. 899)	•	•	•	•
	115 *Portunus depurator.* Swimming Crab Plymouth 25. i. 02 (No. 900)	•	•	•	•
	116 *Maia squinado.* Spider Crab Plymouth 27. ii. 02 (No. 897)	•	•	•	•
Arachnida	117 Spider Serum 10. viii. 03				
	118 Spider Serum vi. 02				
	119 Spider Extract vi. 02				

	Anti-Reptilian Sera												Anti-Arthropod Sera		
	Chelonia				Lacertilia			Ophidia							
Anti-Tortoise (T. ibera)	Anti-Tortoise (T. vicina)	Anti-Turtle	Anti-Alligator	Anti-Lizard (Uromastix)	Anti-Lizard (Varanus)	Anti-Snake serum	Anti-Snake-egg	Anti-Python	Anti-Frog	Anti-Ammocoetes	Anti-Ascidian	Anti-Lobster	Anti-Crab	Anti-Limulus
·	·	·	·	·		·	·	·		·	·	·	·	**+** D
·	·	·	·	·		·	·	·		·	·		* tr	* ?
·	·	·	·	·		·	·	·	·	·	·	**+**	× d	·
·	·	·	·	·		·	·	·	·	·	·	+	+ D	·
·	·	·	·	·		·	·	·	·	·	·	* tr	* d	·
·	·	·	·	·		·	·	·	·	·	·	+ d	* d	·
·	·	·	·	·		·	·	·	·	·	·	+ D	× d	·
·	·	·	·	·		·	·	·	·	·	·		**+** **D**	*
·	·	·	·	·		·	·	·	·	·	·	+ d	× d	·
·	·	·	·	·		·	·	·	·	·	·	+ d	* d	* ?
·	·	·	·	·		·	·	·	·	·	·	·	* tr	·
·	·	·	·	·		·	·	·	·	·	·	·	* tr	·
									·	·				× d
									·	·				× d
									·					*

SERA

Class PISCES

		### *Subclass* PALAEICHTHYES
		#### *Order* I. Chondropterygii
Suborder Plagiostomata	**120**	*Carcharias glaucus.* Common Shark
A. Selachoidei		Sarawak (Hose) 10. iv. 02
Fam. Carchariidae	**121**	*Zygaena malleus* ? Hammerhead Shark
		Trinidad (Tulloch) 15. i. 02 (No. 869)
	122	Glassheaded Shark
		Sarawak (Hose) 10. iv. 02
	123	*Mustelus canis.* Dog-fish
		New Jersey, U.S.A. (Silvester) viii. 01 (No. 870)
	124	*Squalus acanthias.* Dog-fish
		New Jersey, U.S.A. (Silvester) 30. viii. 01 (No. 871)
Fam. Scylliidae	**125**	*Scyllium canicula.* Dog-fish (fluid)
		Plymouth 27. ii. 02 (No. 872)
B. Batoidei		
Fam. Pristidae	**126**	Saw-fish
		Sarawak (Hose) 10. iv. 02
Fam. Rajidae	**127**	*Raja ocellata.* Skate
		New Jersey, U.S.A. (Silvester) viii. 01 (No. 873)
	128	Ray
		Sarawak (Hose) 10. iv. 02
Fam. Trygonidae	**129**	*Dasyatus.* Sting Ray
		New Jersey, U.S.A. (Silvester) viii. 01 (No. 874)
Fam. Myliobatidae	**130**	*Rhinoptera bonasus.* Sting Ray
		New Jersey, U.S.A. (Silvester) 30. viii. 01 (No. 875)
Suborder Holocephala	**131**	*Chimaera monstra*
		Naples 6. xi. 02
		### *Subclass* TELEOSTEI
		#### *Order* II. Acanthopterygii
Division Perciformes	**132**	*Perca fluviatilis.* Common Perch (fluid)
Fam. Percidae		Horsham (S. C.) 15. viii. 02
	133	*Mesoprion*
		Trinidad (Tulloch) 1. i. 02 (No. 893)
Division Cotto-Scomberiformes	**134**	*Naucrates ductor.* Pilot-fish
Fam. Carangidae		Trinidad (Tulloch) 29. v. 02 (No. 876)

378

SERA
Class PISCES

<table>
<tr><td></td><td></td><td style="text-align:center">*Subclass* TELEOSTEI</td></tr>
<tr><td></td><td></td><td style="text-align:center">*Order* II. **Acanthopterygii**</td></tr>
<tr>
<td>*Division*
Cotto-Scomberiformes
Fam. **Scomberidae**</td>
<td>**135**

136

137</td>
<td>*Scomber auxis.* Spanish Mackerel
   Trinidad (Tulloch) 13. ii. 02 (No. 877)
Scomber carangus
   Trinidad (Tulloch) 28. ii. 02 (No. 878)
Scomber scomber. Common Mackerel (muscle extract)
   02</td>
</tr>
<tr>
<td>*Fam.* **Cottidae**</td>
<td>**138**</td>
<td>*Cottus scorpio ?* Bull-head
   vi. 02</td>
</tr>
<tr>
<td>*Division* **Gobiiformes**
Fam. **Gobiidae**</td>
<td>**139**</td>
<td>*Gobius ruthenspari* (muscle extract)
   Plymouth 02</td>
</tr>
<tr>
<td>*Division* **Muciliformes**
Fam. **Sphyraenidae**</td>
<td>**140**</td>
<td>*Sphyraena barracuda.* Barracuda
   Trinidad (Tulloch) 7. iii. 02 (No. 879)</td>
</tr>
<tr>
<td>*Fam.* **Mugilidae**</td>
<td>**141**</td>
<td>*Mugil capito.* Grey Mullet
   Sarawak (Hose) 10. iv. 02</td>
</tr>
<tr>
<td>*Division*
Gastrosteiformes
Fam. **Gastrosteidae**</td>
<td>**142**</td>
<td>*Gastrosteus spinachia.* Stickleback (muscle extract)
   Plymouth 02</td>
</tr>
<tr>
<td>*Division*
Gobiesciformes
Fam. **Gobiesocidae**</td>
<td>**143**</td>
<td>*Lepadogaster*
   Guernsey vi. 02 (No. 880)</td>
</tr>
<tr><td></td><td></td><td style="text-align:center">*Order* III. **Anacanthini**</td></tr>
<tr>
<td>*Division* **Pleuronectoidei**
Fam. **Pleuronectidae**</td>
<td>**144**

145</td>
<td>*Rhombus triacanth.* Butter-fish
   New Jersey, U.S.A. (Silvester) viii. 01
Pleuronectes limanda. Dab (muscle extract)
   15. viii. 02</td>
</tr>
<tr><td></td><td></td><td style="text-align:center">*Order* IV. **Physostomi**</td></tr>
<tr>
<td>*Fam.* **Cyprinidae**</td>
<td>**146**</td>
<td>*Carassius auratus.* Gold-fish
   15. i. 02 (No. 881)</td>
</tr>
<tr>
<td>*Fam.* **Esocidae**</td>
<td>**147**</td>
<td>*Esox lucius.* Pike
   26. xi. 02 (No. 882)</td>
</tr>
</table>

SERA

Class PISCES

		Subclass TELEOSTEI
		Order IV. Physostomi
Fam. Esocidae	148	*Esox lucius.* Pike Ireland (Dillon) 6. v. 01 (No. 883)
Fam. Salmonidae	149	*Salmo fario.* Trout (Leighton) 17. x. 01 (No. 884)
	150	*Salmo salar.* Salmon Ireland (Dillon) 9. v. 01 (No. 885)
	151	*Salmo salar* (fluid) v. 02
Fam. Hyodontidae	152	*Hyodon tergisus ?* Moon-fish Trinidad (Tulloch) 1. iv. 02 (No. 886)
Fam. Muraenidae	153	*Anguilla vulgaris.* Common Eel Sussex (S. C.) 16. viii. 01
		Order VI. Plectognathi
Fam. Sclerodermi	154	*Aluton schoepfii.* File-fish New Jersey, U.S.A. (Silvester) viii. 01 (No. 887)
Fam. Gymnodontes	155	*Tetrodon turgidus.* Toad-fish New Jersey, U.S.A. (Silvester) viii. 01 (No. 888)
		Subclass CYCLOSTOMATA
Fam. Petromyzontidae	156	*Petromyzon branchialis* (Ammocoetes). Larval Lamprey Cambs. (Gaskell) 24. x. 02 + 03
		Also 4 undetermined fish-bloods, 157—160

Most of the above fish bloods were tested with antisera to Emu's, Duck's, Fowl's, Crane's and Snake's egg-albumins, and Tortoise (2), Turtle, Alligator, Lizard (2), Snake, Python, Frog, Ammocoetes, Ascidian, Lobster, Crab, and Limulus (2) sera. No well-marked positive reactions occurred except that between anti-Ammocoetes and Ammocoetes serum.

SECTION IX.

ON THE PRACTICAL APPLICATION OF THE PRECIPITIN REACTIONS IN LEGAL MEDICINE, ETC.

1. *Antisera in the Examination and Identification of Bloods and Blood-Stains.*

THE reactions given by specific haemolysins and agglutinins possess scarcely any value medico-legally for the reason that to make such tests a large number of intact blood corpuscles must be in suspension. The use of the haemolysins medico-legally was first suggested by Deutsch (see p. 41) who claimed that they might even be used upon dried corpuscles. There are certainly grave sources of error in the method, as compared to what we find when using precipitins.

The specific character of precipitins was already indicated by Kraus for bacterio-precipitins and by Bordet, Fich, and Morgenroth, in their work on Lactosera. It was recognized by Ehrlich, who refers in his Croonian Lecture, to still unpublished experiments of Morgenroth upon lactoprecipitins. Although at the time very few data had been collected regarding the precipitins, the assumption seemed justified that they would prove to possess similar specific characters to the haemolysins and agglutinins. Ehrlich's paper was read before the Royal Society, at a meeting which I attended on the 23rd of March, 1900.

Subsequently, Wassermann (18—21 April, 1900)[1] brought the question of specificity into greater prominence, speaking of the reaction as " eine streng specifische Methode " which permits us " über die Stellung verschiedener Eiweisskörper Aufschluss zu geben: nämlich durch die Bildung der *streng* specifischen Agglutinine in Thierkörper die wir seit Jahresfrist kennen, und die mir noch nicht genug gewürdigt scheint." He refers to the work of Bordet on lactosera. Bordet, and

---

[1] I am indebted to Professor Wassermann for a manuscript copy of the paper cited, the original being inaccessible to me. G. H. F. N.

Ehrlich spoke of precipitins as "coagulins," Wassermann termed them "agglutinins[1]." Wassermann referred to experiments of his own upon lactosera. With lactoserum for cows' milk he only obtained a precipitation with cows' milk, not with human or goat milk. "Es handelt sich immer um specifische Körper, je nach der Milchart, welche dem Thier vorher injiciert wurde....Die Eiweisskörper in verschiedenen Thiergattungen sind demnach streng different und verschieden." He added that experiments made with fowl egg-white also gave "in der That specifische Coaguline gegenüber dem Hühnereiweiss." It is therefore fully evident that Wassermann recognized the practical bearings of the discoveries of Tchistovitch and Bordet, and credit is due to him for having suggested the use of precipitins in the differentiation of albumins of different animals.

It will be noticed that Wassermann refers to the precipitins, as affording a means of distinguishing the *albumins* of different animals, irrespective of their being contained in blood or the like, and it is therefore not correct, as he states in a personal letter to me (15. I. '03), strictly speaking, to refer to the precipitin or biological method in *blood* diagnosis. To be certain that a reacting substance is blood, it is still necessary, in forensic practice, to utilize the ordinary tests for blood, such as that for haematin etc. According to Wassermann the reaction demonstrates the presence of albumins belonging to certain animals, without especially telling us which albuminous substance we are dealing with.

The medico-legal use of precipitins was almost simultaneously discovered by Uhlenhuth, Wassermann and Schütze (see p. 162). Uhlenhuth (7. II. 1901) considered that the method might have forensic value in the identification of blood stains. He tested 19 kinds of blood and only obtained a reaction with human blood upon adding anti-human serum to the series of dilutions. He moreover found that human blood which had been dried 4 weeks on a board, could be readily distinguished by means of anti-human serum from the blood of the horse and ox. Wassermann and Schütze (18. II. '01), a few days later, reported having examined 23 bloods, none of which reacted to anti-human serum except human blood and that of a baboon, the reaction in the

---

[1] I prefer to refer to these antibodies as precipitins for the reason that the term conveys a definite meaning with regard to the appearances observed in their reactions. The term agglutinin has been used with a definite meaning hitherto, and it can only lead to confusion to include precipitins under agglutinins even assuming that the antibodies are identical, which is unproved. The same objection holds for the term coagulins (see p. 17).

latter case taking place much more slowly and to a lesser degree than in human blood. Bloods of different species of animals, dried on knives, linen, etc., when dissolved in saline solution after 3 months, and cleared by filtration, gave specific reactions. In making their tests they took about 5—6 c.c. of salt solution for the solution of a dried blood drop of about the size of a sixpence, adding about ·5 c.c. of antiserum to 5—6 c.c. of the blood dilution, the mixture being afterwards placed at 37° C. Under these conditions marked reaction had taken place within 20 minutes. The lists of bloods tested by the above authors will be found on page 162.

Uhlenhuth (25. IV. '01) next reported that his anti-human serum had served to identify human blood which had been dried 3 months. Six samples of blood obtained from human cadavers and from healthy persons, were allowed to putrefy, control tests being made with the putrid bloods of the sheep, pig, horse, donkey, ox, cat, dog, goose, fowl, hare, rabbit, deer. Solutions of these putrid bloods gave specific reactions (see p. 119 *et seq.* also Plate and explanation at the end of this section). All the bloods smelt of sulphuretted hydrogen, the reaction being slightly alkaline. To obtain clear solutions, the bloods were filtered through Berkefeld filters, the sterile filtrate being diluted. A blood which had undergone putrefaction for 3 months still reacted specifically to its antiserum. Faintly alkaline blood solutions containing a small amount of soap, menstrual urine, human blood spots on snow for two weeks (− 10° C.), blood containing CO-haemoglobin, albuminous, and especially pus-containing human urine, all reacted to human antiserum, other bloods not doing so.

Nuttall (11. v. '01), working independently, found antihuman serum to cause specific precipitation in human blood which had undergone putrefaction for 2 months. Bloods dried for 2 months and kept at room temperature, or at 37° C. in the dark, and bloods exposed for a week in the sun, gave specific reactions, as did also blister-fluid, nasal secretion and human tears, the latter however to a slight degree only. On the other hand, old horse serum (31 months in the laboratory, preserved with trikresol) and human pleuritic exudate (6 months in the laboratory) yielded effective antisera when injected into rabbits. He drew attention to the reactions occurring in the bloods of *allied animals* which he had tested. The allied species whose bloods reacted to antiserum for human blood were man and 2 species of monkey, whilst ox and sheep bloods reacted to both anti-ox and anti-sheep sera, the most marked reactions being obtained when an antiserum acted upon its homologous blood.

Ziemke (27. vi. '01) next published an extensive series of tests possessing especial medico-legal interest. Using anti-human serum, he tested bloods which had been dried for years. Human blood dried for 2 years, reacted in 3 hours by clouding, the cloud persisting after 24 hours. Ox blood dating from 1863, 1869, 1876 gave no reaction with anti-human serum either at room temperature or at 37° C. Human bloods dried on various fabrics and dating from 1878 to 1899 (shirting, gauze, linen) were extracted with soda solution and gave reactions after several hours, excepting a sample of the year 1883 which did not go into solution. None of the samples gave more than a clouding which persisted 24 hours. Of the other bloods tested by way of control (sheep, calf, pig, dog, ox, horse, rabbit) and which had been dried 2 months on linen, only the rabbit blood gave a slight "Opazität." Human blood mixed in garden earth since 1898 and 1900, gave a slight but distinct clouding which persisted after 24 hours. This blood was soluble in soda solution, not soluble in saline. As controls he used the bloods of the horse, ox, sheep, calf and pig, which had been kept in garden earth for 8 weeks, but none of these gave a reaction. Soda solution of human blood from a case of CO-poisoning, kept 8 weeks on linen and in earth gave a moderate reaction in one hour, the clouding persisting 24 hours; saline dilutions gave negative results. Human blood dried on instruments such as a rusty knife (1896) and clean axe (1896), gave moderate reactions in 1 hour, clouding persisting after 24 hours. He only used soda dilutions, finding that rust and soda solution alone produced no reaction. Washed human blood-stains, still possessing a pale yellow colour (1883) gave slight clouding in one hour, control tests with rust-spots on linen were negative. Human blood from a white-washed cellar wall (1899), gave rather marked clouding in 3 hours, both in soda and saline dilutions; but he does not state that he controlled the effect of the white-wash. Human blood dried on wooden matches for 1 year, gave moderate reactions in soda dilutions in 3 hours, whilst blood on a tree branch gave a slight clouding in saline solution after 3 hours. Human blood dried 3 months on glass, gave great clouding after one hour at room temperature. Human blood exposed for 7 months on linen in the open, gave marked clouding in soda dilution in 3 hours at room temperature. Human blood dried 10 years on paper, gave slight clouding in 3 hours in soda dilution. Human blood from a cadaver, 3 days old, gave a great clouding in a few minutes and a flocculent precipitum after 24 hours at 37° C. Putrid human blood, diluted in saline and tested at 37° C. reacted, whereas putrid pigeon, fowl, goose, ox and pig blood did not. He gathered the impression that the older the blood, the less intense is the clouding, for he never obtained a regular precipitum with old dried blood.

I have noted elsewhere that there are drawbacks to using soda solution, because of the pseudo-reactions it may give (see pp. 64 and 83), also to the necessity of observing a " time limit," facts which detract from the value of Ziemke's observations. He did not state that controls were used in all his tests, nor that the control bloods were also diluted with soda solution.

Uhlenhuth (25. VII. '01) next reported experiments upon blood-stained articles. He obtained positive reactions with human blood-stains on a stick (1900), blood-stained sand (1896), a stain on cotton (1897), a stain on a coat and pair of trousers (1901), and on a hatchet (1900), upon adding anti-human serum to saline solutions of these bloods. With anti-pig serum he obtained positive reactions with pig blood-stains on linen (anti-human, anti-sheep, anti-horse sera gave no reaction), also with blood dried since 1897. Anti-pig serum gave a reaction with a mixture of sheep and pig blood (dried 1889), also with solutions of the organs of a pig, dried 18 months. I have noted my observations on reactions in mixed bloods (1. VII. '01) on p. 140. Uhlenhuth found that wash-water, containing carbolic acid, sublimate, and soap, gave positive reactions with anti-human serum when it contained human blood, the same being the case with blood containing 3 % borax, and blood-soaked garden earth after the expiration of 3 months.

The observations of Dr Graham-Smith, in this laboratory, do not confirm the results of Uhlenhuth's tests in the presence of carbolic acid, and of soap (see p. 82).

The fact that monkey bloods give similar, though less reactions than human blood upon the addition of anti-human serum, may not be a matter of any importance in most countries, where monkeys are not indigenous. Nevertheless it might happen that a murderous organ-grinder, or perchance a similarly inclined keeper of monkeys, backed by a well-instructed defence, would claim that suspected blood-stains were not human but derived from a monkey. On the other hand, Mr Hankin of Agra informed me, as stated in my paper of 21. XI. '01, of a case in India which had come to his notice, where it appeared essential to make a test to determine if certain blood-stains were caused by human or monkey blood. As I stated at the time, it would be necessary in such cases to be provided with an antiserum for the most prevalent genera or species of monkeys belonging to such a region. My tests with anti-monkey serum (p. 169) have clearly shown that such an antiserum would give a greater reaction with monkey blood, so that by tests conducted with both antisera, there should be no especial difficulty in determining to which primate the blood belongs.

In his papers of later date, Uhlenhuth (11–18 VIII. '02 and '02 a) reports at length upon a large number of tests made for medico-legal purposes, these proving the value of the method.   As it is impossible to give his results in detail, those especially interested will have to refer to the original papers.   The lists of tests in both papers are the same, and include 5 tests mentioned in his paper of 25. VII. '01, already cited, 16 further cases being given.   Leaving out of consideration the 5 cases above referred to, he obtained positive results in the following: (6) Blood stain several years on linen, tested with anti-pig serum, gave a reaction, not so with anti-sheep, anti-horse, anti-human serum. Professor Beumer subsequently informed him that the stain was due to pig blood.   (7) Dried blood (1897) acted similarly, also subsequently stated to be pig blood by Beumer.   (8) Dried blood mixture (1899). reacted to both anti-pig and anti-sheep serum, diagnosis recognized as correct by Beumer, who, as in the other cases, supplied the specimens without letting Uhlenhuth know what they were until after he reported the result of his tests.   (9) Blood-stain on paper found in puddle of blood on a road, reacted to anti-pig serum, a suspicion of murder being thus removed. (10) Blood-stains on penknife and handkerchief, medico-legal case, diagnosis human blood, subsequently confirmed by prisoner, who stabbed a man with the knife, explaining the spots on the handkerchief as due to his own nose having bled.   (11) Blood-stains on trousers and shirt, sent from Landgericht Munich and relating to a case of rape, diagnosis human blood.   (12) Shavings from a blood-stained box, same source as preceding, tests negative with anti-human, anti-sheep, and anti-horse serum, further tests not made, and subsequently discovered that the stains were due to roebuck blood.   (13) Blood-stained waistcoat and trousers, the owner being suspected of having killed some sheep, tests negative with anti-sheep, positive with anti-fowl, and it was subsequently proved that he had killed a fowl a day before the sheep-killing.   (14) Blood-stained wood-shavings from a floor, sent from Braunschweig in connection with a murder case, reaction with anti-human serum, murderer subsequently confirmed this.   (15) Two samples of blood-stained cloth sent by Prof. Minovici (Medico-legal Institute, Bucharest), and 11 other articles, all blood-stained, were correctly diagnosed as subsequently reported by Minovici.   (16) Blood-stained coat, tests negative with anti-human and anti-pig sera, subsequently proved to be roebuck blood-stains, medico-legal case at Marklissa. (17) Dried blood sent from Luxemburg, diagnosed human, subsequently proved to come from a suicide. The blood had been found in front of a house where the suicide lived, the body having been thrown into the Mosel (whence it was recovered) by his relatives who wished to keep the fact hidden that the man had committed suicide. (18) Blood-stains on wool-fragments from waistcoat and basket for carrying wood, diagnosed as human, confirmed by evidence in court. (19) Blood-stained trousers, diagnosis fowl blood. Prisoner suspected of stealing chickens, had claimed the spots to be due to rabbit blood; microscopic examination of the stains had however shown the presence of elliptical corpuscles. Comparative tests made with other avian bloods (goose, duck) showed the reaction to take place much more slowly and feebly with these bloods. (I doubt that such tests to distinguish avian bloods medico-legally can have much value, in view of my results, see p. 200.) The diagnosis confirmed in the course of the trial. (20) Three shirts and a handkerchief in connection with a murder, human blood proved to be spattered on two of the shirts. (21) Blood-stained trousers, shirt, stockings from a murder case at Strassburg Landgericht, diagnosis human blood, the prisoner having claimed that the blood came from a cow which had knocked off a horn. (22) Blood-stains on numerous articles of clothing were diagnosed to be human and from sheep. It was subsequently proved in court that the man had committed a murder, also that he had slaughtered some sheep two weeks before the murder.

Uhlenhuth (18. IX. '02, p. 679) does not attach any particular importance to the weaker or stronger reactions which occur in dilutions of dried blood, depending upon the length of time the blood has been dried. He makes control tests upon material as far as possible of similar age. So as to have such material of different ages on hand for this purpose, he dries sterile bloods of various kinds in Petri dishes, and stores the dried scales in test-tubes, a method by no means as convenient nor as compact as mine, where the blood is allowed to dry on filter-paper which it has saturated (see p. 63).

The generalized reactions obtained with anti-bovine sera with different bovine bloods, will apparently make it unnecessary in most cases to prepare special antisera for the blood of each bovine species, for we have seen that antibovine sera produce marked, in some cases almost equivalent reactions in the blood dilutions of different species of Bovidae. Uhlenhuth reached this conclusion, although he has examined but three bovine bloods in this respect (ox, sheep, goat). The very large number of bloods, both of Bovidae and Cervidae, examined by me and reported upon in part in earlier papers (see pp. 183—192, 252—258) show that either anti-cervine or anti-bovine sera will usually suffice for the identification of a blood as belonging to the group Pecora. I suspect therefore that Uhlenhuth's anti-sheep serum used in Case 12 must have been weak, for I have obtained quite marked reactions with roebuck blood upon the addition of anti-sheep serum, although the reaction was naturally less intense than with the sheep blood.

In medico-legal work, Uhlenhuth states that he would proceed to test the blood dilutions with the different antisera in succession, until a positive reaction is obtained. I should however warn against adding different antisera in succession to the same blood sample. The reasons being that no time limit is observable for the reaction which one or the other antiserum may give, the first antiserum may give a "mammalian reaction" which will be attributed to the last antiserum added, or finally, the increasing serum concentration due to having added several antisera, will mask the reaction.

I have already drawn attention to the fact that over-powerful anti-sera may be a source of error (p. 74) in medico-legal work, as is also stated by Uhlenhuth. Reaction should take place within a few minutes after antiserum and blood-dilution have been mixed. Uhlenhuth properly dwells upon the necessity of every antiserum used being provedly effective, and to insure this he considers, as does also Ziemke, that the preparation and testing of antisera for medico-legal purposes should be under State control. The use of weak antisera, which require a period of 24 hours, or the like, to exert their action should certainly be condemned for medico-legal work. A great many bloods may react to an antiserum after such a lapse of time, and there may be bacterial development (see p. 86). I have not infrequently observed, and this has been confirmed by Ziemke, that quite a marked clouding may occur which does not necessarily lead to a deposit after 24 hours. This result is recorded in my protocols, no deposits being noted at times under bloods which gave even marked cloudings. In my short tables which summarise the contents of those at the end, I found it necessary to give, for instance, equal value to, say a marked clouding, leading to little or no deposit, and to a faint clouding leading to a deposit.

In my investigations, of course, the main thing was to see if any blood-relationship could be established, the question of identification was of secondary importance.

Biondi (1902) obtained reactions with blood stains on rusty knives, on leather, and on fabrics exposed to the rain. Blood stains washed 15 and 30 minutes at 70° and 80° C. respectively gave no reaction, whereas they had to be washed some hours at low temperatures to remove all traces of blood as evidenced by the reaction with antiserum. He found that strong acid, 5 minutes contact with 5% carbolic, and 1 : 1000 sublimate or chloride of lime, caustic potash, soap, and borax destroyed the reacting substance.

Whittier (18. I. '02) and Wood (24. IV. '02) have used the precipitin test in medico-legal cases in America with positive result. Okamato (X. 1902) testing various human blood-stains obtained negative results in about $\frac{1}{4}$ of them, all animal bloods giving negative results when tested by anti-human serum. He found very old and putrid blood generally to give a negative result. The papers of several other authors, bearing indirectly upon this subject, will be found mentioned under the tests with various haematosera p. 161 *et seq.*

In some tests which I conducted together with Mr Sanger and which are mentioned in his Thesis[1] (23. XII. '02) it was found in several instances that blood spots on leather, owing to the varying acidity of leather, at times gave pseudo-reactions, that is, such blood dilutions gave a clouding upon the addition of any serum. Sanger found that it was possible to neutralize this acidity and obtain positive specific reactions, neutralization being effected by the addition of $\cdot 1\%$ sodium carbonate. The studies upon the effect of various agents on blood were continued by Graham-Smith, and are cited fully on pp. 76—86, 390 *et seq.*

Farnum (28. XII. 1901) obtained antisera by injecting semen intra-peritoneally into rabbits, or using testicular emulsions, derived from man, dog, and bull. The injections were made at intervals of 5—6 days, the rabbits receiving 5—8 injections of 5 to 10 c.c. at a time. The antisera were specific, when tested on the semens mentioned. Anti-human serum produced a reaction with human serum dried as long as 34 days.

Layton (1903, p. 220) obtained positive results with anti-human serum tested on fresh human blood-stains on cloth and filter-paper, old, dried stains on cloth, newspapers, filter-paper, putrid blood and blood soaked in earth. Also with human blood mixed with others as in Nuttall's earlier experiments. He does not state the age of the blood-stains he tested.

---

[1] Incorporated in the Paper by Graham-Smith and Sanger, the main results of which are reprinted in this book, having appeared in the *Journal of Hygiene*, vol. III. 1903.

Finally Austin (12. III. '03) in a paper on "the limitations of the Uhlenhuth test for the differentiation of human blood" showed that "other fluids of the human body, like effusions and exudates, were of little value" in the production of antisera. These facts, however, can have no bearing on the test in its medico-legal aspect, and but confirm the observations of others.

Graham-Smith and Sanger (1903, pp. 269—272) at my suggestion examined a number of articles obtained through the courtesy of Mr Henry, Chief of the Criminal Investigation Department of Scotland Yard. The objects in question all possessed forensic interest. They consisted of various weapons and blood-stained fabrics. These authors found that blood which had been dried many years (see also p. 119 *et seq.*) could still be tested by means of precipitins. I quote the following from their publication which appeared in the *Journal of Hygiene.*

A.   *Blood dried on metal.*

"The results of experiments on 17 samples covering a period of 30 years, are arranged according to age in the following table. The number after each specimen refers to the catalogue of the Scotland Yard museum, and a short description of each is given in the appendix.

The reactions of all were neutral.

The following table refers entirely to weapons which had been preserved from rusting by the application of oil to the surface of the metal. This process had caked the blood into black masses, making it frequently difficult to say whether the mass consisted of blood and oil or rust and oil. In the majority of cases however it was possible to make certain of scraping off some blood. The material thus obtained was extracted with distilled water, and subsequently an equal volume of 1·2 % salt solution added to it. If necessary the solution was filtered through filter-paper, and tested in the way described.

Excellent results were obtained from these materials, and showed conclusively that the property of producing a precipitum with its appropriate antiserum is not lost by blood dried on metal even after 30 years have elapsed.

In one case, No. 11, however, no reaction was obtained; the negative result was probably due to little or no blood being present on that part of the knife which was examined. The condition of the weapon was such that it was impossible to be certain that the material scraped from it was blood, but it was thought better to include it in the series,

Instrument from which the blood was obtained	Age	Foam-test*	Result with anti-human serum	Control serum	Remarks
1. Razor	5 months	good	marked cloud, 5 mins.	anti-turtle nil	This had not been preserved with oil.
2. Knife	few months	,,	,,	,,	,,
3. Pocket-knife (554)	9 months	,,	marked cloud, 1 hour	anti-dog nil	,,
4. Razor (550)	10 ,,	,,	,,	,,	,,
5. Hatchet (539)	1 year	,,	cloud, 20 mins.	,,	Much oil. Blood obtained from a crevice in the hatchet.
6. ,, (544)	1 ,,	,,	marked cloud, 1 hour	,,	Thin layer of oil.
7. Pocket-knife (530)	1½ ,,	,,	marked cloud, 10 mins.	,,	No oil.
8. Dagger in sheath (524)	1½ ,,	,,	,,	,,	Very little oil.
9. Two knives (525)	1½ ,,	,,	,,	,,	Much oil. On first occasion no reaction. Instrument again scraped; reaction good.
10. Knife (494)	3 years	slight	,,	,,	Owing to the amount of rust and oil present it was impossible to say whether any blood had been scraped off.
11. Chopper, knives (491), and oil-can	5 ,,	,,	nil	,,	
12. Razor (423)	6 ,,	good	cloud, 20 mins.	anti-turtle nil	See hat, Table p. 393.
13. ,, (357)	10 ,,	,,	cloud, 45 mins.	,,	
14. Pocket-knife (547)	11 ,,	fair	slight cloud, 1 hour	,,	
15. Knife (17)	28 ,,	slight	marked cloud, 15 mins.	anti-dog nil	Tried also with very strong anti-deer serum to see if any ruminant blood present; no reaction.
16. Razor (20)	28 ,,	,,	cloud, 15 mins.	,,	Much blood and oil.
17. ,, (19)	30 ,,	,,	marked cloud, 3 mins.	,,	Thickly smeared with blood; little oil.

* "Foam-test" refers to whether or not the blood dilution foamed on air being blown through it. As Nuttall found, the foam-test is a valuable aid for determining when blood has gone into solution.

so as to point out the possibility of a mistake occurring under such circumstances.

It appears that the effect of oil on blood is to lessen the reaction. This is probably due to the blood being coated with a film of oil, and therefore not so easily passing into solution.

B.   *Blood dried on organic materials.*

The experiments quoted below have been inserted here to show the effects of age on blood dried on organic fabrics, but further experiments (p. 394) indicate some of the fallacies which may arise from the character of the materials.   It happened, however, that in the specimens chosen few were of such a character as to give rise to possibilities of error.

The blood-stained materials tabulated below were all obtained from Scotland Yard, and with them two series of tests were conducted, the antiserum employed in the second being more powerful than that in the first.

In the first series very small quantities were employed, but in the second the amount in each case was slightly greater.   It was, however, not found possible on either occasion to obtain more than very small fragments, and, moreover, none of the specimens, with the exception of No. 11, were markedly encrusted with blood.   The one exception, a specimen of hair, 28 years old, was in some parts thickly plastered, and gave well-marked reactions with each antiserum.

The following table shows that numbers 1, 2, 3, 4, 6, 8, and 11, or 64% of the whole, gave well-marked reactions, their ages varying from 3 to 28 years.   In No. 5 the paper was badly burnt and the capacity for reacting was probably destroyed by the heat.   Nos. 7 and 10 produced alkaline solutions, and in each case the reaction with anti-human serum was very slight.   This was probably due to the retarding influence of the alkali, which has been discussed on p. 79 *et seq*.   At the time these experiments were carried out we were not aware of this action of alkalis. The negative result of No. 9 may have been due to its acidity.   No. 12 failed to react, but we were unable to discover any reason for this. The controls in all cases were negative.

As so little material was available the results may be looked upon as most satisfactory, for it can scarcely be doubted that more distinct reactions would have been obtained had it been possible to make more extensive use of the specimens."

	Material	Age	Foam test	Character of solution	Reaction to litmus	1st series		2nd series	
						Anti-human I.	Normal rabbit	Anti-human II.	Anti-ox
1.	Lining of clothes	3 years	good	clear	neutral	marked cloud, 60 mins.	nil		nil
2.	Felt hat	10 ,,	,,	(cloudy, clear after filtering)	,,			cloud, 15 mins.	,,
3.	Printed paper	11 ,,	,,	clear	alkaline	cloud, 5 mins.	nil	cloud, 60 mins.	,,
4.	Part of same paper	11 ,,	,,	,,	neutral			cloud, 15 mins.	,,
5.	Same scorched	11 ,,	,,	,,	,,	nil	,,	nil	,,
6.	Alpaca dress	11 ,,	,,	,,	,,	immediate cloud	,,	marked cloud, 10 mins.	,,
7.	Braid	11 ,,	,,	,,	alkaline	slight cloud, 60 mins.	? cloud	slight cloud, 60 mins.	,,
8.	Cardigan jacket	11 ,,	fair	(cloudy, clear after filtering)	slightly acid alkaline	slight cloud, 60 mins.	nil	marked cloud, 30 mins.	,,
9.	"Black Rep"	11 ,,	slight	clear	slightly alkaline	slight cloud, 5 mins. no increase	,,	slight cloud, 15 mins.	,,
10.	Cotton fabric, apparently washed	11 ,,	fair	,,	neutral	cloud, 30 mins.	,,	marked cloud, 5 mins.	,,
11.	Hair	28 ,,	good	,,	,,		,,		,,
12.	Wooden handle of chopper	28 ,,	none	,,	,,			nil	,,

Graham-Smith and Sanger (1903, p. 287) further studied the influence of various materials upon the blood with which they had been in contact. They reported upon their results as follows:

Experiments upon the detection of blood dried on fabrics.

(See Plate and explanation p. 402.)

"In order to determine to what extent the composition of different cloths influenced blood which had dried on them we procured a number of samples. Human blood was dropped upon these so as to leave some patches unaffected and others saturated. Subsequently the specimens were allowed to dry under natural conditions and were left undisturbed at room temperature and in the light for at least 30 days; some were not tested for nine months. First a series of control tests were carried out on unstained pieces of cloth in the following way. Small pieces 1×2 cms. were soaked overnight in 2 c.c. of distilled water. In the morning an equal quantity of double normal salt solution was added and the condition and reaction to litmus of the extract recorded. The majority of samples was found to be nearly neutral, some were distinctly alkaline, whilst most of the coarser materials were acid. About ·5 c.c. of each extract, if necessary after filtration, were placed in small test-tubes and 1 drop of serum added. No cloudings were noticed except in the markedly acid specimens. After neutralisation with sodium carbonate these also produced no effect on the serum. Certain solutions, especially the acid ones, were found to be opalescent, or slightly cloudy, before the addition of serum, but it was noticed that neutralisation tended to make these clearer. In all our experiments we have avoided shaking the extracts, as we frequently observed deposits and cloudy precipitates at the bottom of the tubes, which in some cases were very difficult to remove by filtration. After removing the supernatant fluid in solutions containing blood the tubes were, however, shaken to ascertain whether sufficient serum was in solution to produce marked foaming.

In testing for blood, stained patches were treated in the way described above and neutralised if necessary. Two small tubes of each solution were prepared. To one was added one drop of anti-human serum and to the other a drop of anti-ox serum. The results of some of these experiments are given on the opposite page.

Material		Solution	Reaction to litmus	Anti-human serum		Anti-ox serum	
				Immediate	6 hours	Immediate	6 hours
1.	Black dress (1 month)	clear	neutral	marked reaction	large deposit	—	—
2.	Glacé silk "	"	"	"	"	—	—
3.	Serge "	"	"	"	"	—	—
4.	Green fancy serge "	green	"	"	"	—	—
5.	Sateen "	clear	"	"	"	—	—
6.	Merve "	"	"	"	"	—	—
7.	Tweed cloth "	"	"	"	"	—	—
8.	Furniture serge "	red	"	"	"	—	—
9.	Pillow case (9 months)	clear	"	"	"	—	—
10.	Silk handkerchief (1 month)	opalescent	"	"	"	—	—
11.	Blind ticking "	"	"	slight reaction	"	—	—
12.	Green velveteen "	green	alkaline	marked reaction	"	—	—
13.	Dark green cloth "	"	slightly acid	"	"	—	—
14.	Coarse flannel "	clear	acid	"	"	—	—
15.	Felt hat "	"	"	medium reaction	medium deposit	—	—
16.	Coarse duster (9 months)	opalescent	"	"	"	—	—
17.	Brown canvas (1 month)	"	"	"	large deposit	—	—

When acid the solution was neutralised. The sign — indicates that the result of the test was negative.

The detection of blood-stains on leather[1].

Several observations have been made with various samples of leather, which have been placed under a separate heading to more fully bring into prominence their peculiarities. It was found that nearly all gave acid reactions on solution. The degree of acidity, however, varied greatly, chamois leather being alkaline, suède kid glove only slightly acid, and the coarser leathers very decidedly acid. The addition of a drop of serum to the acid solutions produced clouding, and even coagulation with extracts of the coarser leathers. The latter also gave rise, especially if shaken, to bulky deposits in the original solutions.

Nearly all the solutions of leather could be neutralised and the blood test satisfactorily employed. One class of leather was, however, a marked exception, namely, thick polished yellow leather. Solutions of this gave rise to extremely acid yellow fluids, whose colour deepened on the addition of alkali. It was found impossible to obtain the specific test for blood dried on it. At first it was thought possible that the blood was destroyed by the acid after solution, and extracts were made in alkaline salt solution to neutralise this effect. Even under these conditions no positive results could be obtained. Up to the present although many methods have been tried we have been unable to devise one which gives satisfactory results, and are forced to conclude that the mode of preparation of such leathers produces conditions which destroy the blood in contact with them. Under favourable conditions, when blood has been thickly deposited on the surface, it might, however, be possible to scrape it off and obtain a positive reaction. In the following table all solutions when necessary were neutralised, and filtered, before the addition of anti-human serum.

A series of experiments was also made to determine the effects of boot-blacking and polish. Blood-stains blackened over were hard to detect on the boot, but by neutralisation and filtration clear solutions could be obtained, and yielded well-marked reactions. Polish also made no difference to the test.

Experiments with saline solutions of tannin show that it has a very deleterious action on serum, rendering the application of the test when it is present in large quantities impossible. Solutions of 1 in 20 to

---

[1] See also p. 79 *et seq.* regarding the effects of acids on bloods, and Plate and explanation on p. 402.

1 in 500 produce instant coagulation of the serum, and 1 in 1000 produces marked clouding.

Material	Con-dition	Colour	Re-action	Anti-ox				Anti-human	
				Unneutralised		Neutralised		Neutralised	
				15 mins.	24 hours	15 mins.	24 hrs.	15 mins.	24 hours
Chamois leather	clear	clear	neutral	—	—	—	—	medium reaction	medium deposit
Suède kid glove	cloudy	,,	slightly acid	—	—	—	—	good reaction	large deposit
White ,, ,,	clear	yellowish	acid	slight cloud	slight cloud	—	—	,,	,,
Boot	,,	,,	,,	,,	slight deposit	—	—	,,	,,
Leather from in-side shoe	cloudy	,,	very acid	,,	,,	—	—	,,	,,
Patent leather	,,	,,	,,	,,	,,	—	—	,,	,,
Yellow leather	,,	,,	,,	coagu-lation	deposit	cloud	cloud	cloud	cloud

Detection of blood on materials not previously mentioned.

Ten examples of wall paper of various textures and colours, red, brown, yellow, blue, and green, were tested and gave typical reactions. All produced neutral solutions, some of which were tinted.

Extracts of blood dried on various kinds of paper, stones, flint, slate, coal, cork, string, straw, rubber, linoleum, as well as silver and copper coins, yielded satisfactory results.

Although one piece of oak on which blood had been thickly incrusted gave a marked reaction with anti-human serum, we failed to obtain any reaction with blood on two blocks of cedar and pine. The quantity present on each of these was exceedingly small, and the negative result was probably due to this cause.

These experiments demonstrate that many substances in common use give acid solutions. In most instances the acidity is not so marked as to be of importance, but in some, unless recognised and neutralised might be liable to lead to grave error. Extracts of certain substances are sufficiently alkaline to impede the reaction.

The detection of blood in the presence of lime, mortar, and earth.

The wide distribution of these substances rendered it necessary to investigate their action on blood, since in medico-legal practice it might often be necessary to test blood dried on, or mixed with, these materials.

Solutions of earthy salts, mortar, and lime of various strengths were made in salt solution and tested qualitatively with various antisera to determine their action on serum. These actions vary to some extent with the quantity of serum added. In the following table the quantity added was one drop, since this was the unit chosen for qualitative experiments.

Dilutions		Lime	Mortar	Calcium chloride	Calcium phosphate	Chalk	Sodium phosphate	Plaster of Paris	Alum	Caustic soda	Caustic potash
Saturated solution*	30 mins.	*	*	×	*	*	*	*			
	24 hrs.	d	d	d	*	d	d	*			
1 : 10	{	*	*	×?	*?	.	*	.	1:25 **+**	*	*
		*	*	*	.	.	.	.	**D**	*	*
1 : 100	{	.	.	.	.	.	.	.	×	×	×
		.	.	.	.	.	.	.	×	×	×
1 : 1000	{	.	.	.	.	.	.	.	×	*	*
		.	.	.	.	.	.	.	×	.	.
1 : 10,000	{	.	.	.	.	.	.	.	.	*	.
		.	.	.	.	.	.	.	.	.	.

* Where saturated solutions are mentioned the dilutions are 1 : 10 etc. of these. The signs used here are described on p. 338.

The addition of serum to strong solutions of lime resulted in a general clouding, which later gave place to a dense cloud below, which would be hard to distinguish from a positive reaction. Mortar gave rise to a similar but smaller clouding. Calcium chloride and sodium and calcium phosphates caused cloudings in very strong solutions only. The actions of caustic soda and potash in certain solutions are very marked, and are referred to on pp. 83, 85. They are briefly mentioned here owing to their presence in earth.

At this point it should also be noted that strong lime and calcium solutions give rise on standing, even after filtration, to deposits of the salt at the bottom of the tube and a filmy layer on the surface.

It was found, however, that the difficulty in testing due to the presence of lime in mortar, plaster, and earth, could generally be eliminated in the following way. Solutions of all of the above substances were allowed to stand till the excess had settled to the bottom. The supernatant fluid was then pipetted off and filtered. Carbon dioxide gas generated by the action of dilute hydrochloric acid on chalk, and washed by passing through distilled water, was next passed through the fluid and the latter again filtered to free it from the presence of the precipitated calcium carbonate. By this procedure clear filtrates could be obtained in most cases, which remained so for an indefinite period, and produced no cloudings on the addition of sera.

The following formula explains the reaction :

$$Ca\,(OH)_2 + CO_2 = Ca\,CO_3 + H_2O.$$

Too much of the gas must not however be passed into the solution owing to the fact that excess of CO_2 renders the insoluble carbonate again soluble :

$$Ca\,CO_3 + H_2O + CO_2 = Ca\,H_2\,(CO_3)_2.$$

By quantitative experiments it was found that though this process caused a deposition of blood pigment from blood solutions, yet the property of producing precipitation on the addition of appropriate antisera was not in any way affected.

The action of dry and wet lime, etc.

Lime was intimately mixed with human blood and then spread on porcelain and exposed to the action of air for three months. The resulting compound turned a greenish colour. Solutions of this gave an immediate clouding on the addition of serum. After the passage of CO_2, however, and subsequent filtration, no reaction could be obtained with anti-human or other serum. Under these conditions it seems that unslaked lime completely destroys the reacting power of blood in contact with it.

Quantitative experiments over a shorter period bring out the destructive quality of lime and mortar very markedly. To ascertain the action on serum of dry and wet lime, mortar, brick, earth, etc. weighed quantities of one gramme of each were mixed with 1 c.c. of human serum and allowed to act for 4 days. Similar mixtures but with 10 c.c. of water added were also prepared and allowed to stand for 4 days, to determine whether any different action was excited by these materials in the presence of water. At the end of this period all were made up to 1 : 21, by the addition in the former case of 20 c.c. of normal, and in the latter of 10 c.c. of double normal salt solution.

After applying the method of removing lime which has just been described quantitative estimations (Method, p. 315) were made.

Material		Anti-human	Anti-ox	%		Material		Anti-human	Anti-ox	%
1. Control		·0403	—	100	6.	Earth	dry	·0244	tr	61
							wet	·0291	tr	72
2. Chalk	dry	·0367	tr	91	7.	White brick	dry	·0262	tr	65
	wet	·0357	tr	89						
3. Red brick	dry	·0281	tr	70	8.	Mortar	dry	tr	tr	0
	wet	·0347	tr	86			wet	·0028	tr	7
4. Pasteur filter	dry	·0309	tr	77	9.	Lime	dry	tr	tr	0
	wet	·0319	tr	79			wet	tr	tr	0
5. Berkefeld filter	dry	·0291	tr	72						
	wet	·0291	tr	72						

The above table shows that all the materials used in this series produced slight effects on the serum, but that mortar and lime completely destroyed its power of reacting. All had been ground up very finely before the addition of the serum.

The effects of the lime present in ordinary earths.

The next point of importance was to determine whether the amount of lime present in ordinary earths was sufficient to interfere to any serious extent with the reaction. For this purpose 9 samples of analysed earth were obtained, five from a field of gravelly land near Trowse, divided into five plots, and the others from different localities. The results of the analyses of these earths dried at 100° C. are given in the following table, arranged according to the percentage of lime present. We are indebted to Mr T. B. Wood for these analysed earths.

	I. Trowse, plot 1	II. Trowse, plot 2	III. Trowse, plot 3	IV. Trowse, plot 4	V. Wryde clay	VI. Needham salt	VII. Trowse, plot 5	VIII. Bentwick fen	IX. Littleport fen
Total lime	·91	1·01	1·43	1·46	1·48	1·57	1·97	2·95	4·39
Organic matter, loss by ignition	5·62	5·16	5·27	5·52	14·48	5·40	6·31	39·35	50·82
Calcium carbonate	1·16	1·39	1·94	1·98	—	—	3·06	—	—
Total phosphoric acid	·16	·18	·16	·18	·37	·19	·20	·30	·28
Total potash	·13	·14	·13	·14	1·32	·31	·13	·63	·56
Total nitrogen	·15	·11	·14	·53	·24	·24	·13	1·42	1·8

Strong solutions of the above after simple filtration were at first clear but showed a white filmy deposit after standing. The quantity of this increased with the percentage of the lime present. On the addition of anti-human serum a thin cloud spread through the entire solution and gradually deepened, being considerably denser in No. IX than in No. I.

After the passage of CO_2 and filtration every solution was clear, with the exception of V and VIII, and produced no deposit on standing. The two mentioned were opalescent. No clouding occurred on the addition of anti-human serum.

Two sets of quantitative experiments were carried out with these soils. In the first 1 c.c. of finely divided soil was placed in a test-tube with 1 c.c. of human pleuritic exudate and 5 c.c. of water. After 4 days 5 c.c. of double normal salt solution were added to each, making a dilution of 1 in 11 of pleuritic exudate. In the second series 1 c.c. of dry earth was allowed to act for 4 days on 1 c.c. of pleuritic exudate. At the end of this period the specimens were diluted to 1 in 11 with salt solution.

These solutions were treated with CO_2 as described, and the precipita measured quantitatively.

Earth solution	Percentage of lime	Quantity of precipitum, mean of the two observations	Percentage	Control anti-ox
No. I.	·91	·0404	100	—
No. II.	1·01	·0304	97·5	—
No. III.	1·43	·0319	78·9	—
No. IV.	1·46	·0389	96·2	—
No. V.	1·48	·0241	59·6	—
No. VI.	1·57	·0258	63·3	—
No. VII.	1·97	·0389	96·2	—
No. VIII.	2·95	·0314	77·7	—
No. IX.	4·39	·0383	94·8	—

The above table shows that the quantity of precipitum obtained did not decrease in proportion to the increase of the lime, which was apparently never present in sufficient quantity to materially affect the reaction. Excess of potash probably accounts for the low figures obtained in No. V, and possibly No. VIII (p. 83). In neither of these could a clear solution be obtained. Whatever may be the cause of the variations in the quantity of precipitum obtained, these experiments go to show that blood mixed with ordinary earth can be readily detected if present in sufficient quantity and that its specific character remains unaltered.

From our experiments on earth and lime salts we have drawn the following conclusions: (1) that the intimate mixture of lime with blood completely destroys the latter; (2) that a clouding occurs in the solution of earth on the addition of serum; (3) that this is due principally to the presence of lime salts; (4) that the lime can be got rid of and the solution rendered clear, and not liable to clouding, by the passage of CO_2 and subsequent filtration; (5) that the passage of CO_2 in no way interferes with the reaction; (6) that the quantity of lime present in ordinary earth does not materially affect blood mixed with it.

EXPLANATION OF PLATE

(GRAHAM-SMITH AND SANGER).

Fig. 1.   No. 1 shows the precipitum with normal human serum (1 : 21 in salt solution) and anti-human serum (·1 c.c.).   No. 2 with putrid human serum (1 : 21) and antiserum, and No. 3 with normal human serum (1 : 21) and putrid anti-human serum.   No. 4 shows a clear solution of human serum in salt solution (1 : 21).   No. 5 shows the deposit resulting from the solution of human serum in distilled water (1 : 21).   No. 6 the precipitum formed with human serum diluted with distilled water (1 : 21) and anti-human serum.   Nos. 7, 8 and 9 show three capillary tubes such as are used in quantitative measurements, and containing precipitum.

Fig. 2   Shows effects of increasing quantities of NaCl on the formation of precipitum; each tube contains ·5 c.c. of a 1 in 21 dilution of human serum, and ·1 c.c. of anti-human serum.   No. 1 contains ·6 % of salt, and those following 1 %, 2 %, 4 %, 6 %, 8 %, 10 %, 16 %, 18 %, and No. 11 is saturated with salt.   Results of measurements are given on p. 103.

Fig. 3   Shows the specific precipitum in tests for human blood dried for a month on various materials.   The lower series shows controls with anti-ox serum.   The cloudings in the tubes are due to the opalescence of the solutions; and the various solid particles are portions of undescended precipitum.

No. 1 test for blood dried on silk handkerchief, No. 2 on tweed cloth, No. 3 on black dress fabric, No. 4 on dark green cloth, No. 5 on coarse green cloth, No. 6 on coarse red cloth, No. 7 on kid glove, No. 8 on blanket, No. 9 very coarse sack material, No. 10 on flannel.   Nos. 11—20 show control tests with anti-ox serum: all negative.

The solutions whenever necessary were neutralised before testing.

Fig. 4.   Nos. 1 to 5 show the effects on serum of dilutions of Hydrochloric acid in salt solutions of strengths of 1 : 10, 1 : 100, 1 : 1000, 1 : 10,000 and 1 : 100,000.   No. 1 has a dense white cloud, No. 2 a slight cloud at the bottom, No. 3 a marked cloud, and the others are unaffected.   Photographed after 6 hours.

Nos. 6—10 similarly illustrate the action of Tartaric acid, No. 1 (1 : 10) having a slight cloud, No. 2 (1 : 100) a medium cloud, No. 3 (1 : 1000) a marked cloud, Nos. 4 and 5 (1 : 10,000 and 1 : 100,000) are unaffected, the apparent deposit being due to the light.

Nos. 11—15 illustrate the action of Nitric acid.   No. 1 (1 : 10) shows the coagulum, No. 2 (1 : 100) a very faint cloud, No. 3 (1 : 1000) a medium cloud, and Nos. 4 and 5 are unaffected.

Nos. 16—20 show the effects of Acetic acid.   Nos. 1 and 2 (1 : 10 and 1 : 100) have slight clouds, No. 3 (1 : 1000) a medium cloud, and No. 4 (1 : 10,000) a marked cloud.   No. 5 (1 : 100,000) is not affected.

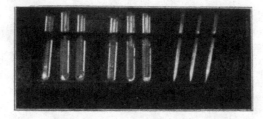

Fig. 1.

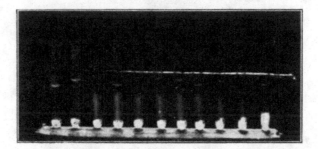

Fig. 2.

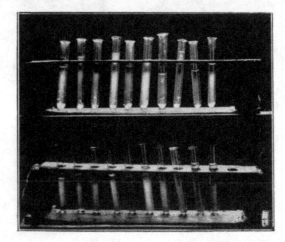

Fig. 3.

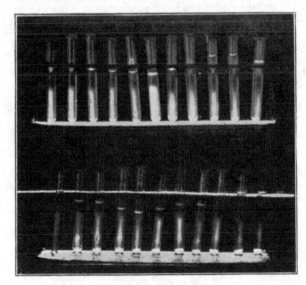

PLATE

1 2 3 4 5 6 7 8 9 10

11 12 13 14 15 16 17 18 19 20

Fig. 4.

1 2 3 4 5 6 7 8 9 10 11

12 13 14 15 16 17 18 19 20 21 22

Fig. 5.

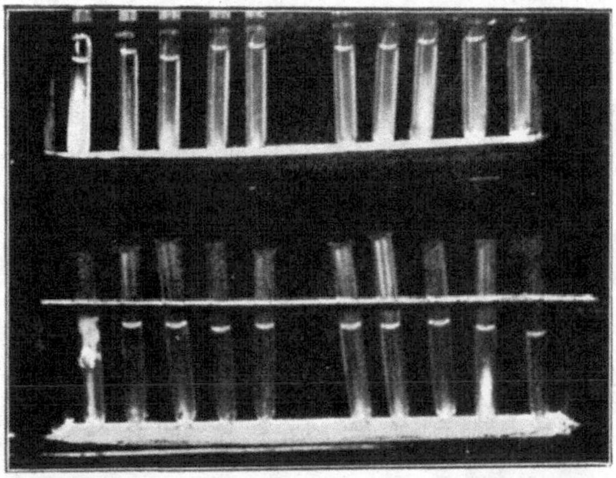

(*Reprinted from Journal of Hygiene, Vol. III. No* 2)

Fig. 5 Illustrates the action of acids and alkalis on the formation of the specific precipitum. All the tubes contain ·5 c.c. of human serum dilution (1 : 21) in salt solution. Nos. 1—5 contain 5 to 1 drops of 1 in 10 Hydrochloric acid. No precipitates have been formed. No. 6 did not receive any acid or alkali. Nos. 7—11 contain 1—5 drops of 1 in 10 sodium carbonate solution ; the quantity of precipitum shows a decrease along the series. Nos. 12—22 have been similarly treated but received drops of 1 in 100 acid and alkali respectively. The precipitum is seen to increase from 12 to 16 and decrease from 18 to 22. The slight clouding above the precipitum in each case is due to bacterial growth, the tubes having stood 48 hours.

We are indebted to Walter Mitchell, our laboratory attendant, for the time and attention he has bestowed on the photographing of these specimens."

Graham-Smith and Sanger (p. 260), as the result of their investigations, reach the following conclusion, which I can but endorse :

"These experiments have led us to the conclusion that with sufficient materials, and due precaution to exclude the various sources of error, there are but few conditions met with in forensic practice under which human could not be readily differentiated from other bloods. By this, however, we do not mean to imply that a considerable acquaintance with the action of precipitating antisera on blood solutions is not necessary in the successful application of this test."

That this conclusion is fully justified is proved by the official recognition of the precipitin method in forensic practice by foreign Governments. Prof. Uhlenhuth (19. IX. '03) informs me that the method has been recommended by the Ministers of Justice in Germany and Austria, and that it has been officially recognized by the Governments of Egypt and Roumania.

I herewith append a copy of the Order issued by the German Authorities, the same having appeared in the *Medicinalblatt für Mediciner und medicinische Unterrichtsangelegenheiten* No. 17 (1. x. '03):—

BERLIN, W. 64,
den 8. *Sept.* 1903.
WILHELMSTRASSE 65.

*Der Minister des geistlichen Unterrichts
und Medicinalangelegenheiten.*

U. I, No. 12328 M.

  Der Justizminister,
   I. No. I, 5283.

    Von dem Stabsarzte Professor Dr. Uhlenhuth in Greifswald ist eine Methode der Blutuntersuchung vermittelt worden welche es ermöglicht, die Art des zu untersuchenden Blutes festzustellen und namentlich Menschenblut mit Sicherheit von Thierblut zu unterscheiden. Bei der Behandlung des zu unter-

26—2

suchenden Blutes mit Serum aus dem Blute von Kaninchen, denen zuvor Blut anderer Thiere oder Menschenblut eingespritzt war, ergeben sich bestimmte Erscheinungen wenn das zu untersuchende Blut von derselben Art ist wie das zuvor den Kaninchen eingespritzte. Es kann desshalb jede Art von Blut, wenn das entsprechende Serum angewendet wird, bestimmt werden. Die wissenschaftliche Deputation für das Medicinalwesen hier hat sich über den Werth der Methode mit Hervorhebung von deren grossen Bedeutung wie folgt geäussert :

"Die Erfahrungen über die Serummethode der Blutuntersuchung sind bereits in Deutschland wie im Auslande so ausgedehnte, die Resultate der Forschungen in wesentlichen so übereinstimmende, dass kein Zweifel mehr darüber bestehen kann, das diese neue biologische Methode in der Mehrzahl der Fälle mit grosser Sicherheit gestattet, frisches sowie angetrocknetes Blut nach seiner Herkunft zu bestimmen, Menschenblut von Thierblut, Blut verschiedener Tierarten zu unterscheiden. Es ist daher dringend geboten, diese vortreffliche Methode, welche natürlich die alten bewährten Methoden des Blutnachweisses nicht verdrängen sondern nur ergänzen u. vervollständigen soll, für die gerichtliche Praxis allgemein nutzbar zu machen."

Als Institute, bei denen diese Methode seit längerer Zeit zur Anwendung gelangt, werden bezeichnet :

Das hygienische Institut der Universität Greifswald.

Das Institut für Infektionskrankheiten in Berlin, Nordufer 39.

Das Institut für Staatsarzneikunde in Berlin.

Das Institut für experimentelle Therapie in Frankfurt a/M.

Diese Institute werden in erster Linie für die Vornahme von Untersuchungen der in Rede stehenden Art empfohlen.

Indem ich auf diese Methode der Blutuntersuchung aufmerksam mache, empfehle ich, in allen geeigneten Fällen die Untersuchungen nach ihr ausführen zu lassen.

Abdrücke dieser Verfügung sind zur weiteren Mitteilung an die Landgerichts-Präsidenten und die Ersten Staatsanwälte des dortigen Bezirks beigefügt.

Im Auftrage,

gez. VITSCH.

An die Herren Vorstandsbeamten des Kammergerichts
　　sowie der sämmtlichen Oberlandesgerichte.

2.　*Antisera in the Examination of Meats.*

From previous observations it seemed natural to conclude that meat extracts would react to corresponding haematosera, for the reason that they contain blood. As was noted on page 385 Uhlenhuth (1901) obtained a positive reaction with anti-pig serum, tested upon the organs of a pig which had been dried for 18 months. Continuing this line of investigation, he obtained positive results with the antisera for pig, sheep, horse, donkey, and cat blood, when these were tested upon the corresponding meats. He found (7. XI. '01) that anti-sheep serum gave almost as much reaction with goat as with sheep, and less reaction with beef extract. The method is of no especial use in the examination of large pieces of meat, for the reason that these can be readily recognized. Tokishige has informed Uhlenhuth that meat is usually sold in small pieces in Japan. Minced meat is sold in most countries. Admixtures of horse, dog, and cat meat can be detected by the precipitin method in minced meat, in sausage, and smoked meat, these of course not having been subjected to cooking. He obtained reactions with pig and horse hams which had been smoked the year before, also with horse sausage (" Pferdemettwurst ").

The *method of examination* consists in scraping the meat and extracting it with water or saline. It takes a long time to extract the meat in some cases. An extract is suitable for testing when it foams on being shaken. Meat can be more rapidly extracted by adding a small amount of chloroform, extraction being then usually effected in a few minutes. The extract is very cloudy, and has to be cleared by repeated filtration through filter-paper or a Berkefeld filter. If extracted with water, an equal volume of double normal salt solution has to be added to the watery extract before testing this. In testing add 10—15 drops of antiserum to 3 c.c. of the saline meat extract.

Von Rigler (1902) has used this method in the study of meat-adulteration. He prepared 20 % watery extracts of the meat of 7 species of animals (roebuck, hare, rabbit, horse, ox, pig, cat) and injected 5—10 c.c. thereof every 3 days into rabbits subcutaneously, during one month. Whereas normal rabbit serum had no effect on the meat extracts, the antisera obtained from the treated rabbits were specific, acting on extracts of mixed meats, as also upon some boiled and roasted meats of an homologous kind.

Nötel (13. III. '02) treated rabbits with horse serum, muscle juice and muscle extract (in ·1 % soda solution) injecting subcutaneously every 2 to 3 days, amounts of 10 c.c., until 10 to 12 doses had been administered, after which 6 days were allowed to elapse before the rabbits were bled. He obtained the least effective antisera from the serum-treated rabbits [1].

To obtain clear solutions of muscle, he found it best to extract by means of ·1 % soda solution, without disturbing the meat mechanically. He found roasted meat, underdone in the centre, as also cold-smoked meat, to give positive reactions. Donkey meat reacted like that of the

---

[1] Valleé and Nicolas (30. VI. '03) have recently confirmed these observations. They speak of sero-precipitins in contradistinction to musculo-precipitins.

horse. He noted that sausage extract alone, when placed at 40° C. for 5 minutes, often becomes clouded in a manner which might lead to mistakes, so it is necessary to use controls. It has not occurred to this author that the soda solution might be responsible for the clouding noted.

3. *Antisera in the Examination of Bones.*

Beumer (1902) has applied the precipitin method to the medico-legal examination of bones, in a case where bone fragments only 1 to 6 cm. long were found in a house which had been destroyed by fire, and it was suspected by the authorities that they might be human. Because of their small size, the fragments could not be determined. The bones were somewhat charred on the outside, remains of soft tissues adhering thereto. After removal of the latter, the bone fragments were placed in saline, where they were allowed to remain 4 days, the fluid foaming then on being shaken. The fluid was now filtered through a Berkefeld filter, and the clear solution thus obtained was tested with anti-human, anti-pig and anti-ox sera, reacting only with anti-ox serum. The conclusion was therefore obvious. Fresh bones were readily determined by the method, best when the marrow or spongy portions were extracted, more slowly and less surely when cortical substance was used. The addition of chloroform facilitated extraction. Cortical substance was best extracted by sawing the bone through and extracting the bone-dust thus obtained. Beumer found bones which had been exposed for several weeks to the air, to be still determinable by the precipitin method. Boiled bones gave a negative result, as did also roasted or incinerated bones.

Schütze (22. I. '03) has made similar experiments, with regard to the identification of bone fragments by means of antisera. He placed the fragments in 0·85 % saline to which ·25 % soda solution had been added, thus obtaining an extract which gave reactions with an homologous antiserum.

4. *Antisera in the Examination of Commercial albuminous Preparations containing egg-white, and in the examination of Honey for adulteration.*

Uhlenhuth (15. XI. 1900) found that the antiserum for the egg-white of the fowl constantly gave negative results with commercial albuminous preparations which did not contain egg-white, the contrary being the case where they contained egg-white. The use of such antisera is therefore suggested where it is desired to prove the presence of egg-white in prepared foods.

von Rigler (1902) treated rabbits with honey, in the manner employed when they are immunified with blood. The anti-honey serum only produced precipitation in dilutions of honey, not in those of grape- or cane-sugar. Normal rabbit serum had no such effect.

5. *Antisera in the Study of Urine.*

In the part referring to precipitins and precipitable substances *in corpore*, I referred to observations on the urine (see p. 133), which I shall not recapitulate. It is obvious that the precipitins may be put to use in the study of assimilation. I would state here that M. Ascoli (11. III. '02) succeeded by means of antiserum for egg-white, in demonstrating the existence of apparently unaltered egg-white in the urine of persons showing albuminuria in consequence of excessive albuminous diet. Linossier and Lemoine (18. IV. '02) made observations on a young man suffering from orthostatic albuminuria, evidently due to malassimilation, the albumen being derived from non-assimilated food. On giving him cow's milk, traces thereof were found in his urine by means of lactoserum for cow's milk.

Conclusions.

In view of the mass of material treated of in this book, it is difficult to draw any detailed conclusions. It has been shown that there are many points of resemblance between the different antibodies. The little work which has been done with the haemolysins bears directly upon the immediate subject of this book, namely, the blood-relationship amongst animals, but it is scarcely to be expected, owing to technical difficulties, that the haemolysins will be of such general use in the study of the problem.

With regard to the precipitins, it is evident that more scientific methods of treating animals for the production of antisera are called for. Powerful antisera may, however, be produced by intravenous injections of much smaller quantities of serum than have hitherto been used. Care should be exercised with regard to the addition of preservatives to antisera, and in the use of solvents other than salt solution for the extraction of dried bloods.

The precipitins and precipitable substances combine quantitatively. The precipitum is soluble in an excess of precipitable substance. Heated antisera cannot be reactivated. The precipitins constitute receptors of the second order (Ehrlich). There is no evidence that the action of precipitins is fermentative. Heated haematoserum (precipitoid) combines with precipitin, and prevents its action upon precipitable substance. There is evidence of the existence of immune-bodies in precipitating antisera. There is no evidence that antisera differ in their ordinary properties from normal sera. The precipitins and precipitable substances are intimately bound up with the globulins in serum, and have not been separated therefrom. The precipitins do not give reactions corresponding to the amount of albumin present in a solution of precipitable substance; the albumin, to be acted upon, must possess certain specific properties. The presence of even small quantities of acids or alkalis, markedly reduces the amount of precipitum formed, but an increase of salt (NaCl) has little effect. The supernatant fluid in a mixture of antiserum and precipitable substance, after precipitation has taken place, may contain an excess of both interacting bodies. There is evidence that the precipitins may exert a special, but not a specific action on different albumins from the same species of animal. It is doubtful if the precipitins and precipitable substance can withstand tryptic digestion, whereas both are certainly destroyed by peptic digestion. The weight of evidence is against the possibility of precipitins being formed for peptones[1].

The rate at which interaction takes place between precipitins and precipitable bodies is markedly influenced by temperature, being retarded at low temperatures, hastened at higher temperatures. The quantity of precipitum formed is not influenced by the temperature (5—37° C.) at which the experiment is made. Undiluted haematosera are but slightly affected by exposure to a temperature of 63° C,; they are inactivated at 68—70°, unaffected at 60° and under. (Bacterio-precipitins are inactivated at 58—60° C.) Undiluted normal sera seem to be rendered non-precipitable at a somewhat lower temperature than that which inactivates haematosera. Both interacting bodies resist desiccation. The precipitins are apparently more unstable than the precipitable substances, which may give reactions even when dried many years. Fluid sera, preserved *in vitro*, may give reactions after being stored 4 or more years; they appear to deteriorate slightly by

---

[1] See Appendix, note 2.

keeping. Fluid antisera, sealed in bulbs and kept cold and in the dark, may at times give good reactions after 6 to 14 months. Putrefaction of serum, or antiserum, does not affect the production of a specific precipitum.

There is evidence that the precipitin-content of the serum *in corpore* undergoes fluctuations during immunization, corresponding to those observed in animals under toxin-treatment. The precipitins begin to disappear about 1 month after treatment has ceased. The precipitins disappear from the blood of animals which have undergone prolonged treatment, the animals having become immune. The precipitins may be present in the *humor aqueus*, and are transmitted to the offspring *in utero*, whilst they are absent from the urine. The precipitin and precipitable substance may co-exist *in corpore*, no precipitation apparently occurring in the body. The presence of foreign precipitable substance in an animal's serum may be readily demonstrated by means of an antiserum for the foreign substance. No explanation can as yet be given of the marked leucocytosis after injection of a foreign albumin into an animal which has been rendered more or less immune to the albumin. The seat of origin of the precipitins is unknown.

The more powerful an antiserum is, the greater is its sphere of action on other bloods. The degree and rate of blood reaction appear to offer an index of the degree of blood-relationship; in other words, closely related bloods react more powerfully (more precipitum) and more rapidly than do distantly related bloods, provided the latter react at all. The interaction of a precipitin with its homologous blood dilution is not impeded through non-homologous bloods of various kinds being present in the blood-dilution in mixture. The amount of reaction may be expressed in terms of the volumetric measurement of the precipitum produced by mixing known quantities of the interacting substances. An antiserum acts upon a higher dilution of homologous than of non-homologous substance. The strength of an antiserum may be expressed either in terms of precipitum-volume, or, by giving the highest dilution of blood with which it reacts, the quantities of the interacting substances being stated. The amount of reaction may be affected by altered blood-concentration in disease, possibly also by the variation in the alkalinity of the blood in disease, or even in health. There is evidence of iso-precipitins, autoprecipitins, and antiprecipitins being artificially formed in the bodies of treated animals. Certain normal sera may contain precipitins, but these do not possess a specific character.

The bacterio-precipitins appear to be distinct from the others hitherto studied. In addition to these, various authors have obtained precipitins for the albumin of yeasts, of higher plants (I have grouped these together as "Phytoprecipitins" in contradistinction to the following "Zooprecipitins"), for the casein of different milks (lactosera), also for different bloods, etc.

To avoid needless repetition, I will refer the reader to pages 156 to 160, where the results of other authors with lactosera are given. The results of tests with different precipitating antisera are summarized on pp. 214, 335, 353, 359, 361, 403. The general bearing of the investigation from a zoological point of view is considered in the Introduction.

In Part II, Section IX., pp. 381—407, evidence of the value of the precipitin reaction in legal-medicine, based upon the work of various authors, is given.

In conclusion I would add that this investigation must necessarily be regarded as preliminary in character. The exhaustive treatment which our present knowledge of the precipitins has received, should prove of use to others, and I hope that the work done will stimulate many to further investigate the many problems which present themselves. Like other lines of investigation, this one appeared relatively simple at first; it is evident however now that the phenomena of precipitation are of an exceedingly complex nature.

Acknowledgments.

The extensive collection of bloods whose examination forms the main subject of this book, was only rendered possible through the generous aid of some seventy gentlemen throughout the world. I take this occasion to thank them very cordially for the friendly assistance they have given me in my work. Of the 900 blood samples collected, nearly 200 were brought together by me personally, some with the aid of our laboratory attendants, 6 by purchase from the Zoological Laboratory at Plymouth; a few have been contributed by Dr Louis Cobbett and Dr Graham-Smith of Cambridge.

My work has been especially furthered by Mr Frank E. Beddard, F.R.S., Prosector of the Zoological Society's Gardens, London, under whose direction 224 blood-samples were collected in the course of about two years by his assistant Mr E. Ockenden. Mr Ockenden has been most painstaking and exact in his attention to my directions with regard to the methods of collecting bloods. It is needless to say that I have obtained an invaluable material from this source.

Especial thanks are also due to the Hon. N. Charles Rothschild, who personally collected 89 samples of blood in Great Britain, Ceylon and Japan, besides stimulating others to supply me with material. Thus Messrs Brazenor Brothers ("B" in the tables), Naturalists at Brighton, have kindly supplied me with 69 specimens, Mr W. J. Clarke, Naturalist at Scarborough, with 19 specimens, a few being also supplied by Mr Head, Naturalist at Scarborough. Mr Rothschild also put me in communication with a number of gentlemen in different parts of the world.

Mr H. M. Phipson, Honorary Secretary of the Bombay Natural History Society, sent me 11 specimens which he had collected. He helped me considerably by circularizing the members of that Society with copies of my letter containing directions for the collection of blood specimens, a task which was continued by Mr W. S. Millard, who assumed his self-imposed duties during Mr Phipson's absence from India. Of the members of this Society, I am indebted to Mr C. H. Donald, of Bhadarwa, Kashmir State, for 29 specimens; Mr Charles M. Inglis, of Tirhut, for 21 specimens; Mr J. Mason, Curator of the Society, for 8 specimens; Captain H. T. Fulton, of Chitral, for 2;

and for one specimen each to Major R. M. Betham, of Akalkote, Deccan; Mr C. Fisher, of Chatrapur; Major G. A. Leslie, of Chitral; Mr A. H. A. Simcox, of Akrani, Khandesh; Captain E. H. Sweet, of Chitral.

In addition, M. Wm. Foster, of Sapucay, Paraguay, supplied me with over 26 specimens, a large number of them from Chiroptera; Herr Kuse, Gamekeeper at Kittendorf in Mecklenburg-Schwerin, sent 20 specimens; Dr John P. Tulloch, of Cronstadt, Trinidad, sent 18 specimens; Prof. E. A. Goeldi, Director of the State Museum of Natural History and Ethnography, at Pará, Brazil, together with his colleague Dr G. Hagmann of the Zoological Gardens at Pará, sent 20 specimens, many from Edentata; Rev. M. C. H. Bird, of Stalham, Norfolk, sent 15; Dr Max Lühe, Privatdocent in Zoology, at Königsberg in Prussia, sent 14; Dr Gerald Leighton, of Pontrilas, Hereford, sent 11 (and over, chiefly Reptilian); Mr Farren, Naturalist at Cambridge, supplied me with 11; Mr F. B. Parkinson, F.R.G.S., etc., of Baviaankrantz, Cape Colony, sent 9 (a number of unique specimens); Dr H. B. Dodds, Medical Officer at Fort Johnson, British Central Africa, sent 8; Dr G. Langmann, of New York City, sent 8 (at the kind suggestion of Prof. T. Mitchell Prudden); Surgeon-Major Leonard Rogers, I.M.S., of Calcutta, sent 8; Dr N. P. Schierbeck, Privatdocent in Hygiene at Copenhagen, sent 8 (several from the Zoological Gardens there); Count Carl Otto von Schlieffen-Schwandt, of Mecklenburg-Schwerin, sent 7 (some collected by his gamekeeper); Prof. R. P. Bigelow, of the Massachusetts Institute of Technology, U.S.A., sent 7 (mostly from fish, collected by Mr C. F. Silvester); Dr Charles Hose, of Baram, Sarawak, sent 6; Kammerherr G. von Oertzen, of Mirow in Mecklenburg-Schwerin, sent 5. Five specimens each were sent by Prof. F. M. Sandwith, of Cairo, Egypt (collected by Mr Littlewood, Chief Veterinary Inspector to the Egyptian Government); Dr J. R. Garrood, of Alconbury, Hunts., and his friend Mr W. F. Beauford; Dr A. S. Grünbaum, of University College, Liverpool (Simian bloods); Dr A. B. Dalgetty, late of Madapore, S. Sylhet, India; Prof. G. Elliott-Smith, of Cairo (Reptilian). Four specimens each were sent by the Hon. R. E. Dillon, of Clonbrock, Co. Galway, Ireland; Prof. E. O. Jordan, Chicago University. Three each were sent by Dr D. E. Salmon, Chief of the Bureau of Animal Industry, U.S. Department of Agriculture, Washington (collected at the Zoological Laboratory of the Bureau); Dr Henry Strachan, Chief Medical Officer at Lagos, W. Africa (Negro bloods); Mr E. G. Wheeler, of Alnwick, Northumberland; Dr Arthur Palmer, of Sydney, N. S. Wales (collected by himself and Dr J. F. Flashman). Two specimens each were supplied by Dr Daniels of the London School of Tropical Medicine (Mongolian and Indian); Major K. C. MacWatt, I.M.S., of Ajmur, Rajputana, India; Mr E. F. Robison, of Evergreen, California; Dr R. J. Scharff, of the Science and Art Museum, Dublin. Finally single specimens were sent me by Dr H. Ainsworth, I.M.S., Bombay; Professor J. Brunchorst, Director of Bergens Museum, Bergen, Norway (fluid whale blood); Mr E. J. Bles, of King's College, Cambridge; Dr John Cropper, of Chepstow, Monmouthshire; Mr William Evans, of Edinburgh; Captain Stanley Flower, Director of the Government Zoological Gardens, Gizeh, Egypt; Mr E. H. Hankin, Director of the Government Laboratory, Agra, India; Dr J. S. Haldane, F.R.S., of Oxford; Mr M. O. Hedley, of Carievale, N.W.T. Canada; Mr W. Kerr, of Singapore; Prof. John MacFadyean, Principal of the Royal Veterinary College, London (donkey serum); Herr G. von Oertzen, Consul-General to the German Empire at Havre, France (fluid porpoise blood); Prof. Sophus Torup, Christiania, Norway (fluid whale serum); Mr I. Ll. Tuckett, of Trinity College, Cambridge.

Some specimens sent by Dr C. Christy (23 bloods) from Kampala, Uganda, by Mr William Foster (2) from Sapucay, Paraguay, by Dr H. Ainsworth (3), by Captain Flower (1), by the Scottish Antarctic Expedition (5), as well as others (11 specimens) received from the Zoological Society's Gardens, London, unfortunately arrived too late to be included in the present work. They will, however, be utilized in due course.

My laboratory attendant, Mr Bertie Clarke, has been of great assistance in the routine work connected with this investigation, where so much depended upon conscientious attention to details, and I feel it is but fair to include him amongst those to whom I am indebted.

The expenses of this investigation have been in part defrayed personally and through two grants, the one from the Government Grants Committee of the Royal Society, the other from the John Lucas Walker Trust, Cambridge.

BIBLIOGRAPHY.

Papers referring to the Precipitins are marked with a star ⁕.

All citations are from the original papers unless otherwise stated. Papers cited second-hand were inaccessible; some of these and others which contained no new matter are only cited in the bibliography, the reference being followed by comments.

ABDERHALDEN, E. (9. X. 1897), Zur quantitativen Analyse des Blutes. *Zeitschr. f. physiol. Chem.*, Bd. XXIII. pp. 521—531.

ACHALME, P. (X. 1901), Recherches sur les propriétés pathogènes de la trypsine et le pouvoir antitryptique du sérum des cobayes neufs et immunisés. *Ann. de l'Inst. Pasteur*, vol. XXV. pp. 737—752.

ARLOING (1898), *Semaine médicale*, p. 261 (cited by van Emden, 1899).

⁕ASCHOFF, L. (1902), Ehrlich's Seitenkettentheorie und ihre Anwendung auf die künstlichen Immunisirungsprozesse. Zusammenfassende Darstellung. *Zeitschr. f. allgem. Physiol.*, Bd. I. pp. 69—248 (Exhaustive bibliography).

⁕ASCHOFF, L. (6. IX. 1902), Note on the Origin of Urine albumin. *Lancet*, vol. II. Reprint.

ASCOLI, G. (9. X. 1902), Ueber haemolytisches Blutplasma. *Deutsche med. Wochenschr.*, Jahrg. XXVIII. pp. 736—738.

⁕ASCOLI, M. (11. III. 1902), Ueber den Mechanismus der Albuminurie durch Eiereiweiss. *München. med. Wochenschr.*, Jahrg. XLIX. pp. 398—401.

⁕— (26. VIII. 1902), Zur Kenntniss der Präcipitinwirkung und der Eiweisskörper des Blutserums. *München. med. Wochenschr.*, Jahrg. XLIX. pp. 1409—1413.

⁕— (1903), Neue Thatsachen und neue Ausblicke in der Lehre der Ernährung. *München. med. Wochenschr.*, No. 5, repr. 9 pages.

⁕AUSTIN, A. E. (12. III. 1903), Limitations of the Uhlenhuth test for the differentiation of human blood. *Boston Med. and Surg. Journ.*, vol. CXLVIII. pp. 279—281.

⁕BAIL, O. (1901), Untersuchungen über die Agglutination von Typhusbakterien. *Prag. med. Wochenschr.*, Bd. XXVI. No. 7 and 12, reprints.

⁕BASHFORD, E. F. (III. 1902), Note on toxic and antitoxic action in vitro and in corpore. *Journ. of Pathol. and Bacteriol.*, vol. VIII. p. 59.

BAUMGARTEN, P. (16. XII. 1901), Microscopische Untersuchungen über Haemolyse im heterogenen Serum. *Berlin. klin. Wochenschr.*, Jahrg. XXXVIII. No. 50, pp. 1241—1244.

— (27. X. 1902), Weitere Untersuchungen über Haemolyse im heterogenen Serum. *Berlin. klin. Wochenschr.*, Jahrg. XXXIX. pp. 997—1000.

BEHRING, E. (1890), Untersuchungen über das Zustandekommen der Diphtherieimmunität bei Thieren. *Deutsche med. Wochenschr.*

BEHRING, E. and KITASATO (1890), Ueber das Zustandekommen der Diphtherieimmunität und der Tetanusimmunität bei Thieren. *Deutsche med. Wochenschr.*

BEHRING, E. and NISSEN, F. (1890), Ueber die bakterienfeindliche Eigenschaften verschiedener Serumarten. *Zeitschr. f. Hygiene*, Bd. VIII. pp. 412—433.

BELFANTI and CARBONE (VII. 1898), Produzione di sostanze tossiche nel siero di animali inoculati con sangue heterogeno. *Giorn. della R. Acad. di Med. di Torino*, No. 8 (cited by Metchnikoff, 1900, p. 371).

*BELJAJEW, W. (1902), Ueber die Bedingungen der Bildung specifischer Kraus'scher Niederschläge. *Russ. Arch. f. Pathol.*, 14 Lief. II. (review by Tchistovitch in *Centralbl. f. Bakteriol.*, Bd. XXXII. p. 629).

BESREDKA (VI. 1900), La leucotoxine et son action sur le système leucocytaire. *Ann. de l'Inst. Pasteur*, vol. XIV. pp. 390—401.

— (25. x. 1901), Les antihémolysines naturelles. *Ann. de l'Inst. Pasteur*, vol. XV. pp. 785—807.

*BEUMER (6. XII. 1902), Die Untersuchung von Menschen- und Thierknochen auf biologischem Wege. *Zeitschr. f. Medicinalbeamte*, Heft XXIII. Repr. 4 pages (read at Med. Verein, Greifswald).

*BINDA (III. and IV. 1901), Sulla diagnosi specifica del sangue per mezzo del siero reattivo. *Giorn. de Medicina Legale* (cited by Biondi, 1902).

*BIONDI (1902), Beitrag zum Studium der biologischen Methode für die specifische Diagnose des Blutes. *Vierteljahresschr. f. gerichtl. Med. u. öffentl. Sanitätswesen*, 3. Folge, Bd. XXIII. Supplementheft, pp. 1—37.

BOERI (28–30. x. 1902), Le sérum névrotoxique. (XI Congr. Soc. ital. di Med. intern.) *Semaine Méd.*, Année XXII. p. 368.

BORDET, J. (VI. 1895), Les leucocytes et les propriétés actives du sérum chez les vaccinés. *Ann. de l'Inst. Pasteur*, vol. IX. pp. 462—506.

*— (x. 1898), Sur l'agglutination et la dissolution des globules rouges par le serum d'animaux injectés de sang défibriné. *Ann. de l'Inst. Pasteur*, vol. XII. pp. 688—695.

*— (III. 1899), Le mécanisme de l'agglutination. *Ann. de l'Inst. Pasteur*, vol. XIII. pp. 225—250.

— (v. 1900), Les sérums hémolitiques, leurs antitoxines et les théories des sérums cytolitiques. *Ann. de l'Inst. Pasteur*, vol. XIV. pp. 257—296.

*BORDET, J. and GENGOU, O. (III. 1901), Recherches sur la coagulation du sang et les sérums anti-coagulants. *Ann. de l'Inst. Pasteur*, vol. XV. pp. 129—144.

— (v. 1901), Sur l'existence de substances sensibilisatrices dans la plupart des sérums antimicrobiens. *Ann. de l'Inst. Pasteur*, vol. XV. pp. 289—302.

BRIEGER and EHRLICH (1893), Beiträge zur Kenntniss der Milch immunisirter Thiere. *Zeitschr. f. Hygiene*, Bd. XIII. pp. 336—346.

BRIOT (1900), Étude sur la présure et l'antiprésure. Sceaux, 1900. *Thèse de la Faculté des Sc. de Paris*, No. 4 (cited by Metchnikoff, 1901, p. 116).

BUCHNER, H. (21. v. 1892), Zur Physiologie des Blutserums und der Blutzellen. *Centralbl. f. Physiol.*, vol. VI. pp. 97—101.

— (1893), Weitere Untersuchungen über die bakterienfeindlichen und globuliciden Wirkungen des Blutserums. *Arch. f. Hygiene*, vol. XVII. pp. 112—178.

*Buchner, H. and Geret, L. (16. vii. 1901), Ueber ein krystallinisches Immunisirungsproduct. *München. med. Wochenschr.*, Jahrg. xlviii. pp. 1163—1164, 1275—1276 (1 figure).

*Butza, J. (18. iv. 1902), Un nouveau moyen pratique pour distinguer le sang de l'homme, d'avec celui des animaux. *Compt. rend. de la Soc. de Biol.*, liv. pp. 406—407.

Calmette, A. and Breton, E. (1. xii. 1902), Sur la formation des anticorps dans le sérum des animaux vaccinés. (Acad. des Sc.) *Semaine méd.*, Année xxii. p. 407.

Camus, L. and Gley, E. (29. i. 1898), De la toxicité du sérum d'anguilles pour des animaux d'espèce différente (lapin, cobaye, hérisson). *Comptes rendus de la Soc. de Biol.*, p. 129.

— (31. i. 1898), De l'action destructive d'un sérum sanguin sur les globules rouges d'une autre espèce animale. Immunisation contre cette action. *Comptes rendus de l'Acad. des Sciences*, T. cxxxvi. p. 428.

— (3. viii. 1898), Sur le mécanisme de l'immunisation contre l'action globulicide du serum d'anguille. *Comptes rendus de l'Acad. des Sciences*, T. cxxvii. p. 330.

— (1898), Recherches sur l'action physiologique du sérum d'anguille. Contribution à l'étude de l'immunité naturelle et acquise. *Arch. internat. de Pharmacodynamie*, vol. v. pp. 247—305.

— (24. vii. 1899), Expériences concernant l'état réfractaire au sérum d'anguille. Immunité cytotoxique. *Comptes rendus de l'Acad. des Sciences*, vol. cxxix. p. 231.

— (x. 1899), Nouvelles recherches sur l'immunité contre le sérum d'anguille. *Ann. de l'Inst. Pasteur*, vol. xiii. pp. 779—787.

*Carrara (1901), Sulla diagnosi specifica di sangue umano. *L' Arte Medica*, No. 34 (cited only by title by Biondi, 1902).

*Castellani, A. (28. vi. 1902), Some experiments on the precipitins. *Lancet*, vol. i. pp. 1827—1829.

*Cheinesse (27. ii. 1901), *Semaine médicale* (not consulted).

*Corin (27. iv. 1901), *Société de méd. legale de Belgique* (cited by Ziemke, 17. x. 1901).

*Corin, G. (1902), Zur praktischen Verwerthung der Serodiagnostik des menschlichen Blutes. *Vierteljahresschr. f. gerichtl. Med. u. öffentl. Sanitätswesen*, 3. F., Bd. xxiii. (review in *Deutsche med. Wochenschr.*, 1902, p. 136).

Courmont, P. (1897), Signification de la réaction agglutinante chez les typhiques. *Thèse de Lyon* (cited by van Emden, 1899).

*Creite (1869), Versuche über die Wirkung des Serumeiweisses nach Injektion in das Blut. *Zeitschr. f. ration. Medizin*, Bd. xxxvi. (cited by Friedenthal, 1902).

Daremberg (1891), *Arch. de méd. expérimentale et d'anatomie pathologique*, p. 720 (cited by Buchner, 1893).

Deutsch, L. (ix. 1899), Contribution à l'étude de l'origine des anticorps typhiques. *Ann. de l'Inst. Pasteur*, vol. xiii. pp. 689—727.

— (25. vii. 1900), Zur Frage der Agglutininbildung. *Centralbl. f. Bakteriol.*, Bd. xxviii. pp. 45—49.

— (9. viii. 1900), Reviews of this author's paper read before the Medico-Legal Section of the Internat. Med. Congress, Paris, to be found under the titles:

Le diagnostic de taches de sang par les sérums hémolitiques Bordet (*Bulletin médical*, Paris, 8. IX. 1900) ; Moyen de reconnaître l'origine des taches de sang (*Revue scientifique*, 20. X. 1900).

DEUTSCH, L. (21. II. 1901), Zur Diagnose menschlicher Blutkörperchen. *Deutsche med. Wochenschr.*, Jahrg. XXVII. p. 127 (telegram addressed to the Editors regarding his priority in the use of haemolytic antisera forensically).

*— (15. V. 1901), Die forensische Serumdiagnose des Blutes. *Centralbl. f. Bakteriol.*, Bd. XXIX. pp. 661—667.

DELEZENNE, C. (2–9. VIII. 1900), Sérum antihépatique. XIII Congr. Internat. de Méd., Paris ; also *Compt. rendus de l'Acad. des Sciences*, 11. VIII. 1900.

— (X. 1900), Sérum névrotoxiques. *Ann. de l'Inst. Pasteur*, vol. XIV. pp. 686—704.

*DIEUDONNÉ, A. (2. IV. 1901), Beiträge zum biologischen Nachweis von Menschenblut. *München. med. Wochenschr.*, Jahrg. XLVIII. pp. 533—534.

DONATH, J. and LANDSTEINER, K. (25. VII. 1901), Ueber antilytische Sera. *Wiener klin. Wochenschr.*, Jahrg. XIV. pp. 713—714.

DUBOIS, A. (25. IX. 1902), Sur la dissociation des propriétés agglutinante et sensibilisatrice des sérums spécifiques. *Ann. de l'Inst. Pasteur*, vol. XVI. pp. 690—693.

VON DUNGERN (15 VIII. 1898), *München. med. Wochenschr.* (cited by Metchnikoff, 1901, and by Morgenroth, 1899).

— (1899, I.), Globulicide Wirkungen des thierischen Organismus. *München. med. Wochenschr.*, Nos. 13 and 14 (review in *Baumgarten's Jahresbericht*).

— (1899, II.), Specifisches Immunserum gegen Epithel. *München. med. Wochenschr.*, p. 1228 (review in *Baumgarten's Jahresbericht*).

— (1900), Beiträge zur Immunitätslehre. *München. med. Wochenschr.*, No. 28 (Reprint).

— (1902), Neue Versuche zur Physiologie der Befruchtung. *Zeitschr. f. allgem. Physiol.*, Bd. I. pp. 34—55.

DZIERGOWSKI (1899), Die Beziehungen der Verdauungsfermente zum Antidiphtherieserum und das Schicksal des letzteren im Magendarmkanal. *Arch. des Sc. biologiques Russes*, p. 337 (cited by Rostoski, 1902, b. p. 60).

EHRLICH, P. (1891), Experimentelle Untersuchungen über Immunität. I. Ueber Ricin ; II. Ueber Abrin. *Deutsche med. Wochenschr.*, Nos. 32 and 44.

*— (22. III. 1900), On immunity with special reference to cell life. Croonian Lecture to the Royal Society, London.

*— (1901), Vorbemerkungen aus der Immunitätslehre. *Nothnagel's Specielle Pathol. u. Therap.*, Bd. VIII. Reprint, 6 pages.

— (1901), Schlussbetrachtungen, zu Erkrankungen des Blutes und der blutbildenden Organe. *Ibid.* Reprint, 25 pages, 1 plate.

— (1901), Die Schutzstoffe des Blutes. *Deutsche med. Wochenschr.*, Nos. 50—52. Reprint, 27 pages.

EHRLICH, P. and MORGENROTH, J. (1899—1901). Ueber Haemolysine (I—VI. Mittheilung). *Berliner klin. Wochenschr.* Reprints.

EISENBERG, P. (17. X. 1901), Ueber Isoagglutinine und Isolysine in menschlichen Seris. *Wiener klin. Wochenschr.* Jahrg. XIV. pp. 1020—1024 (Bibliography).

*— (5. V. 1902), Beiträge zur Kenntniss der specifischen Praecipitationsvorgänge. *Bulletin de l'Acad. des Sc. de Cracovie, Classe des Sc. math. et nat.*, pp. 289—310

N. 27

(Reprint); see also Autoreferat in *Centralbl. f. Bakteriol.*, vol. XXXI. pp. 773—776.

EISENBERG, P. and VOLK, R. (12. XII. 1901), Untersuchungen über die Agglutination. Vorläufige Mittheilung. *Wiener klin. Wochenschr.*, Jahrg. XIV. pp. 1221—1223.

— (6. VIII. 1902), the complete paper appeared in *Zeitschr. f. Hygiene*, vol. XL. pp. 155—195 (Bibliography).

VAN EMDEN, J. E. G. (1899), Ueber die Bildungsstätte der agglutinirenden Substanzen bei der Infektion mit Bacillus aërogenes. *Zeitschr. f. Hygiene*, Bd. XXX. pp. 19—32.

EMMERICH, R. and LOEW, O. (30. XI. 1901), Ueber biochemischen Antagonismus. *Centralbl. f. Bakteriol.*, Bd. XXX. pp. 552—555.

*EWING, J. (II.—III. 1903), Differentiation of monkey and human blood by the serum test. *Proc. New York Pathol. Soc.*, N.S. vol. III. pp. 14—22.

*FALLOISE, A. (25. XI. 1902), Contribution à l'étude des sérums précipitants. *Ann. de l'Inst. Pasteur*, vol. XVI. pp. 833—841.

*FARNUM, C. G. (28. XII. 1901), The biological test for semen. *Journ. American Med. Assoc.*, Reprint, 3 pages.

FERMI, C. and PERNOSSI, L. (1894), Ueber die Enzyme. *Zeitschr. f. Hygiene*, vol. XVIII. pp. 83—127.

*FERRAI, C. (1901), Sulla diagnosi specifica del sangue col metodo biologico in medicina legale. *Bollet. della R. Accad. med. di Genova*, Anno XVI. No. 7 (cited by von Rigler, and by Schütze).

*FICH, C. (1900), Studies on Lactoserum and other cell-sera. *St Louis Courier of Medicine* (cited by Wassermann and Schütze, 18. II. 1901).

FLEXNER, S. and NOGUCHI, H. (II. 1902), Snake venom in relation to haemolysis, bacteriolysis, and toxicity. *University of Pennsylvania Med. Bulletin.* Reprint, 30 pages.

FODOR and RIGLER (4. VI. 1898), Das Blut mit Typhusbacillen inficirter Thiere. *Centralbl. f. Bakteriol.*, Bd. XXIII. pp. 930—934.

FRÄNKEL, C. and SOBERNHEIM (1. I. 1894), Versuche über das Zustandekommen der künstlichen Immunität. *Hygienische Rundschau*, Bd. IV. pp. 97—101, 145—159.

FREUND, E. and STERNBERG, C. (1899), Ueber Darstellung des Heilkörpers aus dem Diphtherieheilserum. *Zeitschr. f. Hygiene*, Bd. XXXI. pp. 429—432.

FRIEDBERGER, E. (16. IX. 1901), Ueber die Bedeutung anorganischer Salze und einiger organischer krystalloider Substanzen für die Agglutination der Bakterien. *Centralbl. f. Bakteriol.*, Bd. XXX. pp. 336—346.

FRIEDENTHAL, H. (1900), Ueber einen experimentellen Nachweis von Blutverwandtschaft. *Arch. f. Anat. u. Physiol.*, Jahrg. 1900, pp. 494—508.

*— (10. VII. 1902), Neue Versuche zur Frage nach der Stellung des Menschen im zoologischen System. *Sitzungsberichte der königl. preussischen Akad. der Wissenschaften zu Berlin*, Bd. XXXV. pp. 830—835.

FRIEDENTHAL, H. and LEWANDOWSKY, M. (28. VII. 1899), Ueber das Verhalten des thierischen Organismus gegen fremdes Blutserum. *Arch. f. Anat. u. Physiol.*, Jahrg. 1899, Physiol. Abtheil., pp. 531—545.

*FULD (1902), Ueber das Bordet'sche Laktoserum. *Hofmeister's Beiträge z. chem.*

Physiol. u. Pathol. Bd. II. pp. 425—429, review in *Centralbl. f. Bakteriol.*, vol. XXXII. p. 404.

FULD and SPIRO (1900—01), Ueber die labende und labhemmende Wirkung des Blutes. *Zeitschr. f. physiol. Chem.*, Bd. XXXI. p. 132 (cited by Rostoski, 1902 b).

FUNCK, M. (26. V. 1900), Das antileukocytäre Serum. *Centralbl. f. Bakteriol.*, I. Abtheil, Bd. XXVII. pp. 670—675.

FORD, W. W. (3. VII. 1902), Beitrag zur Lehre von den Haemagglutininen. *Zeitschr. f. Hygiene*, Bd. XL. pp. 363—372.

*GENGOU (25. X. 1902), Sur les sensibilisatrices des sérums actifs contre les substances albuminoides. *Ann. de l'Inst. Pasteur*, vol. XVI. pp. 734—755.

GESSARD (1902), Tyrosinase et Antityrosinase. *Comptes rendus de la Soc. de Biol.*, vol. LIV. p. 551.

GOUTSCHARNKOW (1901), Ueber die Herstellung eines für die Schilddrüse specifischen Serums. *Centralbl. f. allgem. Pathol. u. pathol. Anat.*, vol. XIII. rev. in *Centralbl. f. Bakteriol.*, Bd. XXXII. p. 250.

*GRAHAM-SMITH, G. S. (29. VII. 1903), The biological or precipitin test for blood considered mainly from its medico-legal aspect. *Journal of Hygiene*, vol. III. pp. 354—363.

*GRAHAM-SMITH, G. S. and SANGER, F. (2. V. 1903), The biological or precipitin test for blood considered mainly from its medico-legal aspect. *Journal of Hygiene*, vol. III. pp. 258—291.

GRUBER, M. (1896), *Wiener klin. Wochenschr.* (cited by Rath, 1899).

— (12. XI. 1901), Zur Theorie der Antikörper. I. Ueber die Antitoxin-immunität; II. Ueber Bakteriolyse und Haemolyse. *München. med. Wochenschr.*, Jahrg. XLVIII. pp. 1965—1968, 1827—1830.

*GRÜNBAUM, A. S. F. (18. I. 1902), Note on the "blood relationship" of man and the anthropoid apes. *Lancet*, vol. I. p. 143.

*HALBAN, J. and LANDSTEINER, K. (25. III. 1902), Ueber die Unterschiede des fötalen und mütterlichen Blutserums und über eine agglutinations- und fällungshemmende Wirkung des Normalserums. *München. med. Wochenschr.*, Jahrg. XLIX. pp. 473—476.

*HAMBURGER, F. (6. XI. 1902), Zur Frage der Immunisirung gegen Eiweiss. *Wiener klin. Wochenschr.*, Jahrg. XV. pp. 1188—1191.

*HERRSCHER (1901), *Arch. d'Anthropologie criminelle*, No. 93 (cited by Minovici, 1902).

HILDEBRANDT, H. (1893), Weiteres über hydrolitische Fermente, deren Schicksal und Wirkungen, sowie über Fermentfestigkeit und Hemmung der Fermentationen im Organismus. *Virchow's Archiv*, Bd. CXXXI. pp. 5—39.

*HONL, I. (7. VII. 1901), Ueber die biologischen Untersuchungen von verschiedenen Blutarten. *Wiener klin. Rundschau*, No. XXVII. p. 473 (cited by Biondi, 1902).

*IDE, M. (1. IV. 1901), Ueber Antikörper gegen chemisch reine Eiweissstoffe. Kurze vorläufige Mittheilung. *Fortschritte der Med.*, Bd. XIX. p. 234.

*— (27. VII. 1902), Hémolyse et antihémoglobine. *La Cellule*, vol. XX. pp. 263—284.

*JACOBY, M. (1901), Ueber Ricinimmunität. *Beitrag zur chemischen Physiol. u. Pathol.*, I. (cited by Michaëlis, 1902, p. 733, and also Bashford, 1902).

JOOS, A. (1901), Untersuchungen über den Mechanismus der Agglutination. *Zeitschr. f. Hygiene*, Bd. XXXVI. pp. 422—439.

— (24. XII. 1901), Ueber die Bedeutung anorganischer Salze für die Agglutination. *Centralbl. f. Bakteriol.*, Bd. XXX. pp. 853—862.

*KISTER, J. and WOLFF, H. (18. XI. 1902), Zur Anwendbarkeit der serodiagnostischen Blutprüfungsverfahrens. *Zeitschr. f. Hygiene*, Bd. XLI. pp. 410—426 ; also *Zeitschrift f. Medicinalbeamte*, 1902, No. VII. p. 213.

*KLEIN, A. (17. VII. 1902), Zur Frage der Antikörperbildung. *Wiener klin. Wochenschr.*, Jahrg. XV. pp. 746—749.

KOSSEL, H. (14. II. 1898), Zur Kenntniss der Antitoxinwirkung. *Berlin. klin. Wochenschr.*, No. VII. (cited by Camus and Gley, and Tchistovitch, 1899).

*KOWARSKI, A. (4. VII. 1901), Ueber den Nachweis von pflanzlichen Eiweiss auf biologischen Wege. *Deutsche med. Wochenschr.*, Jahrg. XXVII. p. 442.

*KRATTER (1902), Ueber den forensischen Werth der biologischen Methode zur Unterscheidung von Thier und Menschenblut. *Arch. f. Kriminalanthropologie*, Bd. X. (cited by Uhlenhuth, 1902).

*KRAUS, R. (12. VIII. 1897), Ueber specifische Reaktionen in keimfreien Filtraten aus Cholera, Typhus und Pestbouillonculturen, erzeugt durch homologes Serum. *Wien. klin. Wochenschr.*, Jahrg. X. pp. 736—738.

*— (18. VII. 1901), Ueber diagnostische Verwerthbarkeit der specifischen Niederschläge. *Wiener klin. Wochenschr.*, Jahrg. XIV. pp. 693—695.

*— (1. VIII. 1901), Ueber das Vorkommen der Immunagglutinine und Immunhämolysine in der Milch. *Wiener klin. Wochenschr.*, Jahrg. XIV. pp. 737—739.

KRAUS, R. and CLAIRMONT, P. (17. X. 1901), Ueber Bacteriohämolysine und Antihämolysine. *Wiener klin. Wochenschr.*, Jahrg. XIV. pp. 1016—1020.

KRAUS, R. and EISENBERG, P. (27. II. 1902), Ueber Immunisierung mit Immunsubstanzen. *Centralbl. f. Bakteriol.*, Bd. XXXI. pp. 208—213.

*KRAUS, R. and VON PIRQUET, C. (5. VII. 1902), Weitere Untersuchungen über specifische Niederschläge. *Centralbl. f. Bakteriol.*, Bd. XXVII. pp. 60—74.

*LAMB, G. (16. VIII. 1902), On the precipitin for cobra venom : A means of distinguishing between the proteids of different snake venoms. *Lancet*, vol. II. pp. 431—435.

LANDOIS (1875), Zur Lehre von der Bluttransfusion. Leipzig. (Cited by Friedenthal.)

LANDSTEINER, K. (29. IV. 1899), Zur Kenntniss der specifisch auf Blutkörperchen wirkenden Sera. *Centralbl. f. Bakteriol.*, Bd. XXV. pp. 546—549.

— (23. III. 1900), Zur Kenntniss der antifermentativen, lytischen und agglutinirenden Wirkungen des Blutserums und der Lymphe. *Centralbl. f. Bakteriol.*, Bd. XXVII. pp. 357—362.

— (1901), Beobachtungen über Hämagglutination. *Wiener klin. Rundschau*, No. XL. p. 774. Autoreferat, in *Centralbl. f. Bakteriol.*, Referir-Bd. XXXII. p. 593.

LANDSTEINER, K. and CALVO, A. (18. VII. 1902), Zur Kenntniss der Reaktionen des normalen Pferdeserums. *Centralbl. f. Bakteriol.*, Bd. XXI. pp. 781—786.

LAVERAN, A. and MESNIL, F. (25. IX. 1901), Recherches morphologiques et expérimentales sur le Trypanosome de rats. (Tr. Lewisi, Kent.) *Ann. de l'Inst. Pasteur*, vol. XV. pp. 673—713.

*LAYTON, E. N. (9. II. 1903), The medico-legal test of blood-stains. *Trans. of the Chicago Pathol. Soc.*, vol. V. pp. 217—222.

*LEBLANC, A. (31. V. 1901), Contribution à l'étude de l'immunité acquise. *La Cellule*, vol. XVIII. fasc. II. pp. 337—382.

*LECLAINCHE, E. and VALLÉE, H. (25. I. 1901), Note sur les anticorps albumineux. *Compt. rend. de la Soc. de Biologie*, vol. LIII. No. 3 ; also *La Semaine médicale*, 1901, No. 4.

*LEVENE, P. A. (21. XII. 1901), On the biological relationship of proteids. (Preliminary Communication.) *Medical News* (New York), vol. LXXIX. pp. 981 —982.

*LIEPMANN, W. (18. XII. 1902), Ueber ein für menschliche Plazenta specifisches Serum. *Deutsche med. Wochenschr.*, Jahrg. XXVIII. pp. 911—912.

*— (29. I. 1903), II. Mittheilung. *Ibid.*, Jahrg. XXIX. pp. 80—81.

LINDEMANN, W. (II. 1900), Sur le mode d'action de certains poisons rénaux. *Ann. de l'Inst. Pasteur*, vol. XIII. pp. 49—59. 1 plate.

*LINOSSIER (25. III. 1902), Sur la recherche medico-légale de l'origine du sang à l'aide des sérums precipitants. (Acad. de Méd.) *Semaine méd.*, Année XXII. p. 104.

*LINOSSIER, G. and LEMOINE, G. H. (25. I. 1902), Sur les substances précipitantes des albumines (précipitines) contenus dans certains sérums spécifiques. *Compt. rendus de la Soc. de Biol.*, vol. LIV. pp. 85—88.

*— (8. III. 1902), Sur la spécificité des sérums précipitants. *Ibid.* pp. 276—279.

*— (21. III. 1902), Sur quelques conditions de l'action des sérums précipitants. *Ibid.* pp. 320—322.

*— (28. III. 1902), Sur la spécificité des sérums précipitants *Ibid* pp. 360—372.

*— (18. IV. 1902), Utilisation des sérums précipitants pour l'étude de certaines albuminuries. *Ibid.* pp. 415—417.

LIPSTEIN, A. (15. IV. 1902), Die Komplementablenkung bei bactericiden Reagenzglasversuchen und ihre Ursache. *Centralbl. f. Bakteriol.*, Bd. 31, pp. 460—468.

*LISLE, J. (13. IX. 1902), The identification of human blood. *New York Med. Journ.*, vol. LXXXI. p. 456. (Nothing new in this paper.)

*DE LISLE, J. (XI. 1902), A study of eel-serum, and the production of an antitoxin in a cold-blooded animal. A contribution to the study of immunity. *Journ. of Med. Research*, vol. VIII. pp. 396—407.

LONDON, E. S. (5. VII. 1902), Der gegenwärtige Stand der Lehre von den Cytolysinen und die cytolitische Theorie der Immunität. *Centralbl. f. Bakteriol.*, Bd. XXXII. pp. 48—60, 147—160. (Extensive bibliography.)

MALVOZ, E. (25. VIII. 1902), Contribution à l'étude des fixateurs du sérum normal de chien. *Ann. de l'Inst. Pasteur*, vol. XVI. pp. 625—632.

MANKOWSKI, A. (1902), Zur Frage nach den Zellgiften (Cytotoxinen). Thyreotoxine. *Russ. Arch. f. Pathol.*, Bd. XIV. Lief. 1 (review by Tchistovitch in *Centralbl. f. Bakteriol.*, Referir-Bd. XXXII. p. 593).

MARKL (13. VI. 1901), Zur Agglutination des Pestbacillus. *Centralbl. f. Bakteriol.*, Bd. XXIX. pp. 810—814.

*MERTENS, V. E. (14. III. 1901), Ein biologischer Beweis für die Herkunft des Albumen in Nephritisharn aus dem Blute. *Deutsche med. Wochenschr.*, Jahrg. XXVII. pp. 161—162.

MERTENS, V. E. (13. VI. 1901), Beiträge zur Immunitätsfrage. *Deutsche med Wochenschr.*, Jahrg. XXVII. pp. 381—383.

*— (1902), Die neuen biologischen Methoden des Menschenblutnachweises. *Wiener klin. Rundsch.*, No. 9, Reprint, 9 pages.

METALNIKOFF, S. (IX. 1900), Études sur la spermatoxine. *Ann. de l'Inst. Pasteur*, vol. XIV. pp. 577 —589.

— (18. IV. 1901), Ueber haemolytisches Serum durch Blutfütterung. *Centralbl. f. Bakteriol.*, Bd. XXIX. pp. 531—533.

METCHNIKOFF, E. (X. 1899), Étude sur la résorption des cellules. *Ann. de l'Inst. Pasteur*, vol. XIII. pp. 737—769. 2 plates.

— (VI. 1900), Sur les cytotoxines. *Annales de l'Inst. Pasteur*, vol. XIV. pp. 369—377.

— (1901), L'Immunité dans les Maladies Infectieuses. Paris (Masson et Cie.) 600 pages.

MEYER, F. and ASCHOFF, L. (7. VII. 1902), Ueber die Receptoren der Milcheiweisskörper. Ein Beitrag zur Specifitätsfrage der Immunkörper. *Berliner klin. Wochenschr.*, Jahrg. XXXIX. pp. 638—639.

*MICHAËLIS, L. (1901-02), Untersuchungen über Eiweisspräcipitine. *Verhandl. des Vereins f. innere Medicin*, p. 479 (cited by Michaëlis and Oppenheimer, 1902).

*— (25. IX. 1902), Inaktivirungsversuche mit Präcipitinen. *Centralbl. f. Bakteriol.*, Bd. XXXII. pp. 458—460.

*— (9. X. 1902), Untersuchungen über Eiweisspräcipitine. *Deutsche med. Wochenschr.*, Jahrg. XXVIII. pp. 733—736.

*MICHAËLIS, L. and OPPENHEIMER, C. (1902), Ueber Immunität gegen Eiweisskörper. *Arch. f. Anat. u. Physiol.* (Physiol. Abtheil., Supplement), pp. 336—366.

*MINOVICI, S. (12. VI. 1902), Ueber die neue Methode zur Untersuchung des Blutes mittels Serum. *Deutsche med. Wochenschr.*, Jahrg. XXVIII. pp. 429—431.

*MIRTO (1901), Sul valore del metodo biologico per la diagnosi specifica del sangue nelle varie contingenze della pratica medico-legale. *Riforma medica*, No. 222 —223 (title only cited by Biondi, 1902).

MODICA (31. V. 1901), Azione ematolitica delle orine de animali trattati con sangue eterogeneo ed influenza di agenti esterni ed interni sulla reazione. *Giornale R. Accad. di Med. di Torino*, Aug.—Sept. 1901 ; also Policlinico, *Sezione pratica*, 1901, p. 1131 (cited by Biondi, 1902).

MOLL (1902), Ueber die Antiurase. Hofmeister's *Beiträge z. chem. Physiol.*, Bd. II. pp. 344—354 (review by Jacoby, in *Centralbl. f. Bakteriol.*, Bd. XXXII. p. 439).

MORGENROTH, J. (10. X. 1899), Ueber den Antikörper des Labenzyms. *Centralbl. f. Bakteriol.*, Bd. XXVI. pp. 349—359.

— (9. VI. 1900), Zur Kenntniss der Labenzyme und ihrer Antikörper. *Centralbl. f. Bakteriol.*, Bd. XXVII. pp. 721—724.

MORGENROTH, J. and SACHS, H. (7. VII. 1902), Ueber die Completirbarkeit der Amboceptoren. *Berliner klin. Wochenschr.*, Jahrg. XXXIX. pp. 631—633.

*MORO, E. (31. X. 1901), Biologische Beziehungen zwischen Milch und Serum. *Wiener klin. Wochenschr.*, Jahrg. XIV. pp. 1073—1077. 2 figures.

MOSSO, A. (1888), Un venin dans le sang des Murénides. *Arch. italiennes de biol.*, vol. X. pp. 141—169 ; also in *Rendiconti della R. Accad. dei Lincei*, 1888, p. 665 ; also

Mosso, A. (1888), Die giftige Wirkung des Serums der Muräniden. *Arch. f. experiment. Pathol. u. Pharmacol.*, Bd. xxv. pp. 111—135.

Mosso, U. (1889), Recherches sur la nature du venin qui se trouve dans le sang de l'anguille. *Arch. ital. de biol.*, vol. xi. pp. 229—236 ; also *Rendiconti della R. Accad. dei Lincei*, 2. vi. 1889, p. 804.

Moxter (25. i. 1900), Ueber ein specifisches Immunserum gegen Spermatozoen. *Deutsche med. Wochenschr.*, Jahrg. xxvi. pp. 61—64.

Müller, P. T. (21. ii. 1901), Ueber Antihaemolysine. *Centralbl. f. Bakteriol.*, Bd. xxix. pp. 175—187.

— (24. vi. 1901), Ueber die Antihaemolysine normaler Sera. *Centralbl. f. Bakteriol.*, Bd. xxix. pp. 860—873.

*— (18. ii. 1902), Vergleichende Studien über die Gerinnung des Caseins durch Lab und Lactoserum. *Münchener med. Wochenschr.*, Jahrg. xlix. p. 272.

— (12. viii. 1902), Ueber die Erzeugung haemolytischer Amboceptoren durch Seruminjektion. *Münchener med. Wochenschr.*, Jahrg. xlix. pp. 1330—1332.

*Myers, W. (14. vii. 1900), On Immunity against Proteids. *Lancet*, vol. ii. pp. 98—100 ; also *Centralbl. f. Bakteriol.*, Bd. xxviii. pp. 237—244.

*Nedriagailoff, V. I. (11. viii. 1901), The Serotoxins and their application to distinguish human blood from that of other animals. *Vratch*, vol. xxii. No. 32 (review in *Philadelphia Med. Journ.*, vol. xiii. p. 870).

Néfédieff, N. (25. i. 1901), Sérum néphrotoxique. *Ann. de l'Inst. Pasteur*, vol. xv. pp. 17—35.

Neisser, M. and Wechsberg, F. (1901), Ueber die Wirkungsart bactericider Sera. *Münchener med. Wochenschr.*, Jahrg. xlviii. No. 18, Reprint, 14 pages.

Neufeld, F. (2. v. 1902), Ueber die Agglutination der Pneumokokken und über die Theorien der Agglutination. *Zeitschr. f. Hygiene*, vol. xl. pp. 54—72.

Nicolle, C. (iii. 1898), Recherches sur la substance agglutinée. *Ann. de l'Inst. Pasteur*, vol. xii. pp. 161—191.

*Noguchi, H. (xi. 1902), A study of immunization haemolysins, agglutinins, precipitins, and coagulins in cold-blooded animals. *Univ. of Pennsylvania Med. Bulletin*, 1902. Reprint, 18 pages.

*— — The interaction of the blood of cold-blooded animals, with reference to haemolysis, agglutination, and precipitation. *Ibid.* Reprint, 18 pages.

*Nolf, P. (v. 1900), Contribution à l'étude des sérums antihématiques. *Ann. de l'Inst. Pasteur*, vol. xiv. pp. 296—330.

*Nötel (13. iii. 1902), Ueber ein Verfahren zum Nachweis von Pferdefleisch. *Zeitschr. f. Hygiene*, vol. xxxix. pp. 373—378.

Nuttall, G. H. F. (5. vii. 1888), Experimente über die bacterienfeindlichen Einflüsse des thierischen Körpers. *Zeitschr. f. Hygiene*, vol. iv. pp. 353—394. 1 plate.

— (1890), Beiträge zur Kenntniss der Immunität. Inaugural-Dissertation zur Erlangung der philosophischen Doctorwürde, Göttingen, 1890, 54 pages.

*— (and Dinkelspiel, E. M.), (11. v. 1901), Experiments upon the new specific test for blood. Preliminary Note. *Brit. Med. Journ.*, vol. i. p. 1141.

*— (and Dinkelspiel, E. M.), (1. vii. 1901), On the formation of specific antibodies in the blood following upon treatment with the sera of different animals, together with their use in legal medicine. *Journ. of Hygiene*, vol. i. pp. 367—387.

424　　　　　　　　*Bibliography*

*NUTTALL, G. H. F. (14. IX. 1901), A further note upon the biological test for blood and its importance in zoological classification. *Brit. Med. Journ.*, vol. II. p. 669.

*— (21. XI. 1901), The new biological test for blood in relation to zoological classification. *Proceedings of the Royal Society*, London, vol. LXIX. pp. 150—153.

*— (16. XII. 1901), The new biological test for blood—its value in legal medicine and in relation to zoological classification. (Abstract of a lecture delivered 28. XI. 1901 at the London School of Tropical Medicine.) *Journ. of Tropical Med.*, vol. IV. pp. 405—408.

*— (XI. 1901), On the formation of specific antibodies in the blood, following upon treatment with the sera of different animals. (Abstract of foregoing.) *American Naturalist*, vol. XXXV. No. 419, pp. 927—932.

*— (20. I. 1902), Further observations upon the biological test for blood. *Trans. of the Cambridge Philos. Soc.*, vol. XI. pp. 334—336.

*— (5. IV. 1902), Progress report upon the biological test for blood as applied to over 500 bloods from various sources, together with a preliminary note upon a method of measuring the degree of reaction. *Brit. Med. Journ.*, vol. I. pp. 825—827.

*OBERMAYER, F. and PICK, E. P. (1902), Biologisch-chemische Studie über das Eiklar. *Wiener klin. Rundschau*, No. 15, Reprint, 6 pages.

*OGIER (1901), Société de médecine légale. (Review in *Deutsche med. Wochenschr.*, 1901, No. 26.)

*OKAMOTO (X. 1902), Untersuchungen über den forensisch-praktischen Werth der serumdiagnostischen Methode, etc. *Vierteljahresschr. f. gerichtl. Med.*, vol. XXIV. p. 207 (review in *Biochemisches Centralbl.*, vol. I. p. 29).

*OPPENHEIMER, C. and MICHAËLIS, L. (18. VII. 1902), Mittheilungen über Eiweisspräcipitine (Physiol. Gesellsch., Berlin). *Deutsche med. Wochenschr.*, Jahrg. XXVIII. p. 245, Vereinsbeilage.

*ORLOVSKY, V. F. (XI. 1901), (A contribution to the biology of the blood). Vratch (rev. in *Philadelphia Med. Journ.* 1902).

*PATEK, A. J. and BENNETT, W. C. (6. IX. 1902), The new antiserum method of differentiating human from other blood, and its medico-legal aspect. Preliminary report of experiments. *American Medicine*, vol. IV. pp. 374—377. (Application of known methods medico-legally.)

PFEIFFER, R. (1894), Weitere Untersuchungen über das Wesen der Choleraimmunität und über specifische baktericide Prozesse. *Zeitschr. f. Hygiene*, vol. XVIII. pp. 1—16.

— and MARX (1898), Die Bildungsstätte der Choleraschutzstoffe. *Zeitschr. f. Hygiene*, vol. XXVII. pp. 272—297.

*PHILIPPSON, M. (1902), Sur les propriétés spécifiques et génériques des sérums sanguins et leur importance au point de vue zoologique. *Recueil des Travaux du Laboratoire de Physiologie* (Instituts Solvay), vol. V. fasc. 1, Reprint, 8 pages. (Bruxelles, Librairie médicale de H. Lamertin.)

*PICK, E. P. (1902), Zur Kenntniss der Immunkörper : I. Versuche zur Isolierung von Immunkörpern des Blutserums. II. Ueber die bei der Agglutination und der specifischen Niederschlagsbildung (Kraus) beteiligten Substanzen. III. Ueber die Einwirkung chemischer Agentien auf die Serum-Koaguline, Agglutinine, sowie auf den Vorgang der specifischen Niederschlagsbildung und der Agglutination. *Beiträge zur chem. Physiol. u. Pathol.*, Bd. I. Reprint, 121 pages.

PIORKOWSKI (13. VI. 1902), Die specifischen Sera. Eine zusammenfassende Ueber-sicht der bis Anfang 1902 erschienenen diesbezüglichen Arbeiten. *Centralbl. f. Bakteriol.*, Referir-Bd. XXXI. pp. 553—560. (Only a short review.)

RADZIEVSKY, A. (1900), Beitrag zur Kenntniss des Bacterium coli. *Zeitschr. f. Hygiene*, vol. XXXIV. pp. 369—453.

RATH, D. (29. IV. 1899), Ueber den Einfluss der blutbildenden Organe auf die Entstehung der Agglutinine. *Centralbl. f. Bakteriol.*, vol. XXV. pp. 549—555.

RINGER (1902), Einfluss der Verdauung auf das Drehungsvermögen von Serum-globulinlösung. *Verhandl. der physik.-med. Gesellsch. zu Würzburg*, N.F. Bd. XXXV. (cited by Rostoski, 1902, b. p. 60).

VON RIGLER, G. (1901), Das Schwanken der Alkalicität des Gesammtblutes und des Blutserums bei verschiedenen gesunden und kranken Zuständen. *Centralbl. f. Bakteriol.*, vol. XXX. pp. 823—830, 862—875, 913—931, 948—989.

*— (1902), Die Serodiagnose in der Untersuchung der Nahrungsmittel. *Oester-reichische Chemiker-Zeitung*, No. 5, Reprint, 3 pages.

RITCHIE, J. (1. IV. to 1. X. 1902), A review of current theories regarding immunity. *Journ. of Hygiene*, vol. II. pp. 215—285, 452—464. (Bibliography.)

*ROBIN, A. (20. XII. 1902), A note on the employment of the hanging-drop method in the study of hemoprecipitins. *Philadelphia Med. Journ.*, vol. X. pp. 1019—1020.

RÖMER, P. (1902), Experimentelle Grundlagen für klinische Versuche einer Serum-therapie der Ulcus corneae serpens nach Untersuchungen über Pneumokokken-immunität. *Gräfes Archiv f. Ophthalmologie*, Bd. LIV. p. 127 (cited by Rostoski, 1902 b).

*ROSTOSKI (1902 a), Zur Kenntniss der Präcipitine. *Verhandl. der Phys.-Med. Gesellsch. zu Würzburg*, N.F. Bd. XXXV. pp. 15—65.

*— (1902 b), Ueber den Werth der Präcipitine als Unterscheidungsmittel für Eiweisskörper. *Münchener med. Wochenschr.*, Jahrg. XLIX. No. 18. Reprint, 3 pages.

ROUX, E. and VAILLARD, L. (II. 1893), Contribution à l'étude du tétanos. Préven-tion et traitement par le sérum antitoxique. *Ann. de l'Inst. Pasteur*, vol. VII. pp. 65—140.

RUFFER, A. and CRENDIROPOULO, M. (5. IV. 1902), A contribution to the study of the presence and formation of agglutinins in the blood. *Brit. Med. Journ.*, vol. I. pp. 821, 825.

— (24. I. 1903), Note on a new method of producing haemolysins. *Brit. Med. Journ.*, vol. I. pp. 190—191.

SACHS (1902), Ueber Antipepsin. *Fortschritte der Medizin*. (Cited by Michaëlis and Oppenheimer, 1902, p. 339.)

SALOMONSEN, C. J. and MADSEN, Th. (IV. 1897), Recherches sur la marche de l'immunisation active contre la diphthérie. *Ann. de l'Inst. Pasteur*, vol. XI. pp. 315—331.

— (XI. 1898), Sur la reproduction de la substance antitoxique après des fortes saignées. *Ann. de l'Inst. Pasteur*, vol. XII. pp. 763—773.

*SANGER, F. (23. XII. 1902), Biological test for blood from its medico-legal aspect.

Thesis (unpublished) for the Degree of Doctor of Medicine in the University of Cambridge.

SCHATTENFROH, A. (30. VII. 1901), Ueber specifische Blutveränderungen nach Harninjektionen. *Münchener med. Wochenschr.*, Jahrg. XLVIII. p. 1239.

*SCHIROKICH, M. A. (21. VII. 1901), (The new method of medico-legal detection of human blood and that of animals, and a few remarks concerning it.) Vratch, No. 29 (review in *Philadelphia Med. Journ.*, vol. VIII. p. 811).

SCHÜTZE, A. (5. VII. 1900), Beiträge zur Kenntniss der zellenlösenden Sera. *Deutsche med. Wochenschr.*, Jahrg. XXVI. pp. 431—434.

* — (29. I. 1901), Ueber ein biologisches Verfahren zur Differenzirung der Eiweissstoffe verschiedener Milcharten. *Zeitschr. f. Hygiene*, vol. XXXVI. pp. 5—8. (See also Wassermann and Schütze.)

* — (12. XII. 1901), A short note only entitled "Specifische Serumreaktion, Isopräcipitin" (Gesellschaft der Charitéärzte, Berlin). *Deutsche med. Wochenschr.*, Jahrg. XXVIII. p. 4, Vereinsbeilage.

*— (6. XI. 1902), Ueber weitere Anwendungen der Präcipitine. *Deutsche med. Wochenschr.*, Jahrg. XXVIII. pp. 804—806.

*— (22. XI. 1902), Weitere Beiträge zum Nachweis verschiedener Eiweissarten auf biologischem Wege. *Zeitschr. f. Hygiene*, vol. XXXVIII. pp. 487—494.

*— (22. I. 1903), Ueber die Unterscheidung vom Menschen- und Thierknochen mittels der Wassermann'schen Differenzirungsmethode. *Deutsche med. Wochenschr.*, Jahrg. XXIX. pp. 62—64.

SENG, W. (1899), Ueber die qualitativen und quantitativen Verhältnisse der Eiweisskörper im Diphtherieheilserum. *Zeitschr. f. Hygiene*, vol. XXXI. pp. 513—532.

SHIBAYAMA, A. (5. XII. 1901), Einige Experimente über Hämolysine. *Centralbl. f. Bakteriol.*, Bd. XXX. pp. 760—764.

*SIERADZKI (1901), Ueber sogenannte Hämotoxine und andere ihnen verwandte Körper nebst deren Bedeutung im Allgemeinen und für die gerichtliche Medicin im Besonderen. *Przeglad lekarski*, Nos. 25—26 (review in *Deutsche med. Wochenschr.*, No. XXVIII. L. B. p. 175).

SOBERNHEIM (1899), Weitere Untersuchungen über Milzbrandimmunität. *Zeitschr. f. Hygiene*, Bd. XXXI. pp. 89—132.

*STERN, R. (28. II. 1901), Ueber den Nachweis menschlichen Blutes durch ein "Antiserum." *Deutsche med. Wochenschr.*, Jahrg. XXVII. p. 135.

*STOCKIS, E. (v. 1901), Le diagnostic du sang humain en médecine légale. *Ann. de la Soc. méd.-chir. de Liège* (review by Wesenberg, in *Centralbl. f. Bakteriol.* Bd. XXXI. p. 23).

*STRUBE, G. (12. VI. 1902), Beiträge zum Nachweis von Blut und Eiweiss auf biologischem Wege. *Deutsche med. Wochenschr.*, Jahrg. XXVIII. pp. 425—429.

SWEET, J. E. (XII. 1902), A study of an hemolytic complement found in the serum of the normal rabbit. *University of Pennsylvania Medical Bulletin* (Philadelphia), Reprint, 92 pages.

SZCZAWINSKA, W. (28. XI. 1902), Sérum cytotoxique pour les globules du sang d'un invertébré. *Comptes rendus de la Soc. de Biol.*, vol. LIV. pp. 1303—1304.

*TARCHETTI (10. v. 1901), Di un nuovo metodo per differenziere il sangue umano da quello di altri animali. *Gazzetta degli Ospedali*, No. LX. p. 631 (cited by Biondi, 1902).

*TCHISTOVITCH, Th. (v. 1899), Études sur l'immunisation contre le sérum d'anguilles. *Ann. de l'Inst. Pasteur*, vol. XIII. pp. 406—425 (see especially p. 413).

UHLENHUTH (1897), Zur Kenntniss der giftigen Eigenschaften des Blutserums. *Zeitschr. f. Hygiene*, Bd. XXVI. pp. 384—397.

*— (15. XI. 1900), Neuer Beitrag zum specifischen Nachweis von Eiereiweiss auf biologischem Wege. *Deutsche med. Wochenschr.*, Jahrg. XXVI. pp. 734—735.

*— (1. XII. 1900), Neuer Beitrag zum specifischen Nachweis von Eiereiweiss auf biologischem Wege (Vortrag a. d. med. Verein zu Greifswald), review in *München. med. Wochenschr.*, 1901, No. VIII. p. 315.

*— (7. II. 1901), Eine Methode zur Unterscheidung der verschiedenen Blutarten, in besondern zum differentialdiagnostischen Nachweise des Menschenblutes. *Deutsche med. Wochenschr.*, Jahrg. XXVII. pp. 82—83.

*— (25. IV. 1901), Weitere Mittheilungen über meine Methode zum Nachweise von Menschenblut. *Deutsche med. Wochenschr.*, Jahrg. XXVII. pp. 260—261.

*— (5. VI. 1901), Naturwissenschaftl. Verein, Greifswald. Reprint, 3 pages.

*— (25. VII. 1901), Weitere Mittheilungen über die praktische Anwendung meiner forensischen Methode zum Nachweis von Menschen- und Thierblut. *Deutsche med. Wochenschr.*, Jahrg. XXVII. pp. 499—501.

*— (7. XI. 1901), Die Unterscheidung des Fleisches verschiedener Thiere mit Hilfe specifischer Sera und die praktische Anwendung der Methode in der Fleischbeschau. *Deutsche med. Wochenschr.*, Jahrg. XXVII. pp. 780—781.

*— (5. VII. 1902), Neue Ergebnisse von weiteren Untersuchungen über die Unterscheidung der verschiedenen Blutarten (Vortrag a. d. med. Verein zu Greifswald). *München. med. Wochenschr.*, 1902, p. 1548.

*— (11. IX. 1902), Praktische Ergebnisse der forensischen Serodiagnostik des Blutes. *Deutsche med. Wochenschr.*, Jahrg. XXVIII. pp. 659—662, 679—681.

*— (6. XII. 1902), Lactoserum und Dotterantiserum (Vortrag a. d. med. Verein zu Greifswald), notice in *Deutsche med. Wochenschr.*, 29. I. 1903, No. V. Vereinsbeilage, p. 39.

*— (1902 a), Bemerkungen zu dem Aufsatz von Kratter (siehe Kratter). *Arch. f. Kriminal-Anthropologie*, Bd. X. pp. 210—224. Reprint.

*— (1903), Zur historischen Entwickelung meines forensischen Verfahrens zum Nachweis von Blut und Fleisch mit Hülfe spezifischer Sera. *Deutsche tierärztliche Wochenschr.*, Jahrg. XI. No. XVI. Reprint, 10 pages.

*— (and BEUMER), (1903), Praktische Anleitung zur gerichtärztlichen Blutuntersuchung vermittelst der biologischen Methode. *Zeitschr. f. Medizinalbeamte*, Jahrg. 1903, Nos. V—VI. Reprint, 21 pages.

*UMBER, F. (14. VII. 1902), Zur Chemie und Biologie der Eiweisskörper. *Berliner klin. Wochenschr.*, Jahrg. XXXIX. pp. 657—659.

*VALLÉE, H. and NICOLAS, E. (30. VI. 1903), Les sérums précipitants. Leur spécificité et leur mode de préparation. *Bulletin de la soc. centr. de méd. vétér.* XXI. N. S. pp. 293—297.

WALKER, E. W. A. (I. 1903), On some factors in bacteriolytic action. *Journ. of Hygiene*, vol. III. pp. 52—67.

WASSERMANN (2. III. 1899), Pneumococcenschutzstoffe. *Deutsche med. Wochenschr.*, Jahrg. XXV. pp. 141—143.

*WASSERMANN (18—21. IV. 1900), (Discussion zu dem Vortrage von A. Magnus-Levy : "Ueber die Bence-Jones'schen Eiweisskörper"). *Verhandlungen des Congresses für innere Medicin in Wiesbaden*, p. 501.

*— (and SCHÜTZE, A.), (2. VII. 1900), Neue Beiträge zur Kenntniss der Eiweissstoffe verschiedener Milcharten (Verein f. innere Med., Berlin), short reference in *Deutsche med. Wochenschr.*, Jahrg. XXVI., Vereinsbeilage, 26. VII. 1900, p. 178 ; see also von Leyden's *Festschrift*, 1902.

— (3. I. 1901), Ueber die Ursachen der natürlichen Wiederstandsfähigkeit gegenüber gewissen Infektionen. *Deutsche med. Wochenschr.*, Jahrg. XXVII. pp. 4—6.

*— (and SCHÜTZE, A.), (18. II. 1901), Ueber eine neue forensische Methode zur Unterscheidung von Menschen und Thierblut. *Berliner klin. Wochenschr.*, Jahrg. XXXVIII. pp. 187—190.

*— (and SCHÜTZE, A.), (3. VII. 1902), Ueber die Entwickelung der biologischen Methode zur Unterscheidung von menschlichen und thierischen Eiweiss mittels Präcipitine. *Deutsche med. Wochenschr.*, Jahrg. XXVIII. p. 483. (Treats only of matters of priority.)

— (1. I. 1903), Ueber biologische Mehrleistung des Organismus bei der künstlichen Ernährung von Säuglingen gegenüber der Ernährung mit Muttermilch. *Deutsche med. Wochenschr.*, Jahrg. XXIX. pp. 16—17.

*— (10. II. 1903), Ueber Agglutinine und Präcipitine. *Zeitschr. f. Hygiene*, Bd. XLIII. pp. 267—292.

WEICHARDT, W. (25. XI. 1901), Recherches sur l'Antispermotoxine. *Ann. de l'Inst. Pasteur*, Vol. XV. pp. 832—841.

WELCH, W. H. (11. X. 1902), Recent studies of Immunity with special reference to their bearing on pathology. (Huxley Lecture, Charing Cross Hospital.) *Brit. Med. Journ.*, Vol. II. pp. 1105—1114 ; also *Lancet*, Vol. II. pp. 977—984 ; also *Medical News*, Vol. LXXXI. p. 721.

WENDELSTADT, H. (end 1901), Ueber einen Antikörper gegen Blutegelextract. *Arch. internat. de Pharmakodynamie et de Thérapie*, Vol. IX., pp. 407—421.

— (16. IV. 1902), Ueber die Vielheit der Amboceptoren und Komplemente bei Hämolyse. *Centralbl. f. Bakteriol.*, Bd. XXXI. pp. 469—473.

*WHITNEY, W. F. (24. IV. 1902), Notes on the production of the test-serum in rabbits. *Boston Med. and Surg. Journ.*, vol. CXLVI. p. 429.

*WHITTIER, F. N. (18. I. 1902), Agglutination test for human blood. *American Medicine*, vol. III. p. 96.

*WOLFF (15. IX. 1902), (A paper read at the I. Hauptversammlung des deutschen Medicinalvereins, München), review in *Deutsche med. Wochenschr.*, Jahrg. XXVIII. Vereinsbeilage, p. 306. (Apparently but a summary of known facts relating to the test.)

*WOOD, E. S. (14. XI. 1901), Medico-legal examination of blood-stains. *Boston Med. and Surg. Journ.*, vol. CXLV. pp. 533—537. (A brief review of known methods of testing blood.)

*— (24. IV. 1902), The serum test for blood. *Boston Med. and Surg. Journ.*, vol. CXLVI. pp. 427—429.

*ZIEMKE, E. (27. VI. 1901), Zur Unterscheidung von Menschen- und Thierblut mit Hilfe eines specifischen Serums. *Deutsche med. Wochenschr.*, Jahrg. XXVII. pp. 424—426.

*ZIEMKE, E. (17. x. 1901), Weitere Mittheilungen über die Unterscheidung von Menschen und Thierblut mit Hilfe eines specifischen Serums. *Deutsche med. Wochenschr.*, Jahrg. XXVII. pp. 731—733.

*— (15. IX. 1902), (A short note on a discussion at the I. Hauptversammlung des deutschen Medicinalvereins, München). *Deutsche med. Wochenschr.*, Jahrg. XXVIII., Vereinsbeilage, p. 306.

*ZUELZER, G. (4. IV. 1901), Zur Frage der biologischen Reaktion auf Eiweiss in Blut und Harn. *Deutsche med. Wochenschr.*, Jahrg. XXVII. pp. 219—220.

APPENDIX.

Note 1 (to p. 10). Wright (*Brit. Med. Journ.*, vol. I., p. 1069, 9. v. 1903) observed a decrease ("negative phase") of the bactericidal power of blood serum following inoculation with anti-typhoid vaccine, the antibody subsequently increasing in amount ("positive phase"). The haemolysins and anti-rennet behave in a similar manner.

Note 2 (to p. 112). Regarding *Anti-Peptones*: Dr E. F. Bashford has sent me his Report to the Scientific Assessors of the Worshipful Company of Grocers, London, dated 25. III. 1902, in which he writes:

"In his paper on immunisation against various proteids, the late Dr Myers stated, that the precipitate produced by the reaction between Witte peptone and the serum of an animal immunised against it, did not give the biuret reaction, and therefore was a new product. By the kindness of Professor Liebreich I was enabled to immunise two goats against Witte peptone (albumose), and thereby to obtain the product referred to in large amount. It seemed that this step was likely to throw much light on the nature of the reaction, and on that between toxine and antitoxine. It was, however, found that the reason why the biuret reaction had failed was simply the extreme insolubility of the product. Thoroughly dried in vacuo over phosphoric anhydride along with a specimen of the mother albumose (*i.e.* which had been boiled and filtered as for injection into the immunised animals), both were submitted to elementary analysis, with the following mean result for two analyses:—

$$\text{Product precipitated—}C_{48\cdot0}\ N_{13\cdot29}\ H_{6\cdot84}\ \text{per cent.}$$
$$\text{Mother substance—}C_{53\cdot15}\ N_{13\cdot67}\ H_{7\cdot6}\ \text{per cent.}$$

This did not afford grounds sufficient for speculation as to the nature of the product precipitated, nor as to the mechanism of its production. Many other difficulties, especially the fact that albumoses in presence of other albuminous bodies in neutral solution may yield precipitates made it doubtful if the product analysed had any claim to be considered sufficiently free from admixtures, etc. I have not given up this investigation, on the contrary hope that immunisation against nucleohiston now proceeding at Dr T. H. Milroy's suggestion, may afford means for obtaining a reaction product more distinctly differing in composition from the mother nucleohiston than was found to be the case for the albumoses of Witte peptone."

Note 3. On looking through the *Index Medicus*, a paper by W. d'E. Emery (*Bart's Hosp. Journ.*, London, 1902—1903, x. pp. 34—40) relating to the precipitins was noted, but too late to be considered in this volume.

INDEX TO NAMES.

See in addition under Bibliography, p. 414, and under Acknowledgments, p. 411.

INDEX TO SUBJECTS.

To avoid overloading the Index with the names of the numerous animals whose bloods were tested with precipitins it has been deemed advisable to exclude them. The names will, however, be readily found by reference to the Tables on pp. 220–311, 364–380, where they have been ordered zoologically. In the following Index the reader will find references to the different bloods tested by looking up the *Class or Order* to which the animal belongs, and in addition by referring to the Tables as follows: